PRAISE FOR THE *MANGA GUIDE* SERIES

"Highly recommended."
—CHOICE MAGAZINE ON *THE MANGA GUIDE TO DA̶*

"Stimulus for the next generation of scienti̶"
—SCIENTIFIC COMPUTING ON *THE MANGA GUIDE TO̶*

"A great fit of form and subject. Recommended."
—OTAKU USA MAGAZINE ON *THE MANGA GUIDE TO PHYSICS*

"The art is charming and the humor engaging. A fun and fairly painless lesson on what many consider to be a less-than-thrilling subject."
—SCHOOL LIBRARY JOURNAL ON *THE MANGA GUIDE TO STATISTICS*

"This is really what a good math text should be like. Unlike the majority of books on subjects like statistics, it doesn't just present the material as a dry series of pointless-seeming formulas. It presents statistics as something *fun*, and something enlightening."
—GOOD MATH, BAD MATH ON *THE MANGA GUIDE TO STATISTICS*

"I found the cartoon approach of this book so compelling and its story so endearing that I recommend that every teacher of introductory physics, in both high school and college, consider using it."
—AMERICAN JOURNAL OF PHYSICS ON *THE MANGA GUIDE TO PHYSICS*

"The series is consistently good. A great way to introduce kids to the wonder and vastness of the cosmos."
—DISCOVERY.COM ON *THE MANGA GUIDE TO THE UNIVERSE*

"A single tortured cry will escape the lips of every thirty-something biochem major who sees *The Manga Guide to Molecular Biology*: 'Why, oh why couldn't this have been written when I was in college?'"
—THE SAN FRANCISCO EXAMINER

"Scientifically solid . . . entertainingly bizarre."
—CHAD ORZEL, AUTHOR OF *HOW TO TEACH PHYSICS TO YOUR DOG*, ON *THE MANGA GUIDE TO RELATIVITY*

"A lot of fun to read. The interactions between the characters are lighthearted, and the whole setting has a sort of quirkiness about it that makes you keep reading just for the joy of it."
—HACK A DAY ON *THE MANGA GUIDE TO ELECTRICITY*

"*The Manga Guide to Databases* was the most enjoyable tech book I've ever read."
—RIKKI KITE, LINUX PRO MAGAZINE

"The *Manga Guides* definitely have a place on my bookshelf."
—SMITHSONIAN'S "SURPRISING SCIENCE"

"For parents trying to give their kids an edge or just for kids with a curiosity about their electronics, *The Manga Guide to Electricity* should definitely be on their bookshelves."
—SACRAMENTO BOOK REVIEW

"This is a solid book and I wish there were more like it in the IT world."
—SLASHDOT ON *THE MANGA GUIDE TO DATABASES*

"*The Manga Guide to Electricity* makes accessible a very intimidating subject, letting the reader have fun while still delivering the goods."
—GEEKDAD BLOG, WIRED.COM

"If you want to introduce a subject that kids wouldn't normally be very interested in, give it an amusing storyline and wrap it in cartoons."
—MAKE ON *THE MANGA GUIDE TO STATISTICS*

"A clever blend that makes relativity easier to think about—even if you're no Einstein."
—STARDATE, UNIVERSITY OF TEXAS, ON *THE MANGA GUIDE TO RELATIVITY*

"This book does exactly what it is supposed to: offer a fun, interesting way to learn calculus concepts that would otherwise be extremely bland to memorize."
—DAILY TECH ON *THE MANGA GUIDE TO CALCULUS*

"The art is fantastic, and the teaching method is both fun and educational."
—ACTIVE ANIME ON *THE MANGA GUIDE TO PHYSICS*

"An awfully fun, highly educational read."
—FRAZZLEDDAD ON *THE MANGA GUIDE TO PHYSICS*

"Makes it possible for a 10-year-old to develop a decent working knowledge of a subject that sends most college students running for the hills."
—SKEPTICBLOG ON *THE MANGA GUIDE TO MOLECULAR BIOLOGY*

"This book is by far the best book I have read on the subject. I think this book absolutely rocks and recommend it to anyone working with or just interested in databases."
—GEEK AT LARGE ON *THE MANGA GUIDE TO DATABASES*

"The book purposefully departs from a traditional physics textbook and it does it very well."
—DR. MARINA MILNER-BOLOTIN, RYERSON UNIVERSITY ON *THE MANGA GUIDE TO PHYSICS*

"Kids would be, I think, much more likely to actually pick this up and find out if they are interested in statistics as opposed to a regular textbook."
—GEEK BOOK ON *THE MANGA GUIDE TO STATISTICS*

THE MANGA GUIDE™ TO REGRESSION ANALYSIS

CROISSANT

THE MANGA GUIDE™ TO
REGRESSION ANALYSIS

SHIN TAKAHASHI,
IROHA INOUE, AND
TREND-PRO CO., LTD.

Ohmsha

no starch press

THE MANGA GUIDE TO REGRESSION ANALYSIS.

The Manga Guide to Regression Analysis is a translation of the Japanese original, *Manga de wakaru tōkei-gaku kaiki bunseki-hen*, published by Ohmsha, Ltd. of Tokyo, Japan, © 2005 by Shin Takahashi and TREND-PRO Co., Ltd.

This English edition is co-published by No Starch Press, Inc. and Ohmsha, Ltd.

First printing

20 19 18 17 16   1 2 3 4 5 6 7 8 9

ISBN-10: 1-59327-728-8
ISBN-13: 978-1-59327-728-4

MIX
Paper from responsible sources
FSC® C014174

Publisher: William Pollock
Author: Shin Takahashi
Illustrator: Iroha Inoue
Producer: TREND-PRO Co., Ltd.
Production Editor: Serena Yang
Developmental Editors: Liz Chadwick and Tyler Ortman
Technical Reviewers: James Church, Dan Furnas, and Alex Reinhart
Compositor: Susan Glinert Stevens
Copyeditor: Paula L. Fleming
Proofreader: Alison Law
Indexer: BIM Creatives, LLC.

For information on distribution, translations, or bulk sales, please contact No Starch Press, Inc. directly:
No Starch Press, Inc.
245 8th Street, San Francisco, CA 94103
phone: 415.863.9900; info@nostarch.com; http://www.nostarch.com/

Library of Congress Cataloging-in-Publication Data
Names: Takahashi, Shin. | Inoue, Iroha. | Trend-pro Co.
Title: The manga guide to regression analysis / by Shin Takahashi, Iroha Inoue, and Trend-Pro Co., Ltd.
Other titles: Manga de wakaru tåokeigaku. Kaiki bunsekihen. English
Description: San Francisco : No Starch Press, [2016] | Includes index.
Identifiers: LCCN 2016000594 (print) | LCCN 2016003356 (ebook) | ISBN 9781593277284 | ISBN 1593277288 | ISBN 9781593277529 (epub) | ISBN 9781593277536 (mobi)
Subjects: LCSH: Regression analysis. | Graphic novels.
Classification: LCC QA278.2 .T34713 2016 (print) | LCC QA278.2 (ebook) | DDC 519.5/36--dc23
LC record available at http://lccn.loc.gov/2016000594

CONTENTS

PREFACE

This book is an introduction to regression analysis, covering simple, multiple, and logistic regression analysis.

Simple and multiple regression analysis are statistical methods for predicting values; for example, you can use simple regression analysis to predict the number of iced tea orders based on the day's high temperature or use multiple regression analysis to predict monthly sales of a shop based on its size and distance from the nearest train station.

Logistic regression analysis is a method for predicting probability, such as the probability of selling a particular cake based on a certain day of the week.

The intended readers of this book are statistics and math students who've found it difficult to grasp regression analysis, or anyone wanting to get started with statistical predictions and probabilities. You'll need some basic statistical knowledge before you start. *The Manga Guide to Statistics* (No Starch Press, 2008) is an excellent primer to prepare you for the work in this book.

This book consists of four chapters:

- Chapter 1: A Refreshing Glass of Math
- Chapter 2: Simple Regression Analysis
- Chapter 3: Multiple Regression Analysis
- Chapter 4: Logistic Regression Analysis

Each chapter has a manga section and a slightly more technical text section. You can get a basic overview from the manga, and some more useful details and definitions from the text sections.

I'd like to mention a few words about Chapter 1. Although many readers may have already learned the topics in this chapter, like differentiation and matrix operations, Chapter 1 reviews these topics in context of regression analysis, which will be useful for the lessons that follow. If Chapter 1 is merely a refresher for you, that's great. If you've never studied those topics or it's been a long time since you have, it's worth putting in a bit of effort to make sure you understand Chapter 1 first.

In this book, the math for the calculations is covered in detail. If you're good at math, you should be able to follow along and make sense of the calculations. If you're not so good at math, you can just get an overview of the procedure and use the step-by-step instructions to find the actual answers. You don't need to force yourself to understand the math part right now. Keep yourself relaxed. However, do take a look at the procedure of the calculations.

We've rounded some of the figures in this book to make them easier to read, which means that some of the values may be inconsistent with the values you will get by calculating them yourself, though not by much. We ask for your understanding.

I would like to thank my publisher, Ohmsha, for giving me the opportunity to write this book. I would also like to thank TREND-PRO, Co., Ltd. for turning my manuscript into this manga, the scenario writer re_akino, and the illustrator Iroha Inoue. Last but not least, I would like to thank Dr. Sakaori Fumitake of College of Social Relations, Rikkyo University. He provided with me invaluable advice, much more than he had given me when I was preparing my previous book. I'd like to express my deep appreciation.

Shin Takahashi
September 2005

PROLOGUE

MORE TEA?

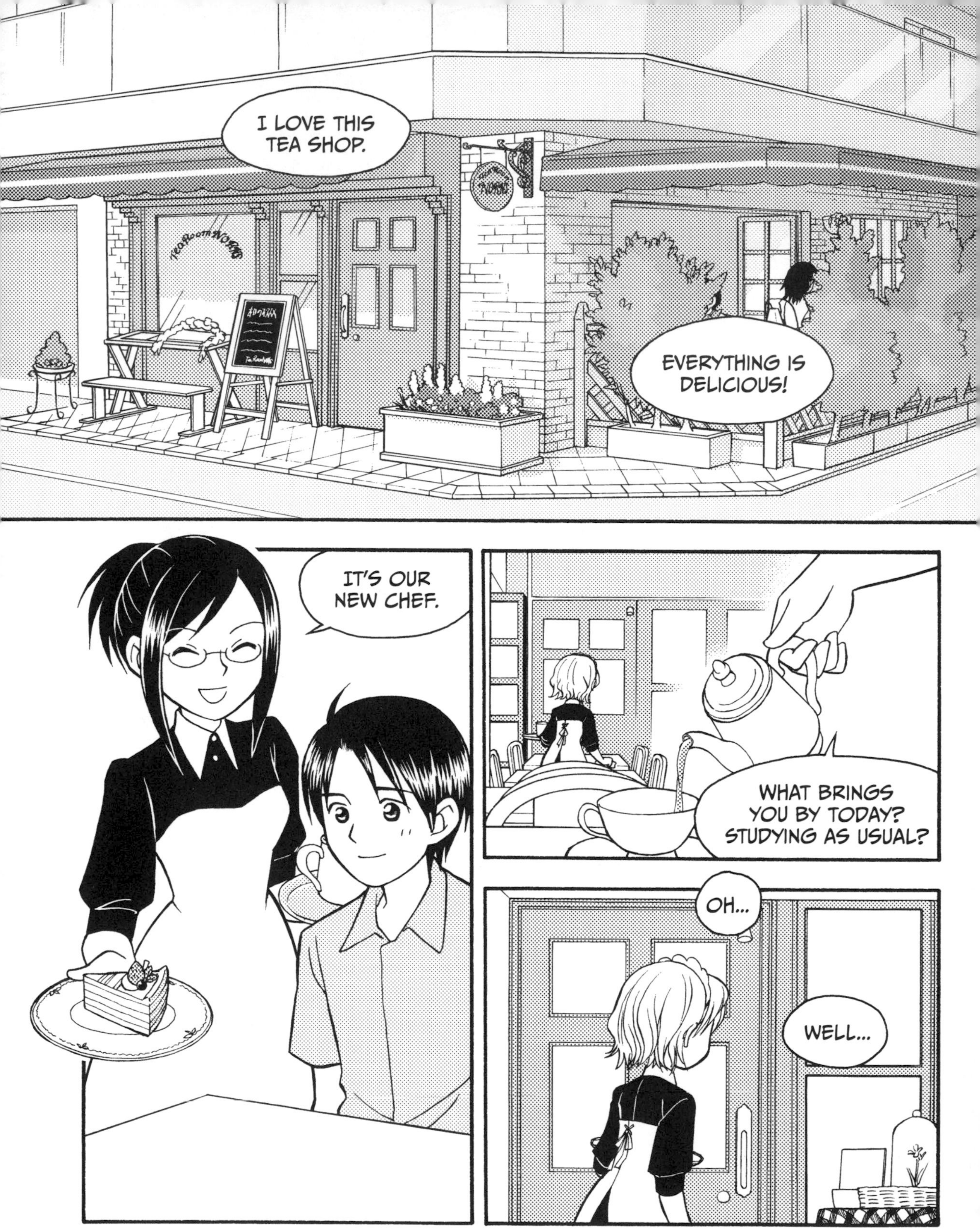
I LOVE THIS TEA SHOP.
EVERYTHING IS DELICIOUS!
IT'S OUR NEW CHEF.
WHAT BRINGS YOU BY TODAY? STUDYING AS USUAL?
OH...
WELL...

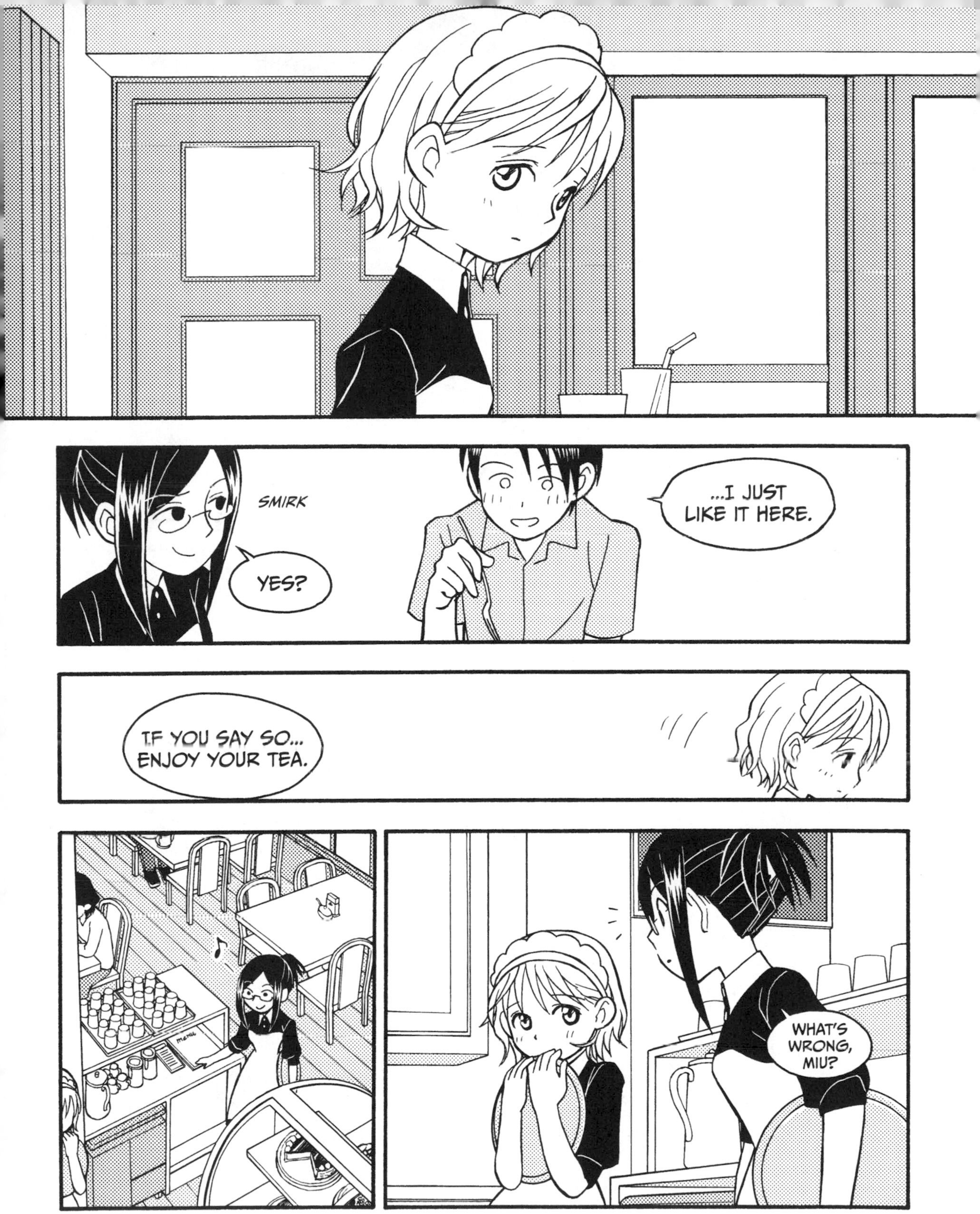
SMIRK
YES?
...I JUST LIKE IT HERE.
IF YOU SAY SO... ENJOY YOUR TEA.
menu
WHAT'S WRONG, MIU?

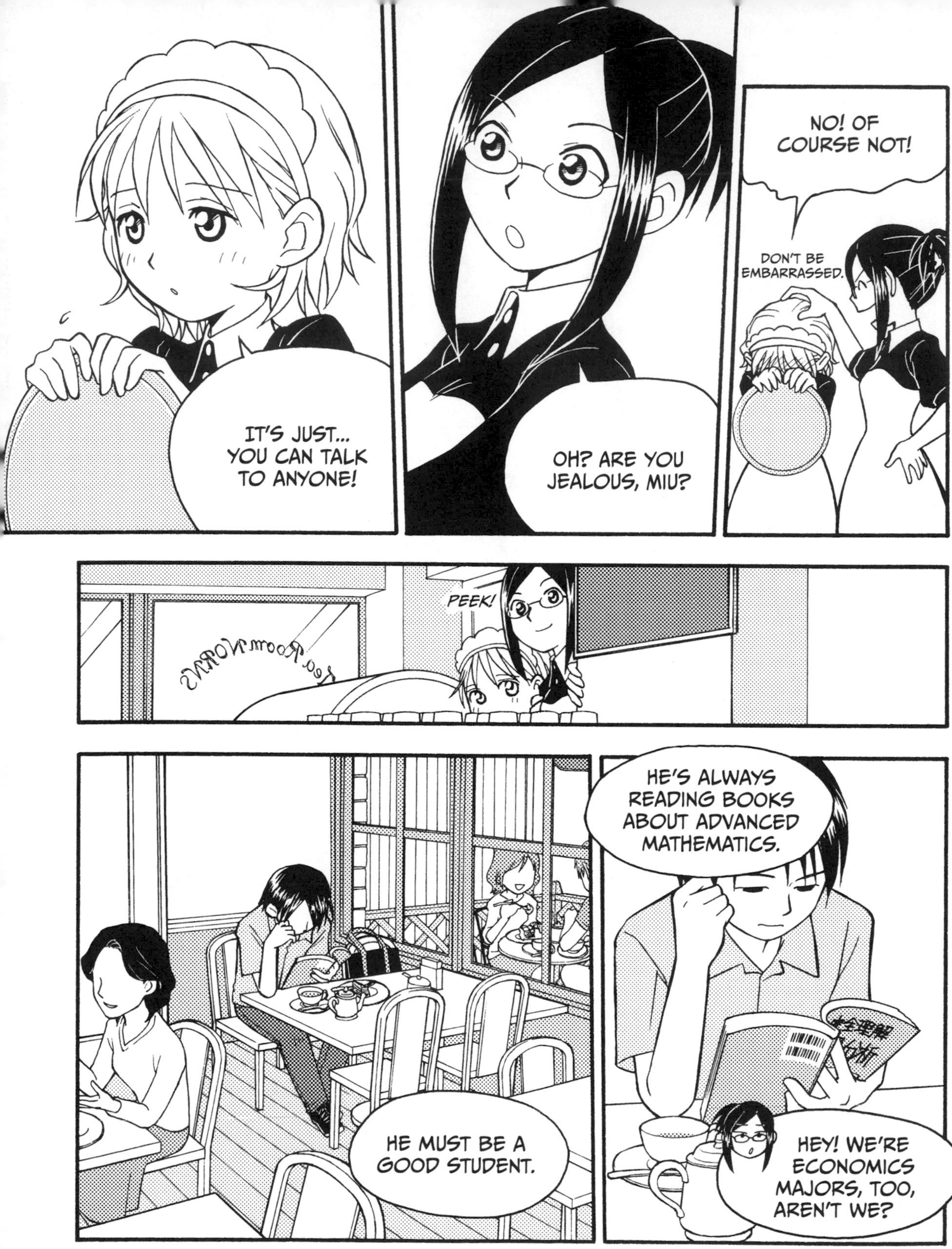
IT'S JUST... YOU CAN TALK TO ANYONE!
OH? ARE YOU JEALOUS, MIU?
NO! OF COURSE NOT!
DON'T BE EMBARRASSED.
PEEK!
HE MUST BE A GOOD STUDENT.
HE'S ALWAYS READING BOOKS ABOUT ADVANCED MATHEMATICS.
HEY! WE'RE ECONOMICS MAJORS, TOO, AREN'T WE?

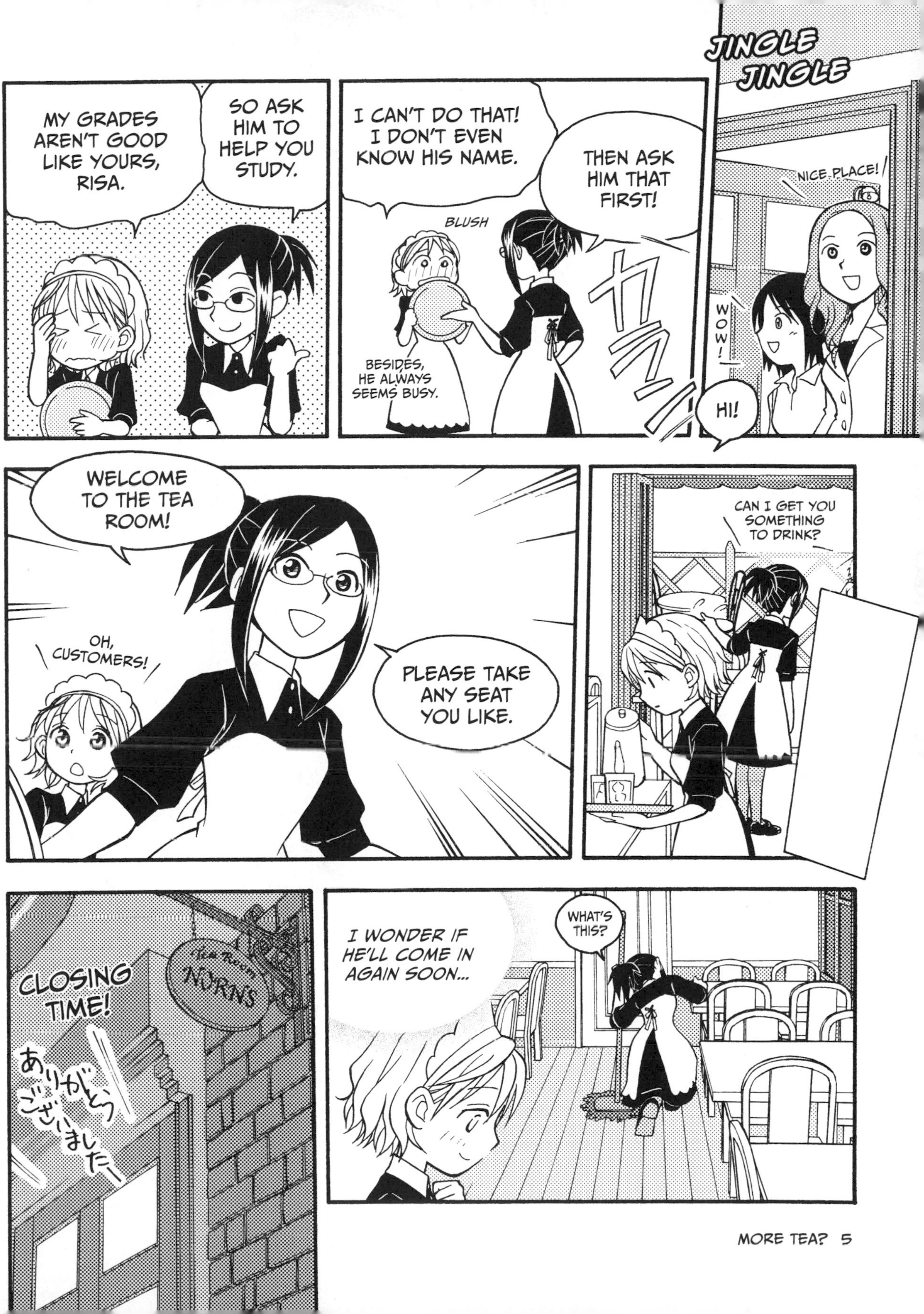
MY GRADES AREN'T GOOD LIKE YOURS, RISA.
SO ASK HIM TO HELP YOU STUDY.
I CAN'T DO THAT! I DON'T EVEN KNOW HIS NAME.
BLUSH
BESIDES, HE ALWAYS SEEMS BUSY.
THEN ASK HIM THAT FIRST!
JINGLE JINGLE
NICE PLACE!
WOW!
HI!
WELCOME TO THE TEA ROOM!
OH, CUSTOMERS!
PLEASE TAKE ANY SEAT YOU LIKE.
CAN I GET YOU SOMETHING TO DRINK?
Tea Room NORN'S
CLOSING TIME!
ありがとうございました〜
I WONDER IF HE'LL COME IN AGAIN SOON...
WHAT'S THIS?

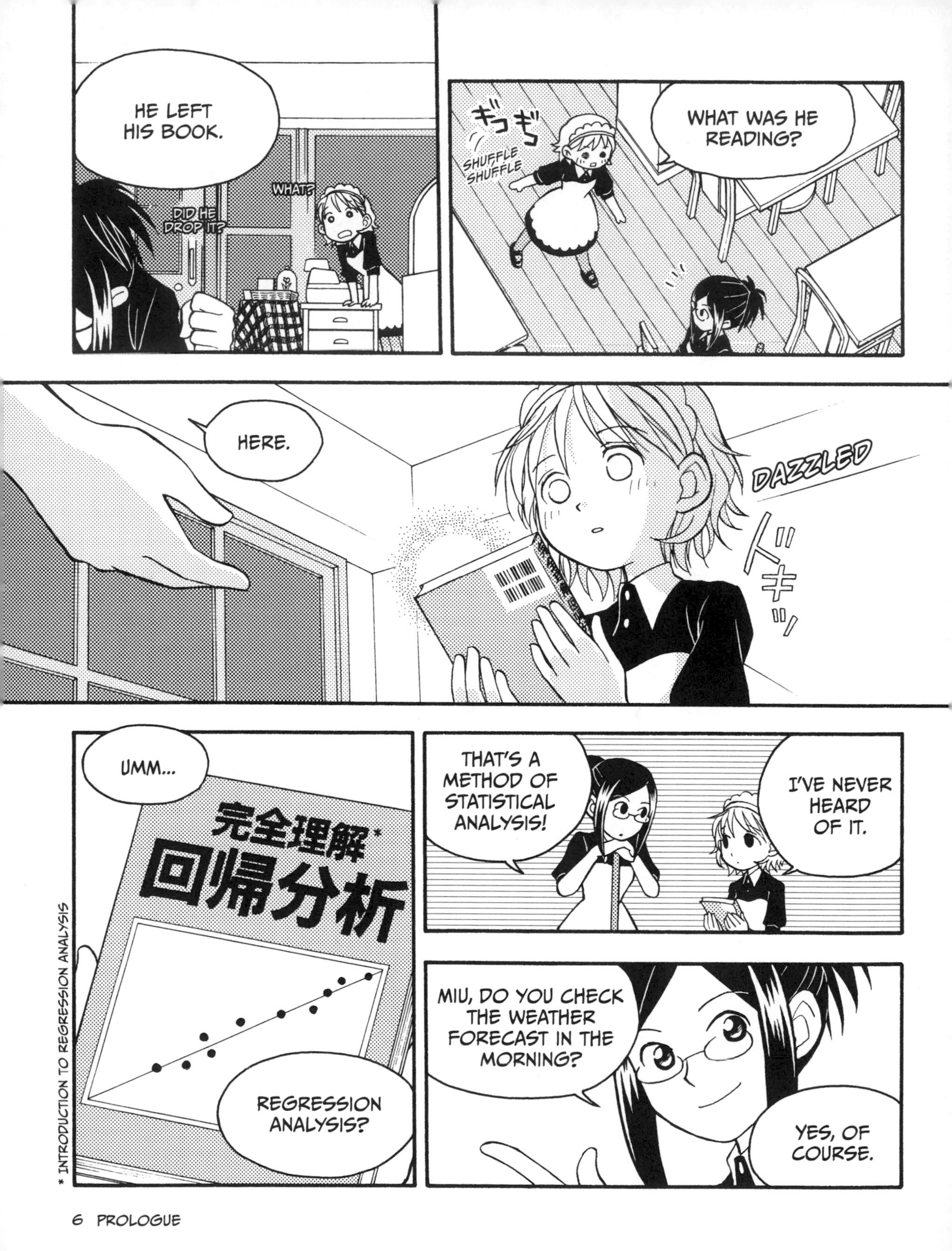
HE LEFT HIS BOOK.
DID HE DROP IT?
WHAT?
SHUFFLE SHUFFLE
WHAT WAS HE READING?
HERE.
DAZZLED
ドキッ
UMM...
完全理解*
回帰分析
REGRESSION ANALYSIS?
THAT'S A METHOD OF STATISTICAL ANALYSIS!
I'VE NEVER HEARD OF IT.
MIU, DO YOU CHECK THE WEATHER FORECAST IN THE MORNING?
YES, OF COURSE.

* INTRODUCTION TO REGRESSION ANALYSIS

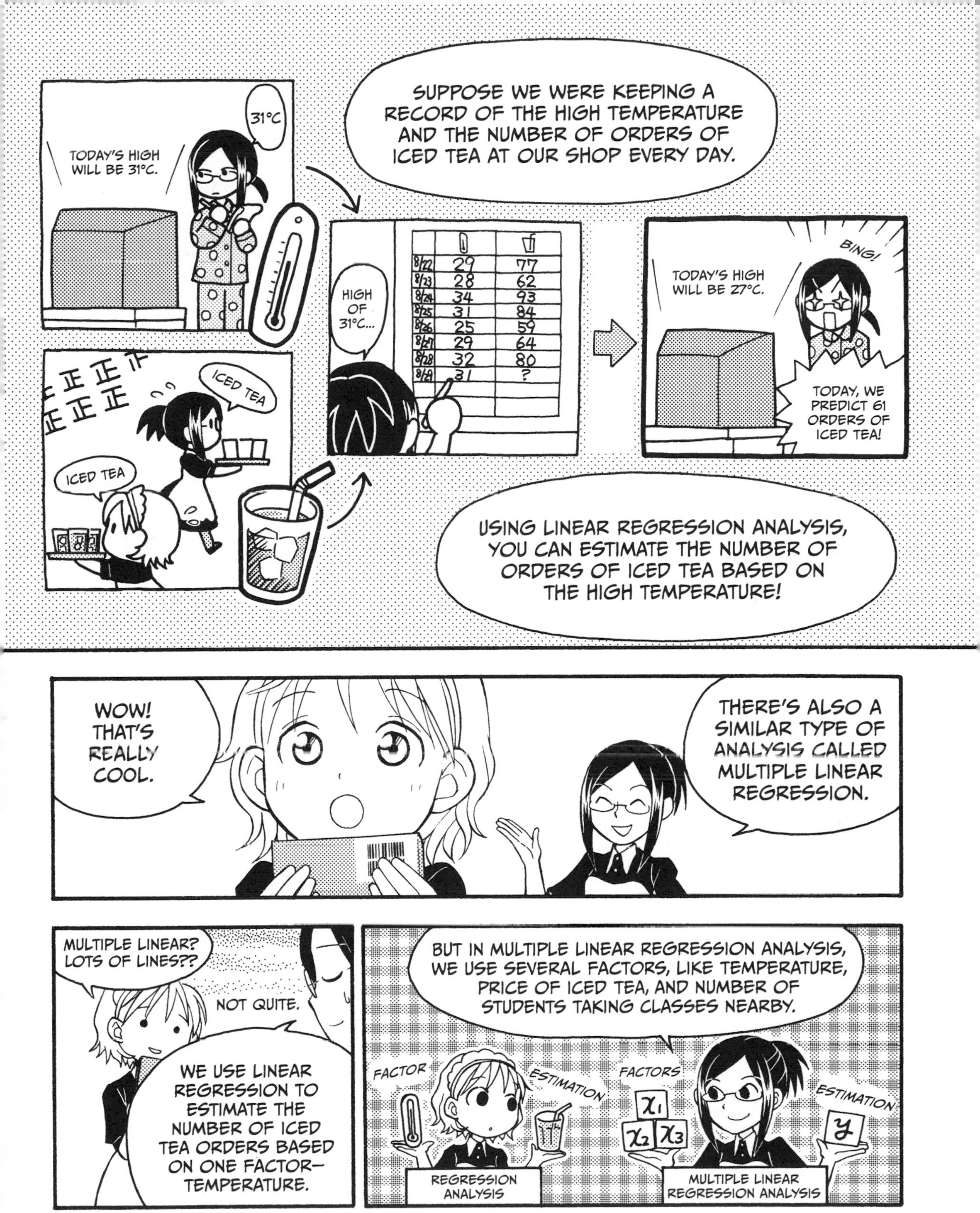
SUPPOSE WE WERE KEEPING A RECORD OF THE HIGH TEMPERATURE AND THE NUMBER OF ORDERS OF ICED TEA AT OUR SHOP EVERY DAY.
TODAY'S HIGH WILL BE 31°C.
31°C
ICED TEA
ICED TEA
HIGH OF 31°C...
8/22 29 77
8/23 28 62
8/24 34 93
8/25 31 84
8/26 25 59
8/27 29 64
8/28 32 80
8/29 31 ?
TODAY'S HIGH WILL BE 27°C.
BING!
TODAY, WE PREDICT 61 ORDERS OF ICED TEA!
USING LINEAR REGRESSION ANALYSIS, YOU CAN ESTIMATE THE NUMBER OF ORDERS OF ICED TEA BASED ON THE HIGH TEMPERATURE!
WOW! THAT'S REALLY COOL.
THERE'S ALSO A SIMILAR TYPE OF ANALYSIS CALLED MULTIPLE LINEAR REGRESSION.
MULTIPLE LINEAR? LOTS OF LINES??
NOT QUITE.
WE USE LINEAR REGRESSION TO ESTIMATE THE NUMBER OF ICED TEA ORDERS BASED ON ONE FACTOR—TEMPERATURE.
BUT IN MULTIPLE LINEAR REGRESSION ANALYSIS, WE USE SEVERAL FACTORS, LIKE TEMPERATURE, PRICE OF ICED TEA, AND NUMBER OF STUDENTS TAKING CLASSES NEARBY.
FACTOR
ESTIMATION
REGRESSION ANALYSIS
FACTORS
x1
x2 x3
ESTIMATION
y
MULTIPLE LINEAR REGRESSION ANALYSIS

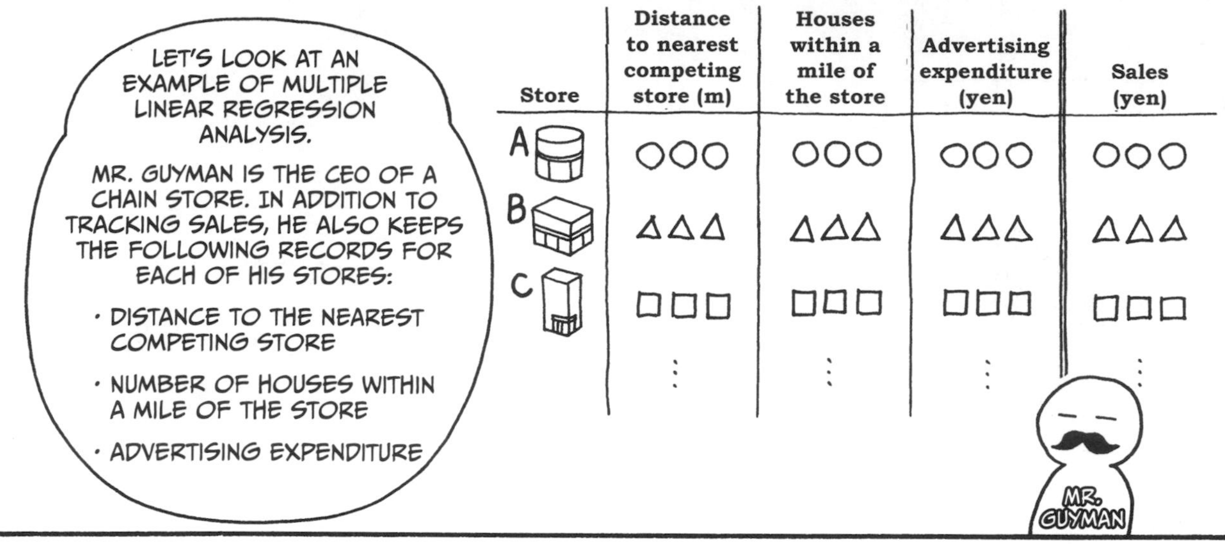
LET'S LOOK AT AN EXAMPLE OF MULTIPLE LINEAR REGRESSION ANALYSIS.
MR. GUYMAN IS THE CEO OF A CHAIN STORE. IN ADDITION TO TRACKING SALES, HE ALSO KEEPS THE FOLLOWING RECORDS FOR EACH OF HIS STORES:
• DISTANCE TO THE NEAREST COMPETING STORE
• NUMBER OF HOUSES WITHIN A MILE OF THE STORE
• ADVERTISING EXPENDITURE
MR. GUYMAN

Store	Distance to nearest competing store (m)	Houses within a mile of the store	Advertising expenditure (yen)	Sales (yen)
A	○○○	○○○	○○○	○○○
B	△△△	△△△	△△△	△△△
C	□□□	□□□	□□□	□□□
	⋮	⋮	⋮	⋮

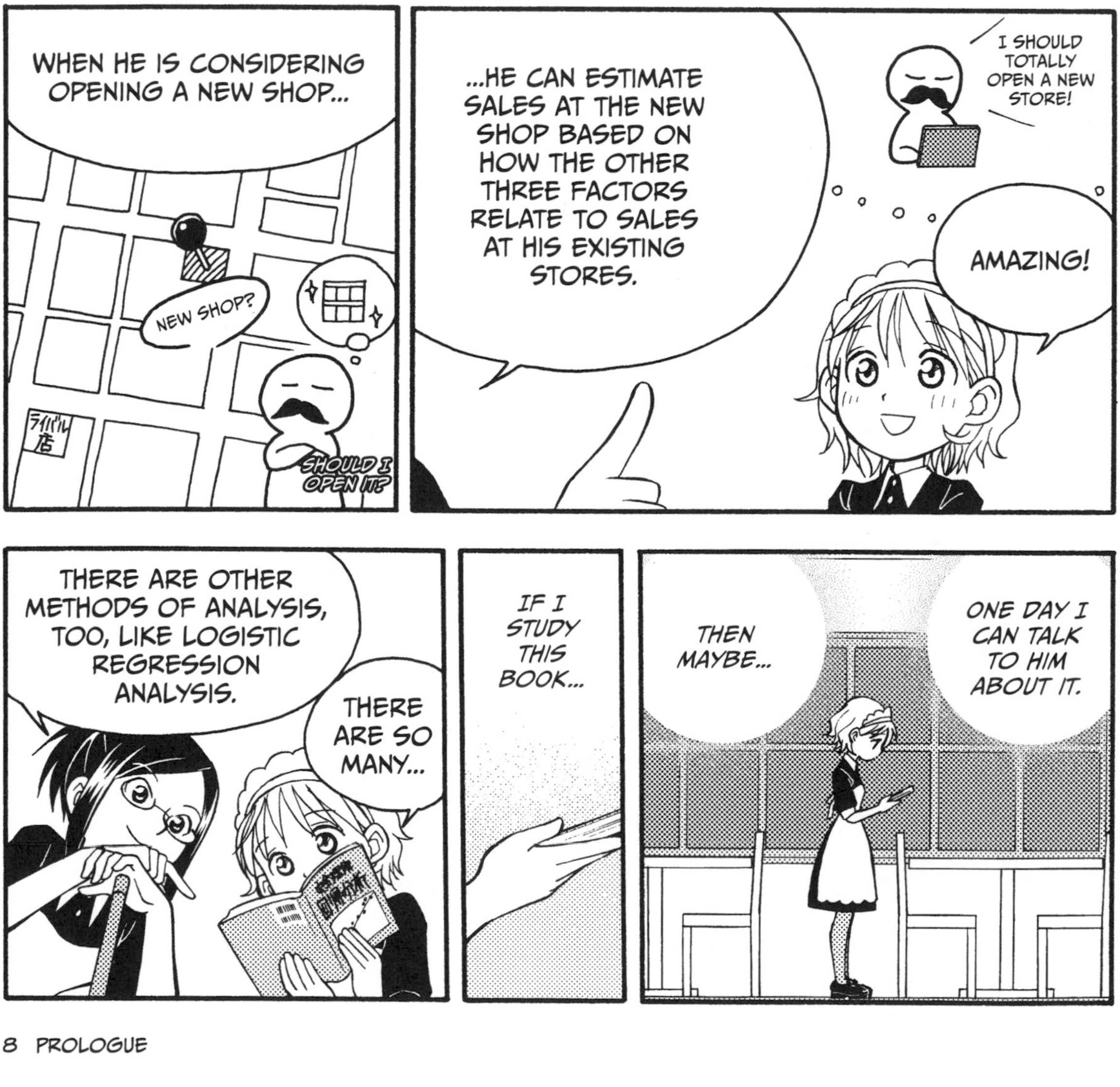
WHEN HE IS CONSIDERING OPENING A NEW SHOP...
NEW SHOP?
ライバル店
SHOULD I OPEN IT?
...HE CAN ESTIMATE SALES AT THE NEW SHOP BASED ON HOW THE OTHER THREE FACTORS RELATE TO SALES AT HIS EXISTING STORES.
I SHOULD TOTALLY OPEN A NEW STORE!
AMAZING!
THERE ARE OTHER METHODS OF ANALYSIS, TOO, LIKE LOGISTIC REGRESSION ANALYSIS.
THERE ARE SO MANY...
IF I STUDY THIS BOOK...
THEN MAYBE...
ONE DAY I CAN TALK TO HIM ABOUT IT.

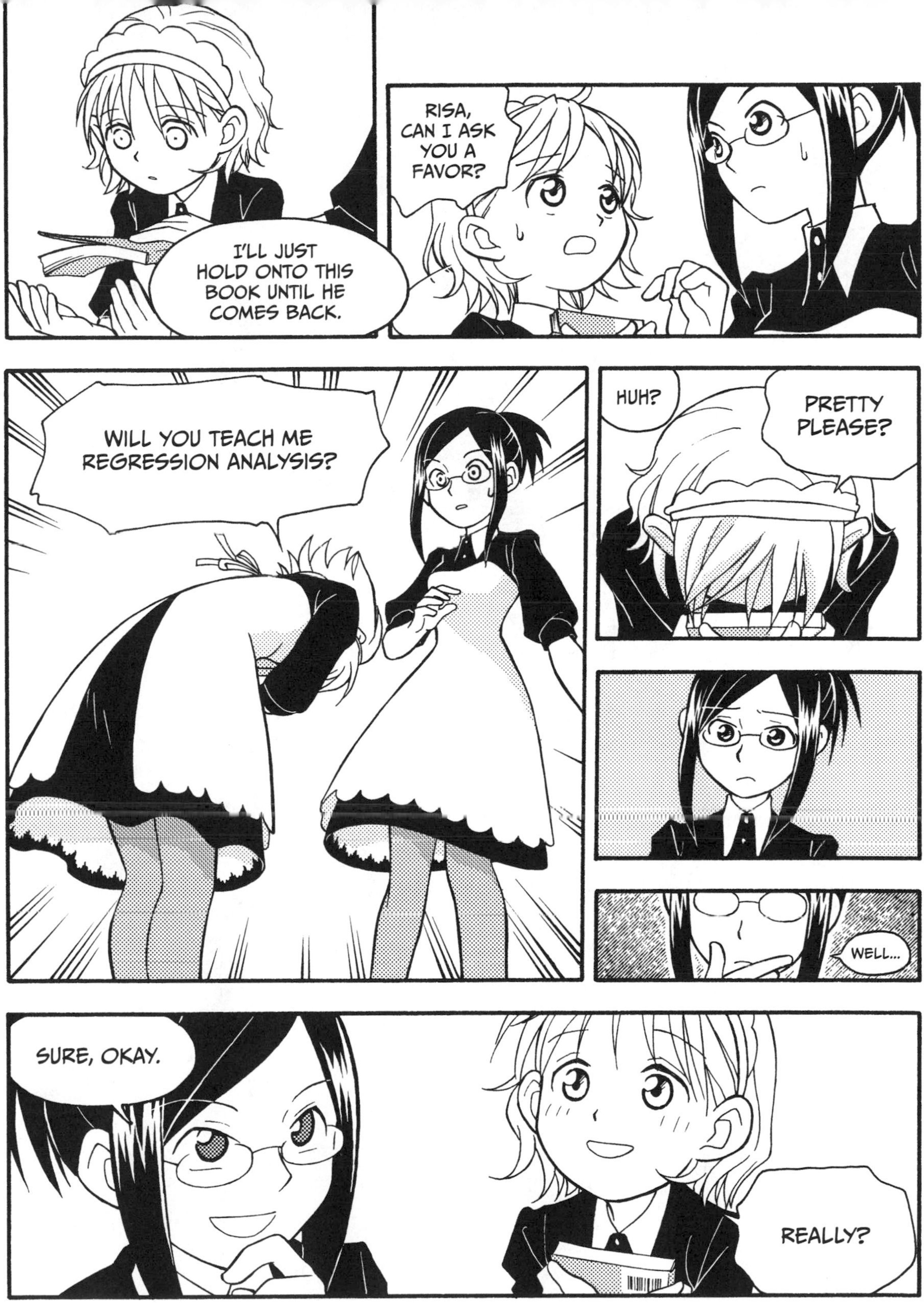
I'LL JUST HOLD ONTO THIS BOOK UNTIL HE COMES BACK.
RISA, CAN I ASK YOU A FAVOR?
WILL YOU TEACH ME REGRESSION ANALYSIS?
HUH?
PRETTY PLEASE?
WELL...
SURE, OKAY.
REALLY?

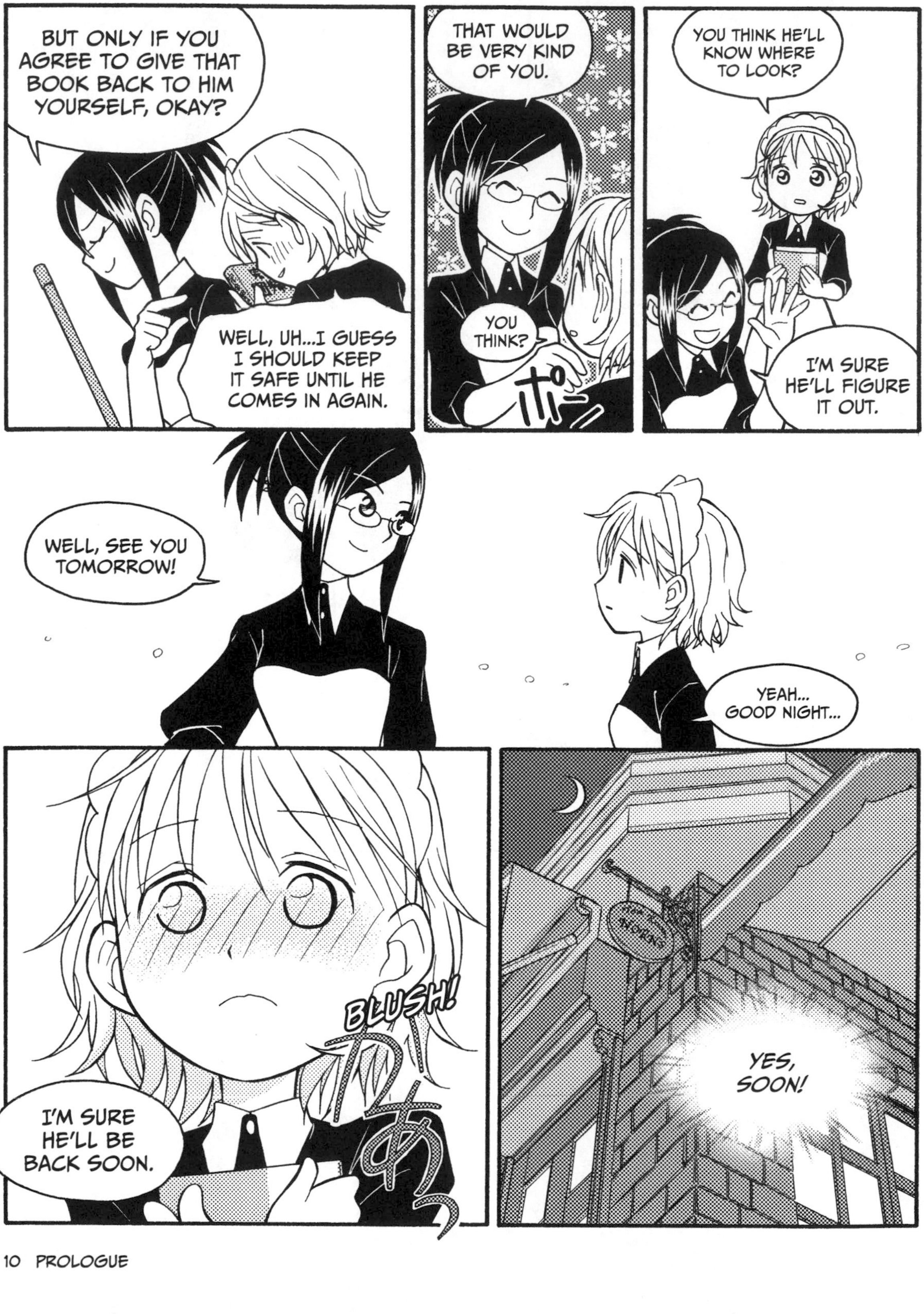
BUT ONLY IF YOU AGREE TO GIVE THAT BOOK BACK TO HIM YOURSELF, OKAY?
WELL, UH...I GUESS I SHOULD KEEP IT SAFE UNTIL HE COMES IN AGAIN.
THAT WOULD BE VERY KIND OF YOU.
YOU THINK?
YOU THINK HE'LL KNOW WHERE TO LOOK?
I'M SURE HE'LL FIGURE IT OUT.
WELL, SEE YOU TOMORROW!
YEAH... GOOD NIGHT...
BLUSH!
I'M SURE HE'LL BE BACK SOON.
YES, SOON!

1
A REFRESHING GLASS OF MATH

BUILDING A FOUNDATION

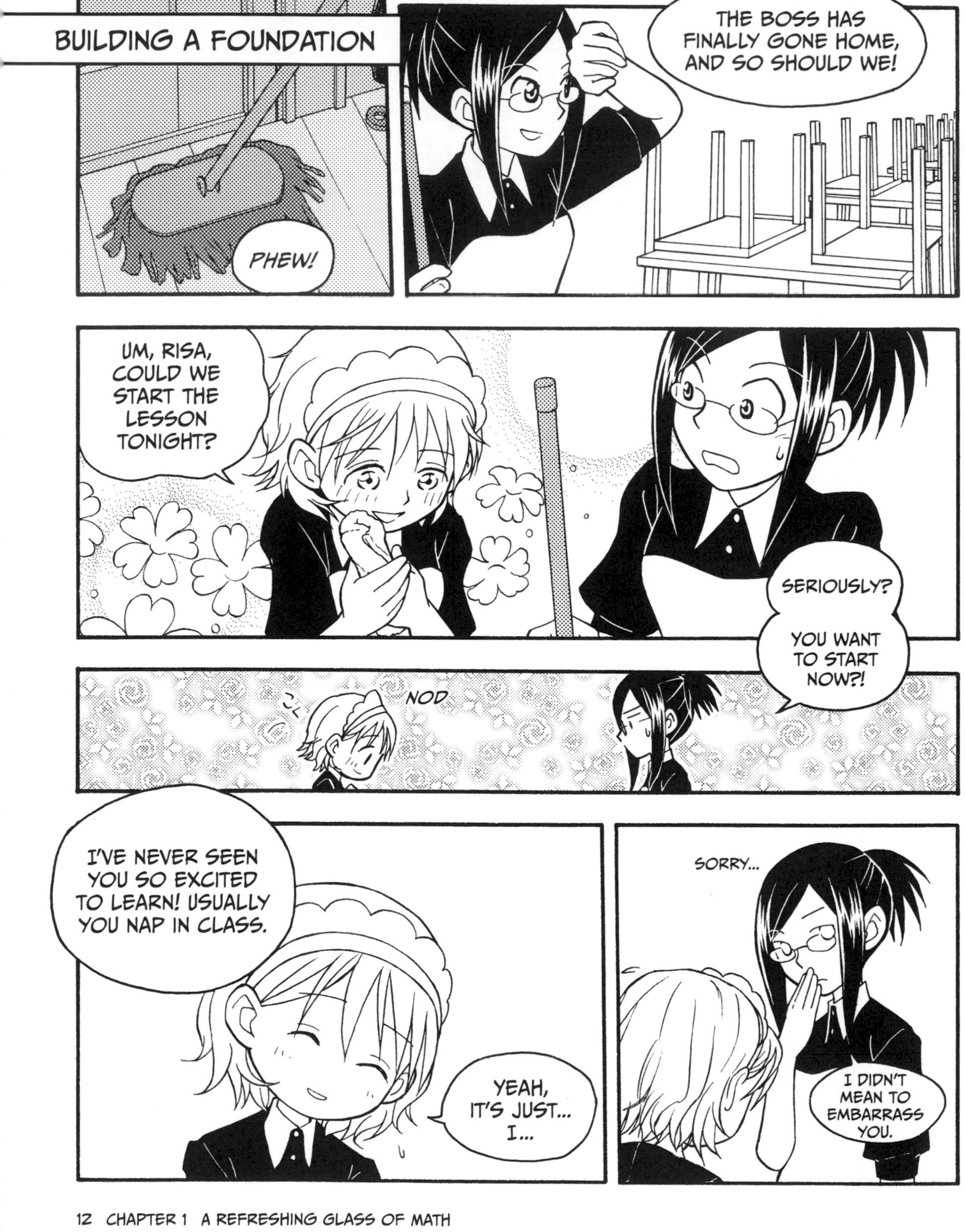
PHEW!
THE BOSS HAS FINALLY GONE HOME, AND SO SHOULD WE!
UM, RISA, COULD WE START THE LESSON TONIGHT?
SERIOUSLY?
YOU WANT TO START NOW?!
NOD
I'VE NEVER SEEN YOU SO EXCITED TO LEARN! USUALLY YOU NAP IN CLASS.
YEAH, IT'S JUST... I...
SORRY...
I DIDN'T MEAN TO EMBARRASS YOU.

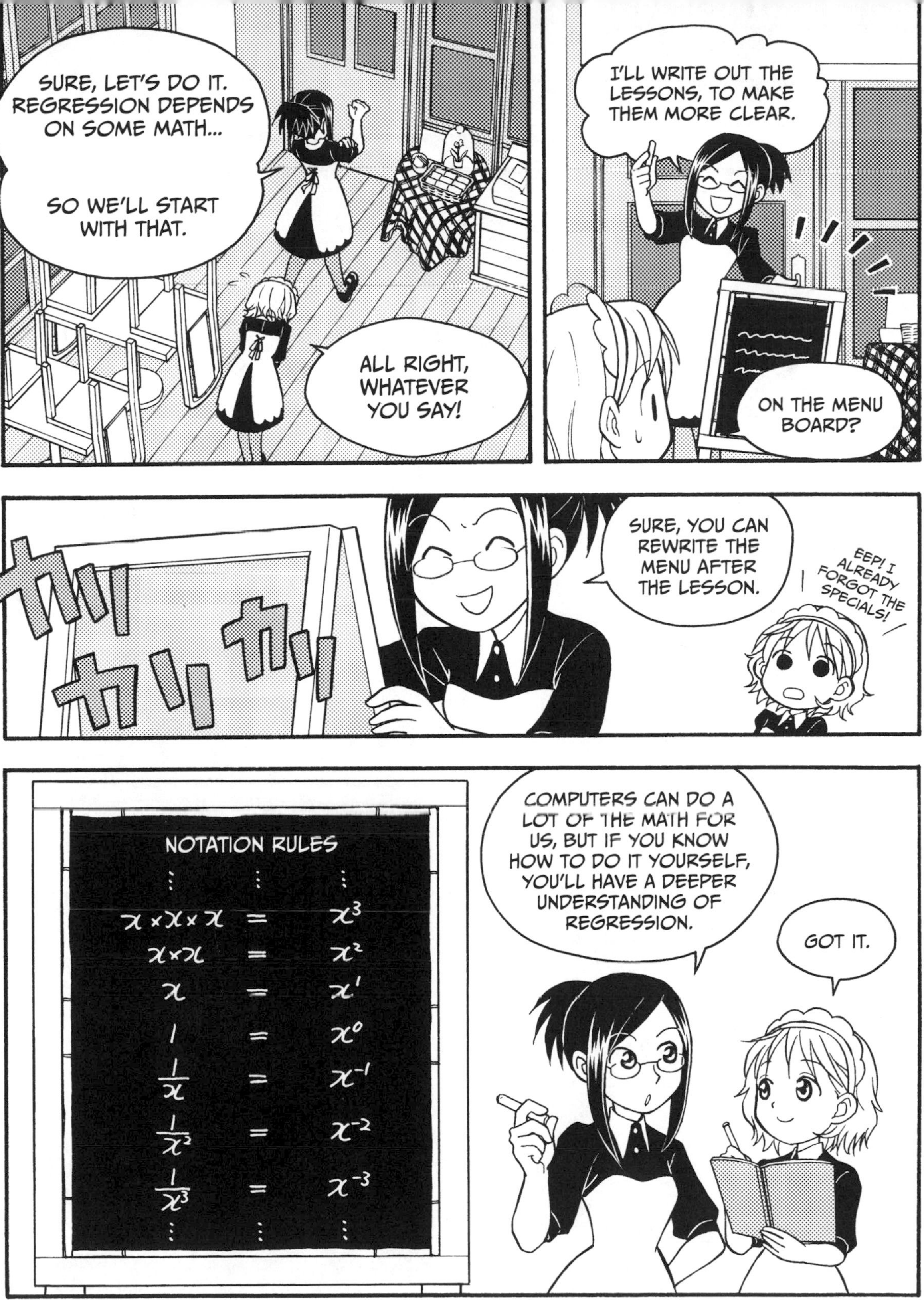

SURE, LET'S DO IT. REGRESSION DEPENDS ON SOME MATH...
SO WE'LL START WITH THAT.
ALL RIGHT, WHATEVER YOU SAY!
I'LL WRITE OUT THE LESSONS, TO MAKE THEM MORE CLEAR.
ON THE MENU BOARD?
カリ カリ カリ
SURE, YOU CAN REWRITE THE MENU AFTER THE LESSON.
EEP! I ALREADY FORGOT THE SPECIALS!
NOTATION RULES
$x \times x \times x = x^3$
$x \times x = x^2$
$x = x^1$
$1 = x^0$
$\frac{1}{x} = x^{-1}$
$\frac{1}{x^2} = x^{-2}$
$\frac{1}{x^3} = x^{-3}$
COMPUTERS CAN DO A LOT OF THE MATH FOR US, BUT IF YOU KNOW HOW TO DO IT YOURSELF, YOU'LL HAVE A DEEPER UNDERSTANDING OF REGRESSION.
GOT IT.

INVERSE FUNCTIONS

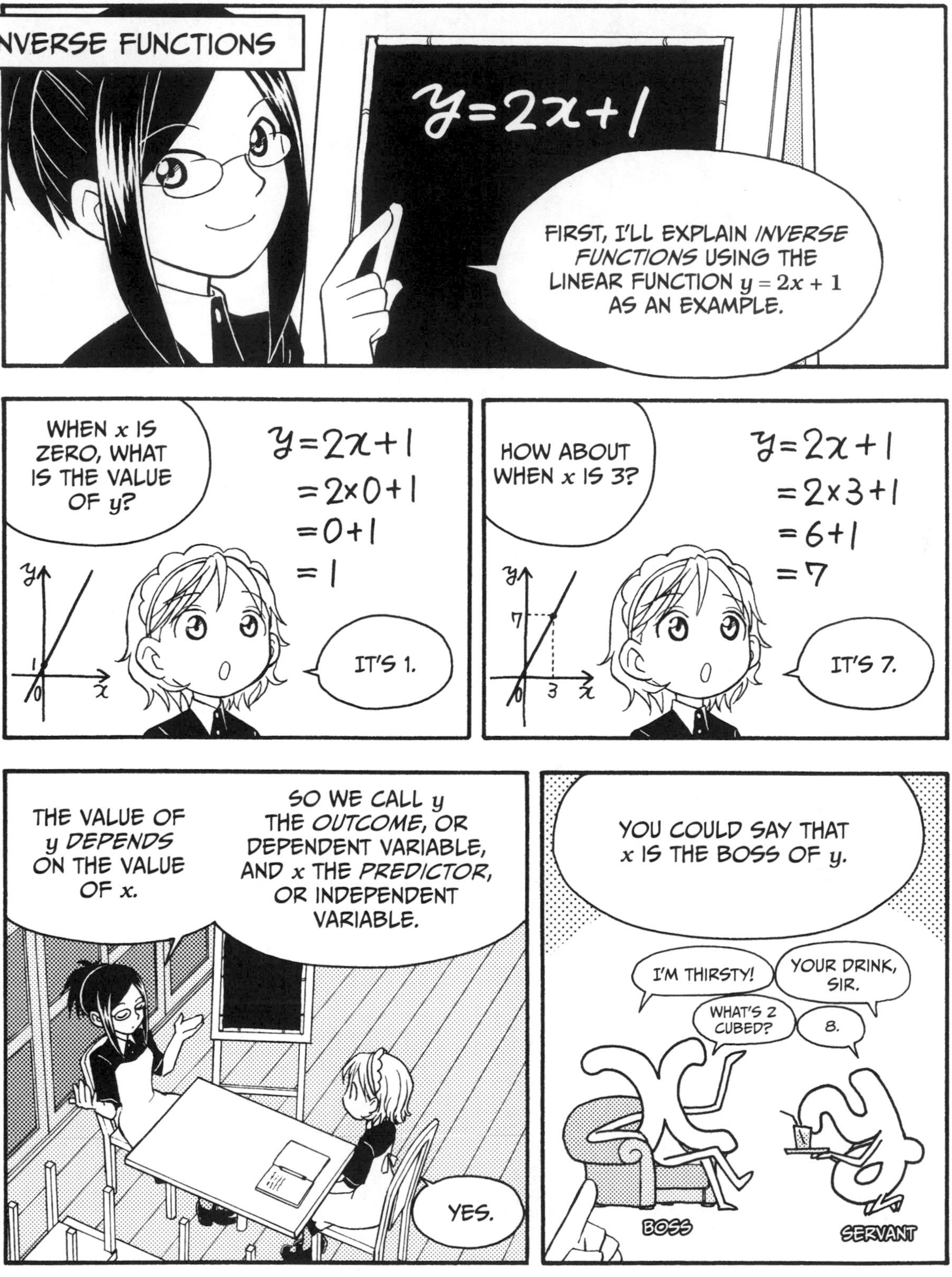
y=2x+1
FIRST, I'LL EXPLAIN *INVERSE FUNCTIONS* USING THE LINEAR FUNCTION $y = 2x + 1$ AS AN EXAMPLE.
WHEN x IS ZERO, WHAT IS THE VALUE OF y?
$y=2x+1$
$=2\times0+1$
$=0+1$
$=1$
y
1
0
x
IT'S 1.
HOW ABOUT WHEN x IS 3?
$y=2x+1$
$=2\times3+1$
$=6+1$
$=7$
y
7
0
3
x
IT'S 7.
THE VALUE OF y *DEPENDS* ON THE VALUE OF x.
SO WE CALL y THE *OUTCOME*, OR DEPENDENT VARIABLE, AND x THE *PREDICTOR*, OR INDEPENDENT VARIABLE.
YES.
YOU COULD SAY THAT x IS THE BOSS OF y.
I'M THIRSTY!
YOUR DRINK, SIR.
WHAT'S 2 CUBED?
8.
BOSS
SERVANT

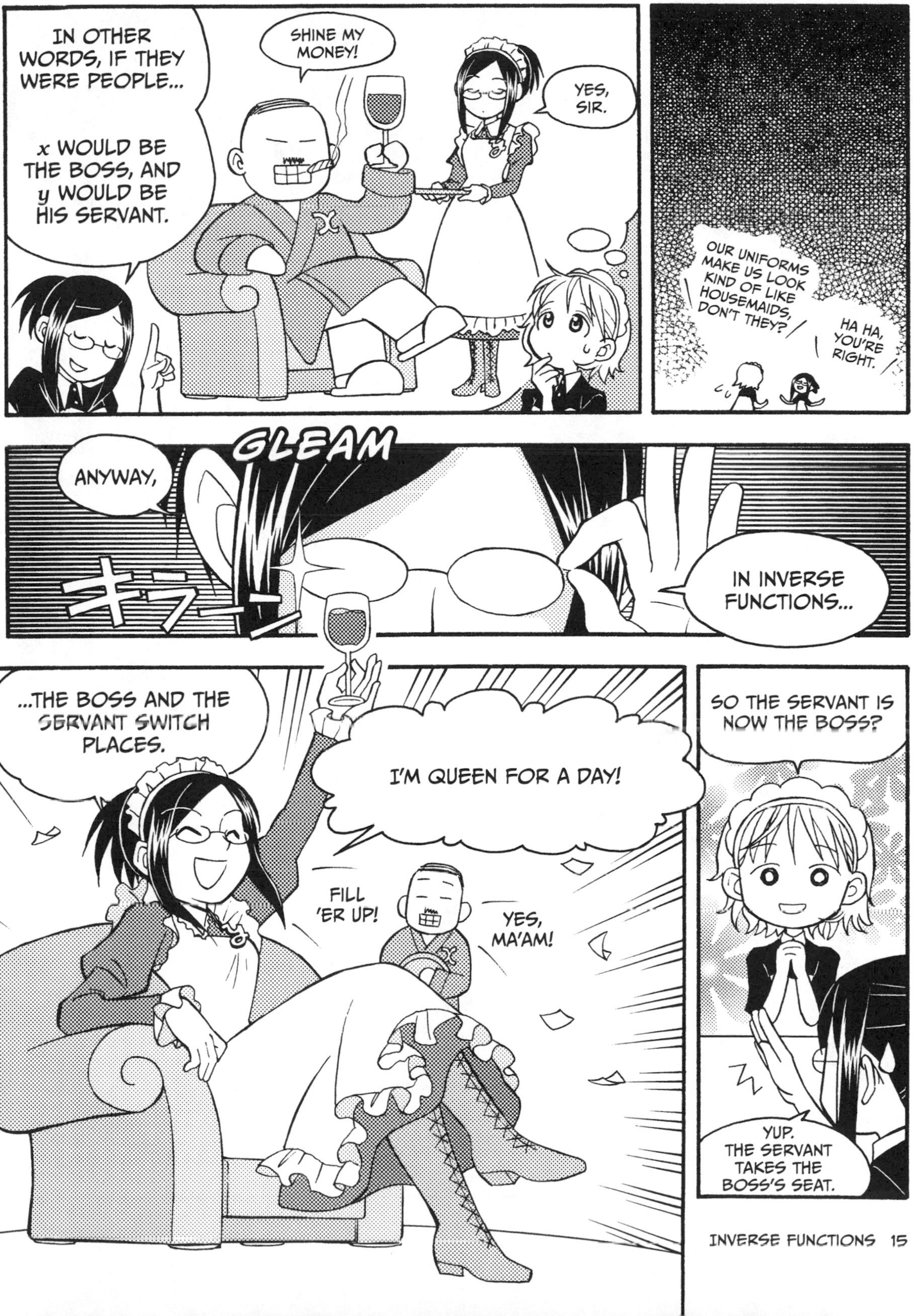
IN OTHER WORDS, IF THEY WERE PEOPLE...
x WOULD BE THE BOSS, AND y WOULD BE HIS SERVANT.
SHINE MY MONEY!
YES, SIR.
OUR UNIFORMS MAKE US LOOK KIND OF LIKE HOUSEMAIDS, DON'T THEY?
HA HA, YOU'RE RIGHT.
GLEAM
ANYWAY,
キラー
IN INVERSE FUNCTIONS...
...THE BOSS AND THE SERVANT SWITCH PLACES.
I'M QUEEN FOR A DAY!
FILL 'ER UP!
YES, MA'AM!
SO THE SERVANT IS NOW THE BOSS?
YUP. THE SERVANT TAKES THE BOSS'S SEAT.

SO FOR THE EXAMPLE $y = 2x + 1$, THE INVERSE FUNCTION IS...

SCRITCH コン コン SCRATCH

$$y = 2x + 1$$

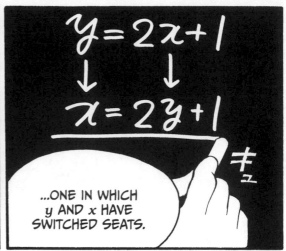

$$y = 2x + 1$$
$$\downarrow \qquad \downarrow$$
$$x = 2y + 1$$

...ONE IN WHICH y AND x HAVE SWITCHED SEATS.

キュ

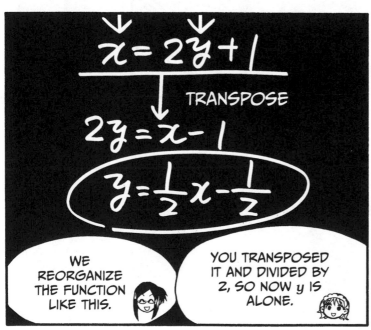

HOWEVER,

WE WANT y ALL BY ITSELF, SO...

$$x = 2y + 1$$
$$\downarrow \quad \text{TRANSPOSE}$$
$$2y = x - 1$$
$$y = \frac{1}{2}x - \frac{1}{2}$$

WE REORGANIZE THE FUNCTION LIKE THIS.

YOU TRANSPOSED IT AND DIVIDED BY 2, SO NOW y IS ALONE.

THAT'S RIGHT. TO EXPLAIN WHY THIS IS USEFUL, LET'S DRAW A GRAPH.

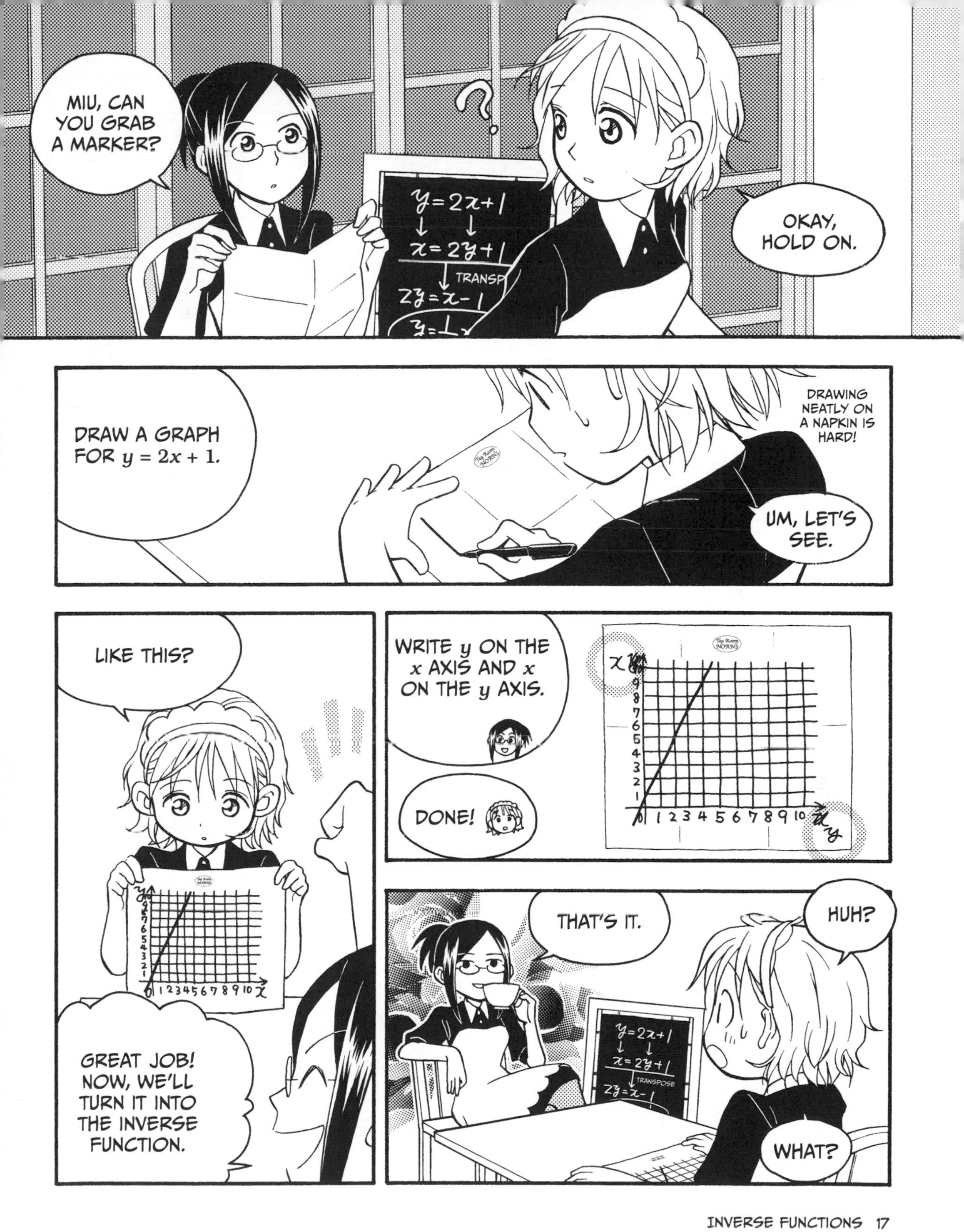
MIU, CAN YOU GRAB A MARKER?
$y=2x+1$
$x=2y+1$
TRANSP
$2y=x-1$
OKAY, HOLD ON.
DRAW A GRAPH FOR $y = 2x + 1$.
DRAWING NEATLY ON A NAPKIN IS HARD!
UM, LET'S SEE.
LIKE THIS?
GREAT JOB! NOW, WE'LL TURN IT INTO THE INVERSE FUNCTION.
WRITE y ON THE x AXIS AND x ON THE y AXIS.
DONE!
THAT'S IT.
HUH?
TRANSPOSE
WHAT?

...
IT'S THE SAME THING! TURN IT AROUND SO THAT x IS ON THE BOTTOM, LIKE IT USUALLY IS.
Tea Room NORNS
YOU MEAN...
...TURN IT LIKE THIS?
YES, BUT NOW ZERO IS ON THE RIGHT. IT SHOULD BE ON THE LEFT.
ERM...
SHOULD I FLIP THE NAPKIN OVER?
THERE WE GO!
NOW IT'S A GRAPH FOR THE FUNCTION $y = \frac{1}{2}x - \frac{1}{2}$.

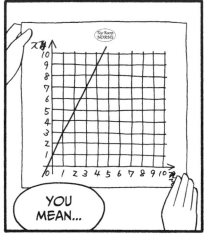

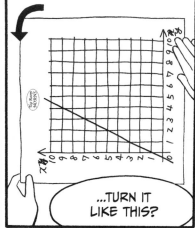

EXPONENTS AND LOGARITHMS

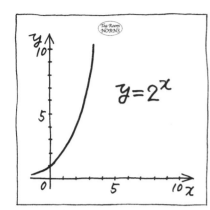

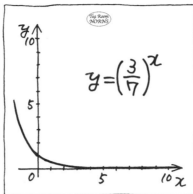

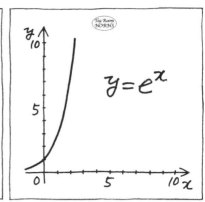

Tea Room NORNS
$y = 2^x$
Tea Room NORNS
$y = \left(\frac{3}{7}\right)^x$
Tea Room NORNS
$y = e^x$
OKAY...
ON TO THE NEXT LESSON. THESE ARE CALLED *EXPONENTIAL FUNCTIONS.*
THEY ALL CROSS THE POINT (0,1) BECAUSE ANY NUMBER TO THE ZERO POWER IS 1.
RIGHT! NOW, HAVE YOU SEEN THIS e BEFORE?
$y = e^x$
THIS e IS THE BASE OF THE NATURAL LOGARITHM AND HAS A VALUE OF 2.7182.
IT'S CALLED *EULER'S NUMBER.*
I'VE HEARD OF IT.
$\log_e y = x$ IS THE INVERSE OF THE EXPONENTIAL EQUATION $y = e^x$.
AH! MORE INVERSE FUNCTIONS!
FLIP!

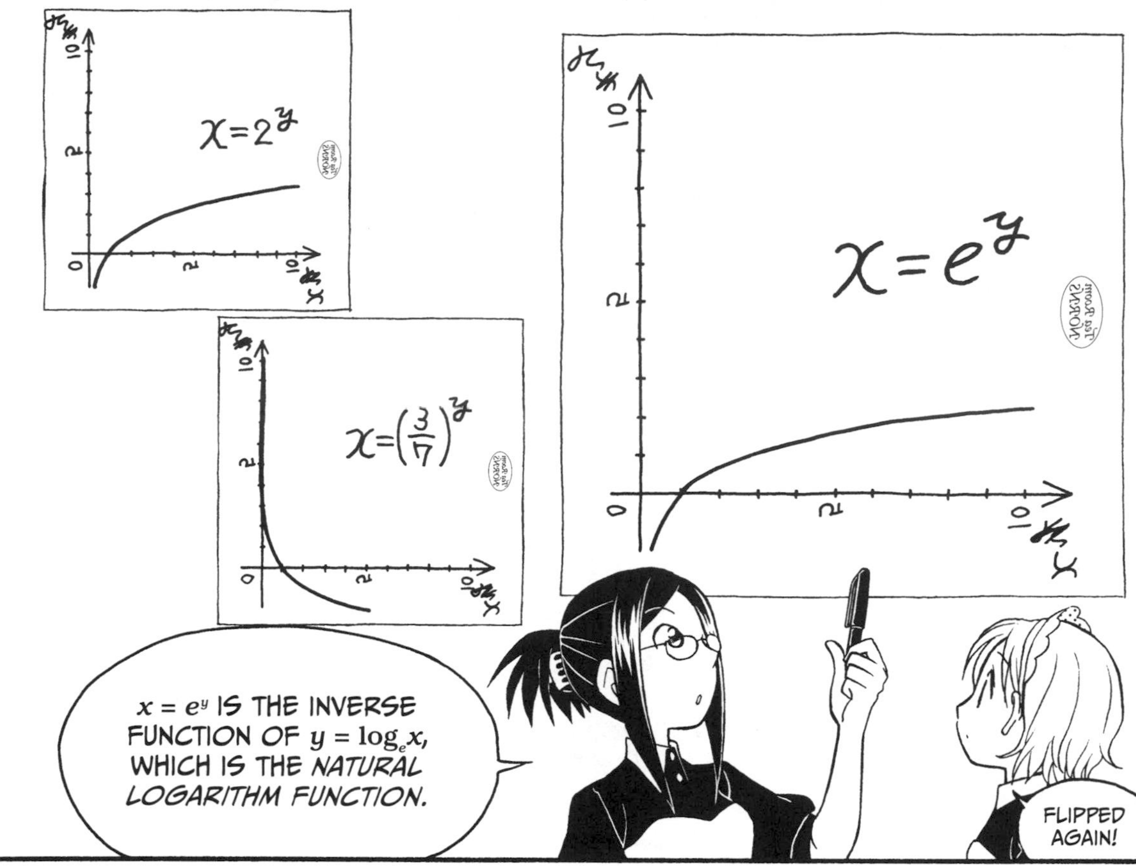

x = 2^y
x = (3/7)^y
x = e^y
x = e^y IS THE INVERSE FUNCTION OF y = log_e x, WHICH IS THE NATURAL LOGARITHM FUNCTION.
FLIPPED AGAIN!

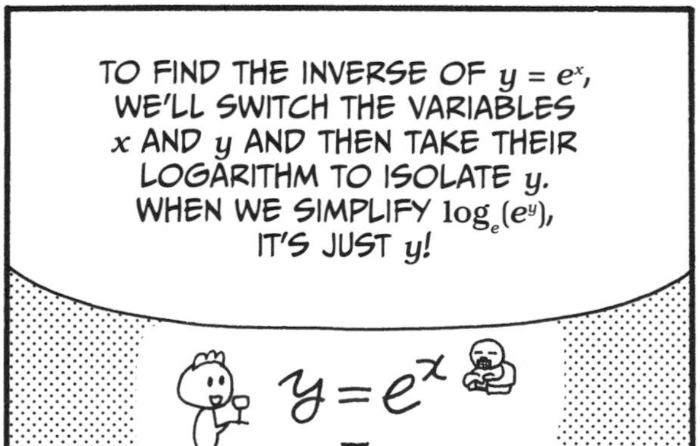

TO FIND THE INVERSE OF y = e^x, WE'LL SWITCH THE VARIABLES x AND y AND THEN TAKE THEIR LOGARITHM TO ISOLATE y. WHEN WE SIMPLIFY log_e(e^y), IT'S JUST y!
y = e^x
SWITCH THE VARIBLES!
x = e^y
y = log_e x
WE FLIPPED THE EQUATION TO PUT y BACK ON THE LEFT.

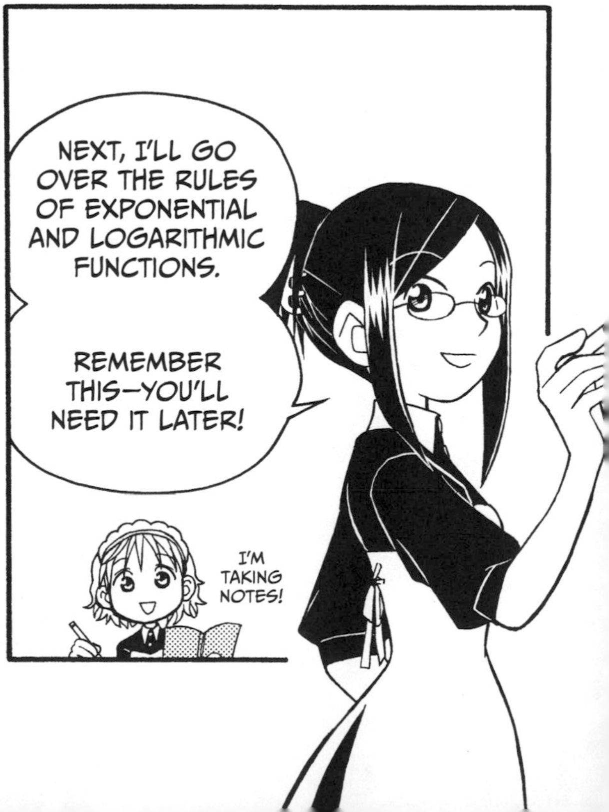

NEXT, I'LL GO OVER THE RULES OF EXPONENTIAL AND LOGARITHMIC FUNCTIONS.
REMEMBER THIS—YOU'LL NEED IT LATER!
I'M TAKING NOTES!

RULES FOR EXPONENTS AND LOGARITHMS

1. POWER RULE $(e^a)^b$ AND $e^{a \times b}$ ARE EQUAL.

Let's try this. We'll confirm that $(e^a)^b$ and $e^{a \times b}$ are equal when $a = 2$ and $b = 3$.

$$\left(e^2\right)^3 = \underbrace{e^2 \times e^2 \times e^2}_{3} = \underbrace{(e \times e) \times (e \times e) \times (e \times e)}_{3} = \underbrace{e \times e \times e \times e \times e \times e}_{6} = e^{2 \times 3}$$

This also means $(e^a)^b = e^{a \times b} = (e^b)^a$.

2. QUOTIENT RULE $\dfrac{e^a}{e^b}$ AND e^{a-b} ARE EQUAL.

Now let's try this, too. We'll confirm that $\dfrac{e^a}{e^b}$ and e^{a-b} are equal when $a = 3$ and $b = 5$.

$$\frac{e^3}{e^5} = \frac{e \times e \times e}{e \times e \times e \times e \times e} = \frac{\cancel{e} \times \cancel{e} \times \cancel{e}}{e \times e \times \cancel{e} \times \cancel{e} \times \cancel{e}} = \frac{1}{e^2} = e^{-2} = e^{3-5}$$

3. CANCELING EXPONENTIALS RULE a AND $\log_e(e^a)$ ARE EQUAL.

As mentioned page 20, $y = \log_e x$ and $x = e^y$ are equivalent. First we need to look at what a logarithm is. An exponential function of base b to a power, n, equals a value, x. The logarithm function inverts this process. That means the logarithm base b of a value, x, equals a power, n.

We see that in $\log_e(e^a) = n$, the base b is e and the value x is e^a, so $e^n = e^a$ and $n = a$.

So $b^n = x$ also means $\log_b x = n$.

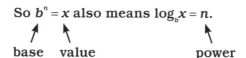

base   value   power

4. EXPONENTIATION RULE $\log_e(a^b)$ AND $b \times \log_e(a)$ ARE EQUAL.

Let's confirm that $\log_e(a^b)$ and $b \times \log_e(a)$ are equal. We'll start by using $b \times \log_e(a)$ and e in the Power Rule:

$$e^{b \times \log_e(a)} = \left(e^{\log_e(a)}\right)^b$$

And since e is the inverse of $\log_e$, we can reduce $e^{b \times \log_e(a)}$ on the right side to just a:

$$e^{b \times \log_e(a)} = a^b$$

Now we'll use the rule that $b^n = x$ also means $\log_b x = n$, where:

$$b = e$$
$$x = a^b$$
$$n = b \times \log_e(a)$$

This means that $e^{b \times \log_e(a)} = a^b$, so we can conclude that $\log_e(a^b)$ is equal to $b \times \log_e(a)$.

5. PRODUCT RULE $\log_e(a) + \log_e(b)$ AND $\log_e(a \times b)$ ARE EQUAL.

Let's confirm that $\log_e(a) + \log_e(b)$ and $\log_e(a \times b)$ are equal. Again, we'll use the rule that states that $b^n = x$ also means $\log_b x = n$.

Let's start by defining $e^m = a$ and $e^n = b$. We would then have $e^m e^n = e^{m+n} = a \times b$, thanks to the Product Rule of exponents. We can then take the log of both sides,

$$\log_e(e^{m+n}) = \log_e(a \times b),$$

which on the left side reduces simply to

$$m + n = \log_e(a \times b).$$

We also know that $m + n = \log_e a + \log_c b$, so clearly

$$\log_e(a) + \log_e(b) = \log_e(a \times b).$$

HERE I HAVE SUMMARIZED THE RULES
I'VE EXPLAINED SO FAR.

RULE 1	$(e^a)^b$ and e^{ab} are equal.
RULE 2	$\dfrac{e^a}{e^b}$ and e^{a-b} are equal.
RULE 3	a and $\log_e(e^a)$ are equal.
RULE 4	$\log_e(a^b)$ and $b \times \log_e(a)$ are equal.
RULE 5	$\log_e(a) + \log_e(a)$ and $\log_e(a \times b)$ are equal.

In fact, one could replace the natural number e in these equations with any positive real number d. Can you prove these rules again using d as the base?

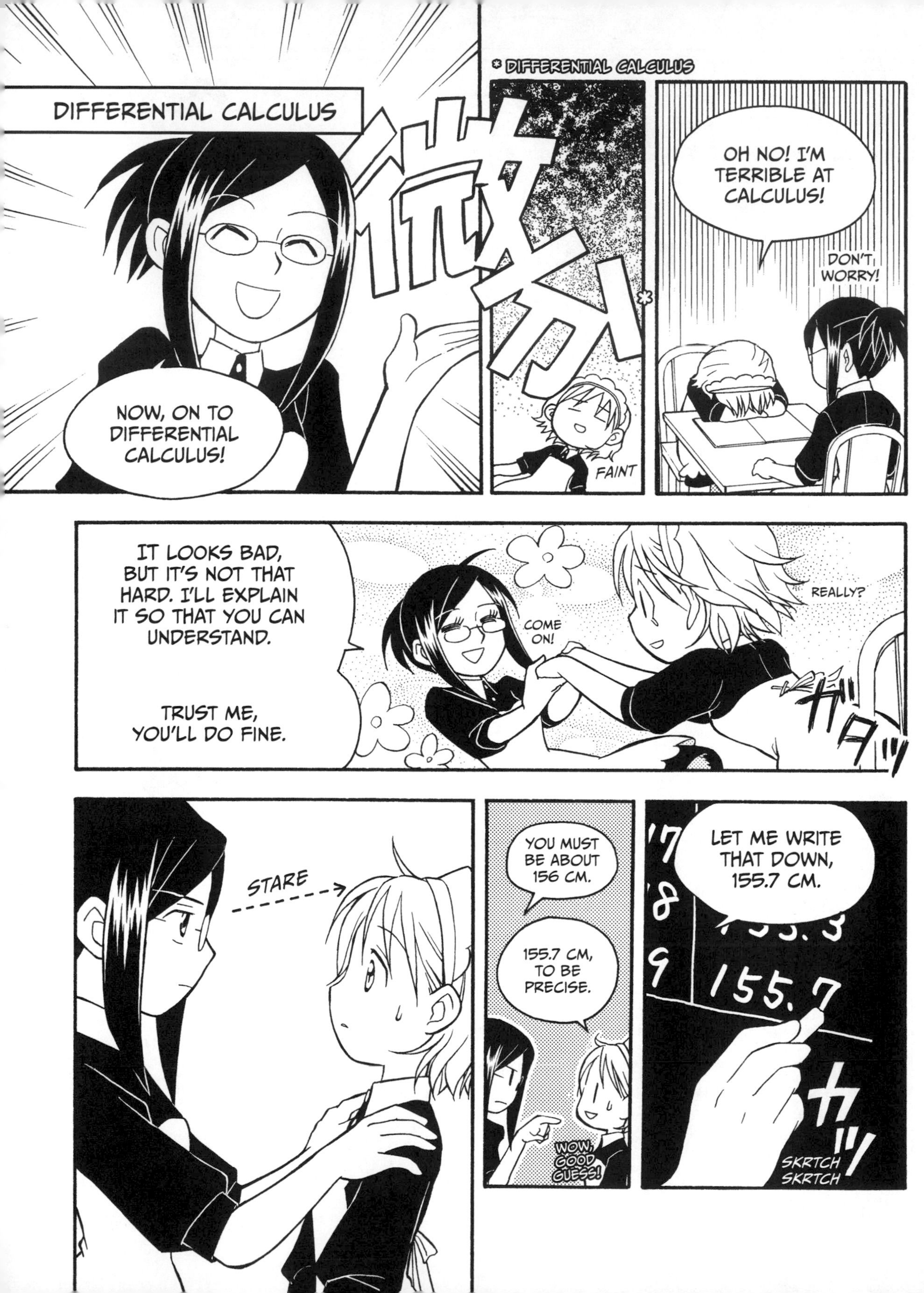
DIFFERENTIAL CALCULUS
微分*
NOW, ON TO DIFFERENTIAL CALCULUS!
* DIFFERENTIAL CALCULUS
FAINT
OH NO! I'M TERRIBLE AT CALCULUS!
DON'T WORRY!
IT LOOKS BAD, BUT IT'S NOT THAT HARD. I'LL EXPLAIN IT SO THAT YOU CAN UNDERSTAND.
TRUST ME, YOU'LL DO FINE.
COME ON!
REALLY?
ガタッ
STARE
YOU MUST BE ABOUT 156 CM.
155.7 CM, TO BE PRECISE.
WOW, GOOD GUESS!
LET ME WRITE THAT DOWN, 155.7 CM.
155.7
カッ
SKRTCH SKRTCH

MIU'S AGE AND HEIGHT

AGE	HEIGHT
4	100.1
5	107.2
6	114.1
7	121.7
8	126.8
9	130.9
10	137.5
11	143.2
12	149.4
13	151.1
14	154.0
15	154.6
16	155.0
17	155.1
18	155.3
19	155.7

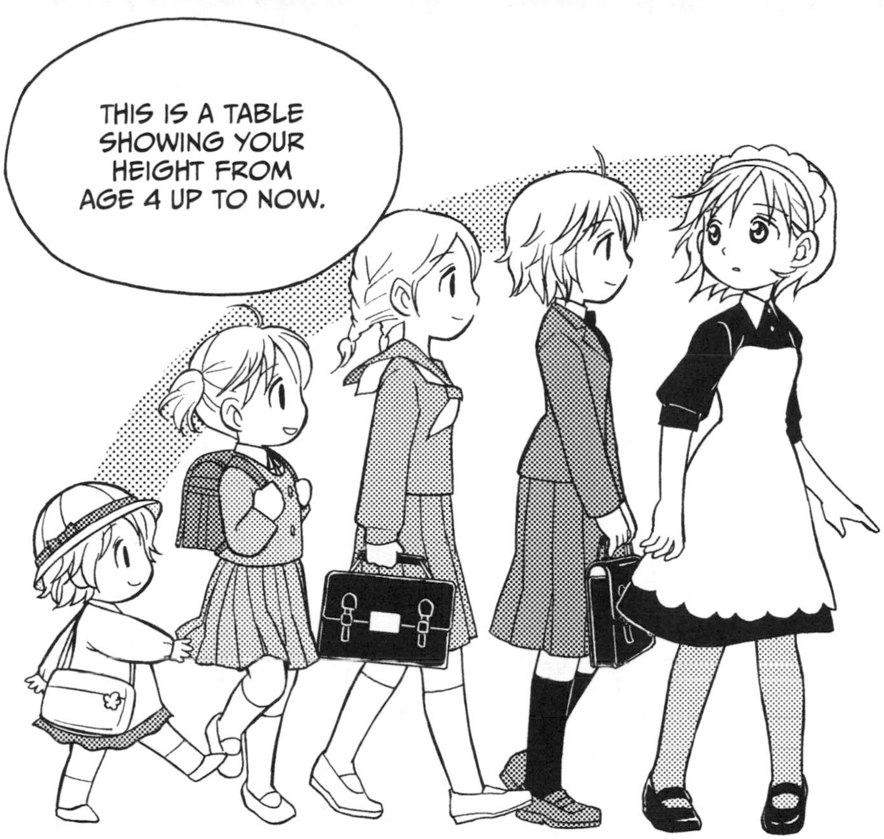
THIS IS A TABLE SHOWING YOUR HEIGHT FROM AGE 4 UP TO NOW.

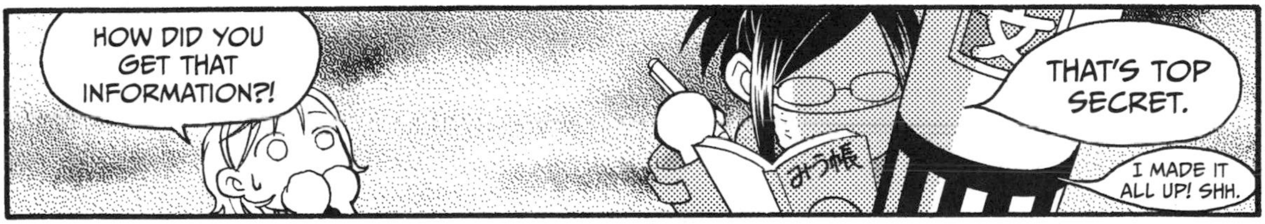
HOW DID YOU GET THAT INFORMATION?!
THAT'S TOP SECRET.
みう帳
I MADE IT ALL UP! SHH.

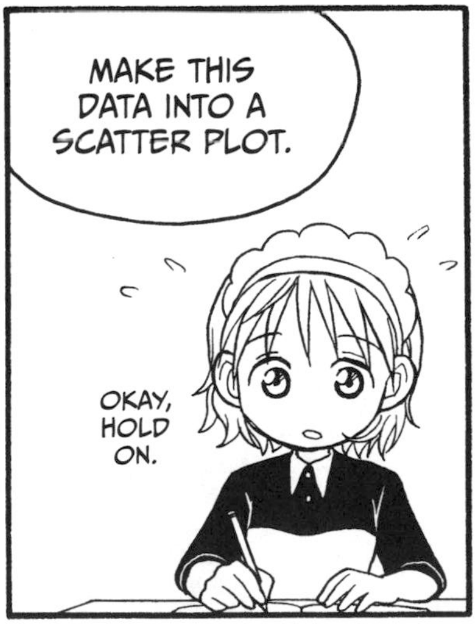
MAKE THIS DATA INTO A SCATTER PLOT.
OKAY, HOLD ON.

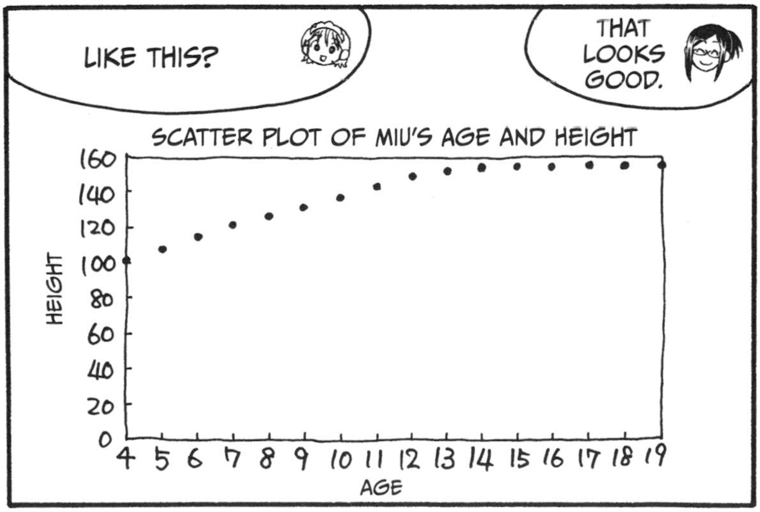
LIKE THIS?
THAT LOOKS GOOD.
SCATTER PLOT OF MIU'S AGE AND HEIGHT
HEIGHT
0
20
40
60
80
100
120
140
160
AGE
4 5 6 7 8 9 10 11 12 13 14 15 16 17 18 19

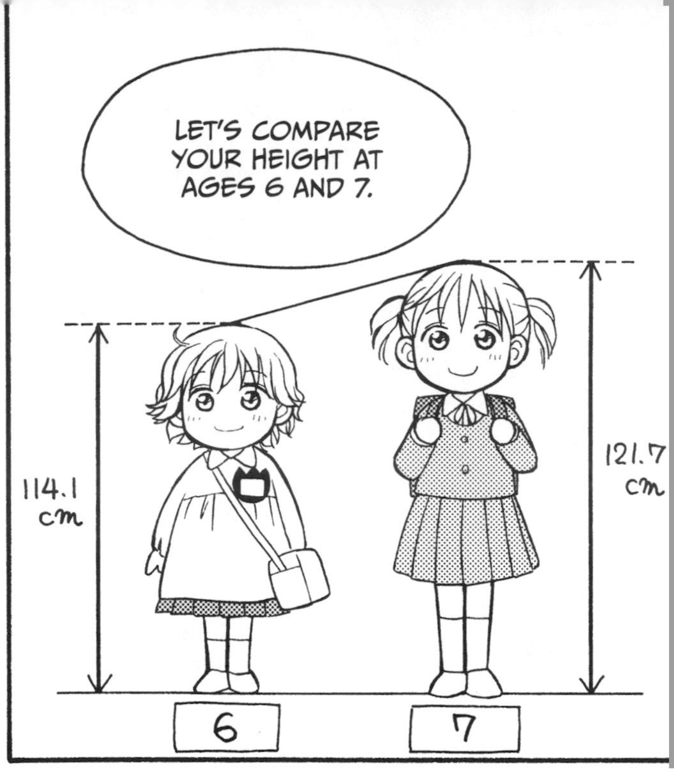
LET'S COMPARE YOUR HEIGHT AT AGES 6 AND 7.
114.1 cm
121.7 cm
6
7

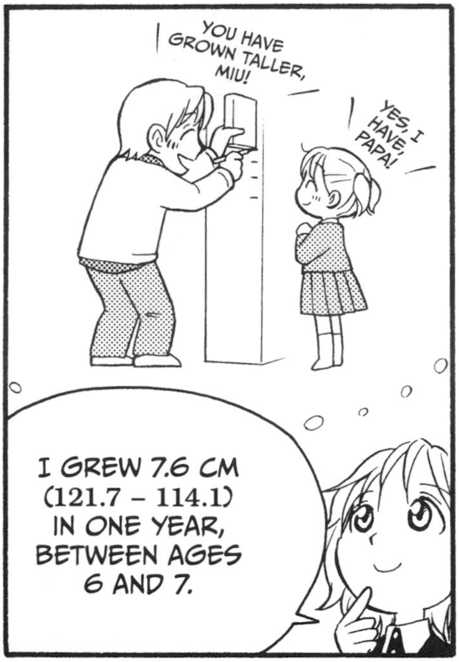
YOU HAVE GROWN TALLER, MIU!
YES, I HAVE, PAPA!
I GREW 7.6 CM (121.7 – 114.1) IN ONE YEAR, BETWEEN AGES 6 AND 7.

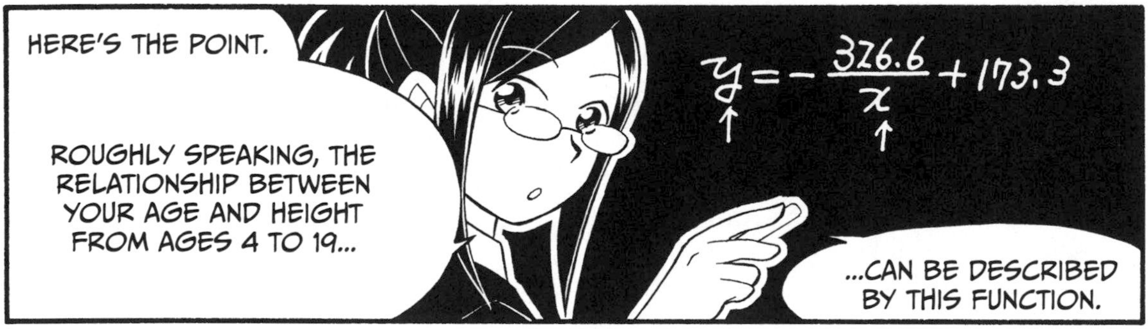
HERE'S THE POINT.
ROUGHLY SPEAKING, THE RELATIONSHIP BETWEEN YOUR AGE AND HEIGHT FROM AGES 4 TO 19...
$y = -\frac{326.6}{x} + 173.3$
...CAN BE DESCRIBED BY THIS FUNCTION.

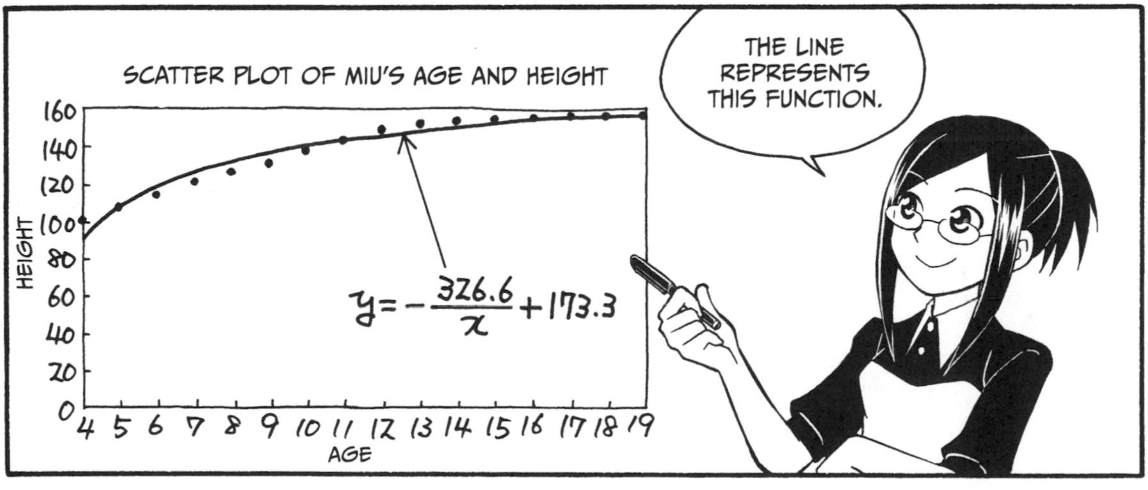
SCATTER PLOT OF MIU'S AGE AND HEIGHT
HEIGHT
0
20
40
60
80
100
120
140
160
AGE
4 5 6 7 8 9 10 11 12 13 14 15 16 17 18 19
$y = -\frac{326.6}{x} + 173.3$
THE LINE REPRESENTS THIS FUNCTION.

WHERE DID THAT
$$y = -\frac{326.6}{x} + 173.3$$
FUNCTION COME FROM?!

SHOCKED

HEE HEE

THAT IS A REGRESSION EQUATION! DON'T WORRY ABOUT HOW TO GET IT RIGHT NOW.

JUST ASSUME IT DESCRIBES THE RELATIONSHIP BETWEEN YOUR AGE AND YOUR HEIGHT.

OKAY, RISA.

FOR NOW, I'LL JUST BELIEVE THAT THE RELATIONSHIP IS
$$y = -\frac{326.6}{x} + 173.3.$$

GREAT.

NOW, CAN YOU SEE THAT "7 YEARS OLD" CAN BE DESCRIBED AS "(6 + 1) YEARS OLD"?

WELL YEAH, THAT MAKES SENSE.

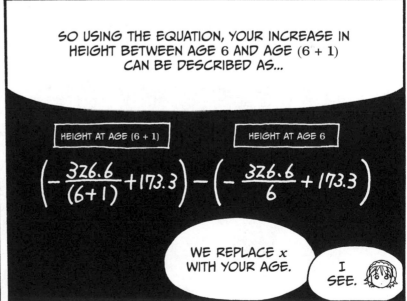

SO USING THE EQUATION, YOUR INCREASE IN HEIGHT BETWEEN AGE 6 AND AGE (6 + 1) CAN BE DESCRIBED AS...

HEIGHT AT AGE (6 + 1)

HEIGHT AT AGE 6

$$\left(-\frac{326.6}{(6+1)} + 173.3\right) - \left(-\frac{326.6}{6} + 173.3\right)$$

WE REPLACE x WITH YOUR AGE.

I SEE.

WE CAN SHOW THE RATE OF GROWTH AS CENTIMETERS PER YEAR, SINCE THERE IS ONE YEAR BETWEEN THE AGES WE USED.

$$\frac{\left(-\frac{326.6}{(6+1)}+173.3\right)-\left(-\frac{326.6}{6}+173.3\right)}{1} \text{ CM/YEAR}$$

OH! YOU DIVIDED THE PREVIOUS FORMULA BY 1 BECAUSE THE INTERVAL IS ONE YEAR.

NEXT, LET'S THINK ABOUT THE INCREASE IN HEIGHT IN HALF A YEAR.

6   6½   7

WHAT IS AGE SIX AND A HALF IN TERMS OF THE NUMBER 6?

LET ME SEE... (6 + 0.5) YEARS OLD?

CORRECT!

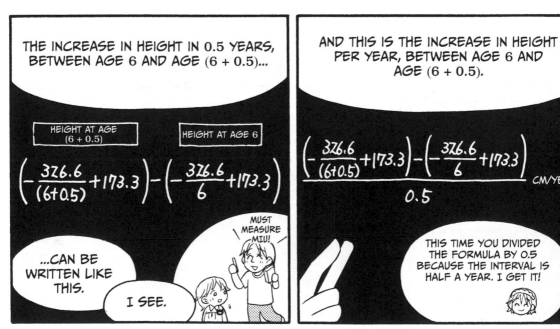

THE INCREASE IN HEIGHT IN 0.5 YEARS, BETWEEN AGE 6 AND AGE (6 + 0.5)...

HEIGHT AT AGE (6 + 0.5)   HEIGHT AT AGE 6

$$\left(-\frac{326.6}{(6+0.5)}+173.3\right)-\left(-\frac{326.6}{6}+173.3\right)$$

...CAN BE WRITTEN LIKE THIS.

I SEE.

MUST MEASURE MIU!

AND THIS IS THE INCREASE IN HEIGHT PER YEAR, BETWEEN AGE 6 AND AGE (6 + 0.5).

$$\frac{\left(-\frac{326.6}{(6+0.5)}+173.3\right)-\left(-\frac{326.6}{6}+173.3\right)}{0.5} \text{ CM/YEAR}$$

THIS TIME YOU DIVIDED THE FORMULA BY 0.5 BECAUSE THE INTERVAL IS HALF A YEAR. I GET IT!

FINALLY...

LET'S THINK ABOUT THE HEIGHT INCREASE OVER AN *EXTREMELY* SHORT PERIOD OF TIME.

MUST MEASURE, MUST MEASURE, KEEP MEASURING!

OH, MIU, YOU ARE GROWING SO FAST!

P-PAPA?

DELTA

Δ

IN MATHEMATICS, WE USE THIS SYMBOL Δ (DELTA) TO REPRESENT CHANGE.

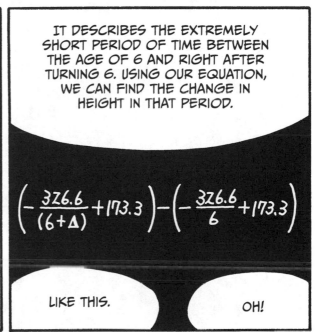

IT DESCRIBES THE EXTREMELY SHORT PERIOD OF TIME BETWEEN THE AGE OF 6 AND RIGHT AFTER TURNING 6. USING OUR EQUATION, WE CAN FIND THE CHANGE IN HEIGHT IN THAT PERIOD.

$$\left(-\frac{326.6}{(6+\Delta)}+173.3\right)-\left(-\frac{326.6}{6}+173.3\right)$$

LIKE THIS.

OH!

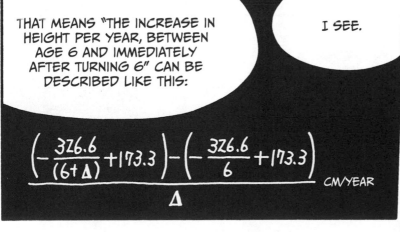

THAT MEANS "THE INCREASE IN HEIGHT PER YEAR, BETWEEN AGE 6 AND IMMEDIATELY AFTER TURNING 6" CAN BE DESCRIBED LIKE THIS:

I SEE.

$$\frac{\left(-\frac{326.6}{(6+\Delta)}+173.3\right)-\left(-\frac{326.6}{6}+173.3\right)}{\Delta}$$ CM/YEAR

FOLLOW ME AS I REARRANGE THIS EQUATION IN A SNAP!

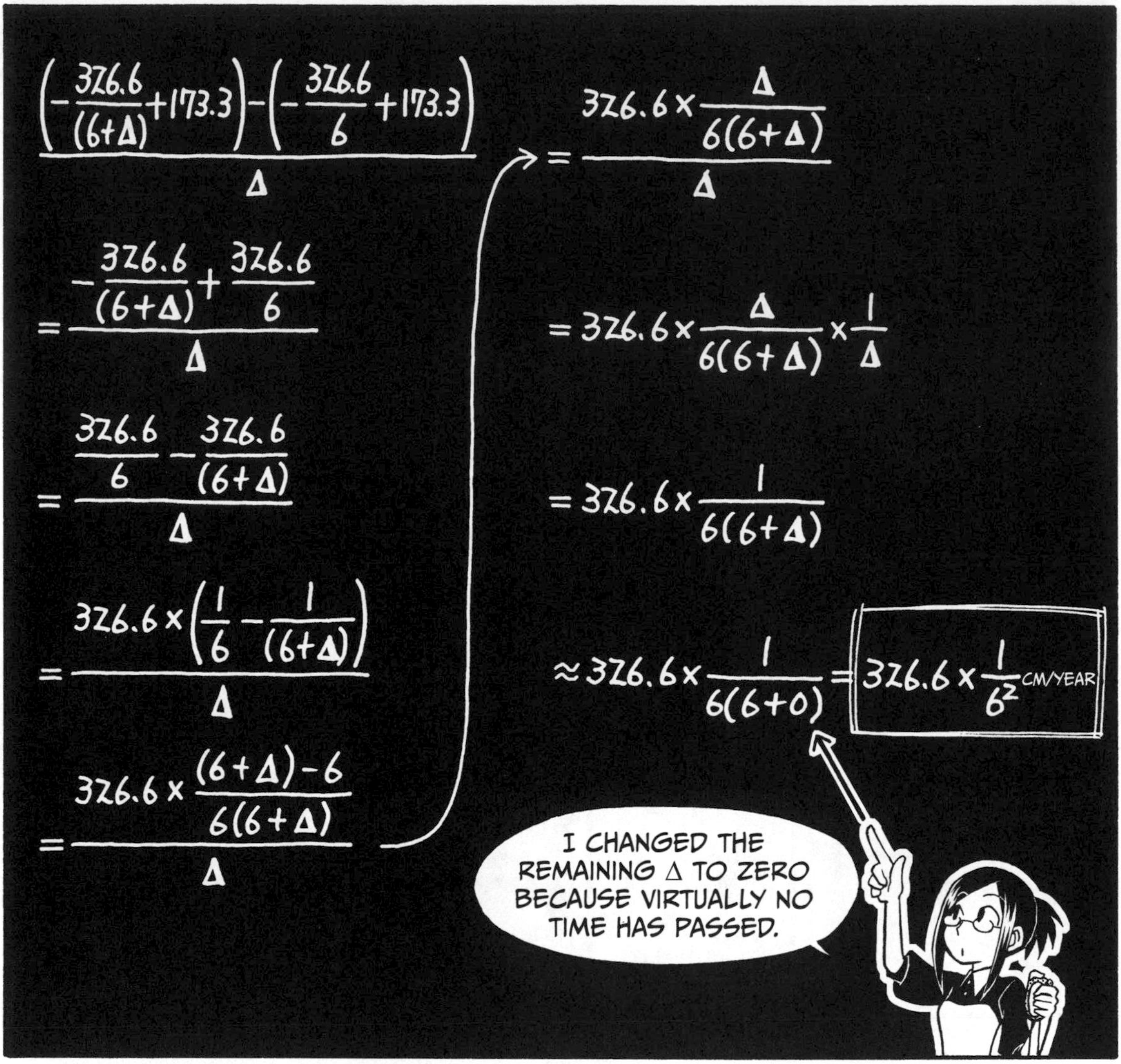
$\dfrac{\left(-\dfrac{326.6}{(6+\Delta)}+173.3\right)-\left(-\dfrac{326.6}{6}+173.3\right)}{\Delta}$
$=\dfrac{-\dfrac{326.6}{(6+\Delta)}+\dfrac{326.6}{6}}{\Delta}$
$=\dfrac{\dfrac{326.6}{6}-\dfrac{326.6}{(6+\Delta)}}{\Delta}$
$=\dfrac{326.6\times\left(\dfrac{1}{6}-\dfrac{1}{(6+\Delta)}\right)}{\Delta}$
$=\dfrac{326.6\times\dfrac{(6+\Delta)-6}{6(6+\Delta)}}{\Delta}$
$=\dfrac{326.6\times\dfrac{\Delta}{6(6+\Delta)}}{\Delta}$
$=326.6\times\dfrac{\Delta}{6(6+\Delta)}\times\dfrac{1}{\Delta}$
$=326.6\times\dfrac{1}{6(6+\Delta)}$
$\approx 326.6\times\dfrac{1}{6(6+0)}=326.6\times\dfrac{1}{6^2}$ CM/YEAR
I CHANGED THE REMAINING Δ TO ZERO BECAUSE VIRTUALLY NO TIME HAS PASSED.

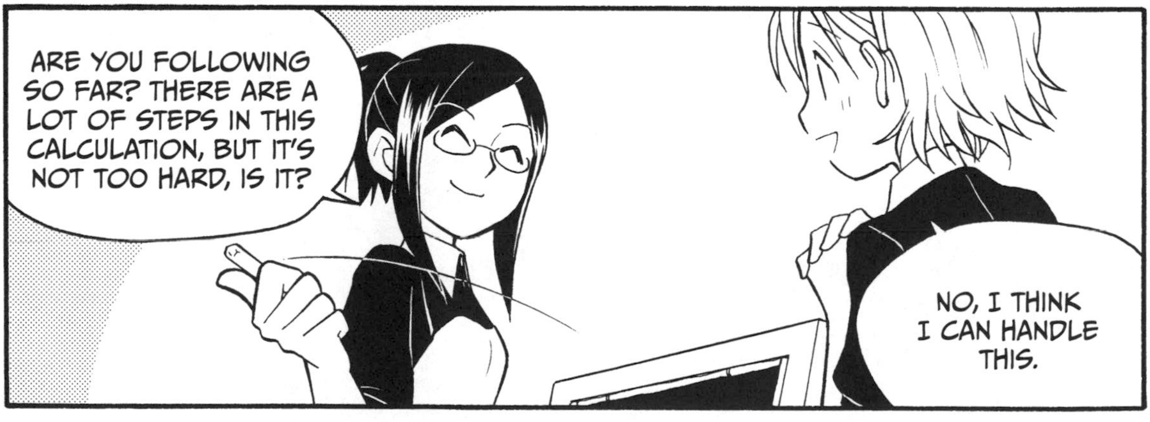
ARE YOU FOLLOWING SO FAR? THERE ARE A LOT OF STEPS IN THIS CALCULATION, BUT IT'S NOT TOO HARD, IS IT?
NO, I THINK I CAN HANDLE THIS.

GREAT, THEN SIT DOWN AND TRY THIS PROBLEM.

CAN YOU DESCRIBE THE INCREASE IN HEIGHT PER YEAR, BETWEEN AGE x AND IMMEDIATELY AFTER AGE x, IN THE SAME WAY?

LET ME SEE...

IS THIS RIGHT?

$$\frac{\left(-\dfrac{326.6}{(x+\Delta)}+173.3\right)-\left(-\dfrac{326.6}{x}+173.3\right)}{\Delta} \quad \text{CM/YEAR}$$

YES, INDEED!

NOW, SIMPLIFY IT.

WILL DO.

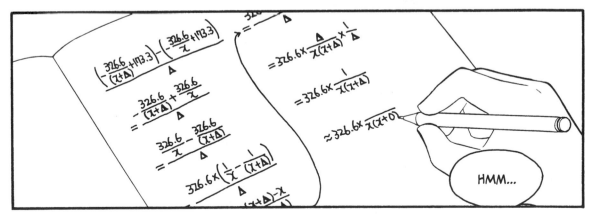

$$\frac{\left(-\dfrac{326.6}{(x+\Delta)}+173.3\right)-\left(-\dfrac{326.6}{x}+173.3\right)}{\Delta}$$

$$=\frac{-\dfrac{326.6}{(x+\Delta)}+\dfrac{326.6}{x}}{\Delta}$$

$$=\frac{\dfrac{326.6}{x}-\dfrac{326.6}{(x+\Delta)}}{\Delta}$$

$$=\frac{326.6\times\left(\dfrac{1}{x}-\dfrac{1}{(x+\Delta)}\right)}{\Delta}$$

$$=326.6\times\frac{\Delta}{x(x+\Delta)}\times\frac{1}{\Delta}$$

$$=326.6\times\frac{1}{x(x+\Delta)}$$

$$\approx 326.6\times\frac{1}{x(x+0)}$$

HMM...

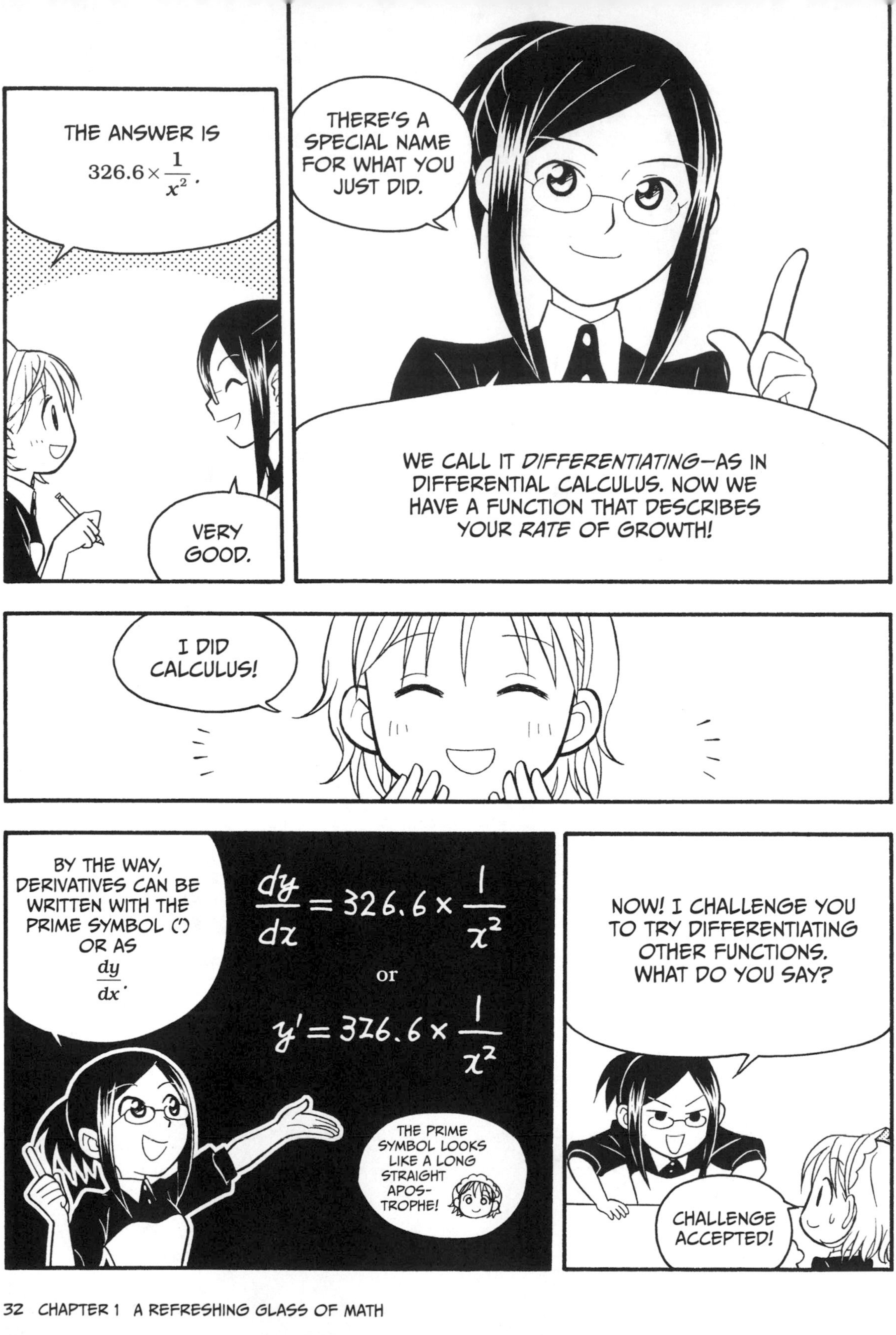
THE ANSWER IS $326.6 \times \frac{1}{x^2}$.
VERY GOOD.
THERE'S A SPECIAL NAME FOR WHAT YOU JUST DID.
WE CALL IT *DIFFERENTIATING*—AS IN DIFFERENTIAL CALCULUS. NOW WE HAVE A FUNCTION THAT DESCRIBES YOUR *RATE* OF GROWTH!
I DID CALCULUS!
BY THE WAY, DERIVATIVES CAN BE WRITTEN WITH THE PRIME SYMBOL (') OR AS $\frac{dy}{dx}$.
$\frac{dy}{dx} = 326.6 \times \frac{1}{x^2}$
or
$y' = 326.6 \times \frac{1}{x^2}$
THE PRIME SYMBOL LOOKS LIKE A LONG STRAIGHT APOSTROPHE!
NOW! I CHALLENGE YOU TO TRY DIFFERENTIATING OTHER FUNCTIONS. WHAT DO YOU SAY?
CHALLENGE ACCEPTED!

DIFFERENTIATE $y = x$ WITH RESPECT TO x.

$$\frac{(x + \Delta) - x}{\Delta} = \frac{\Delta}{\Delta} = 1 \qquad \text{SO} \quad \frac{dy}{dx} = 1$$

IT'S A CONSTANT RATE OF CHANGE!

DIFFERENTIATE $y = x^2$ WITH RESPECT TO x.

$$\frac{(x + \Delta)^2 - x^2}{\Delta} = \frac{x^2 + 2x\Delta + \Delta^2 - x^2}{\Delta} = \frac{(2x + \Delta)\Delta}{\Delta} = 2x + \Delta$$

$$\approx 2x + 0 = 2x \qquad\qquad \text{SO} \quad \frac{dy}{dx} = 2x$$

DIFFERENTIATE $y = \dfrac{1}{x}$ WITH RESPECT TO x.

$$\frac{\dfrac{1}{x + \Delta} - \dfrac{1}{x}}{\Delta} = \frac{\dfrac{x - (x + \Delta)}{(x + \Delta)x}}{\Delta} = \frac{-\Delta}{(x + \Delta)x} \times \frac{1}{\Delta} = \frac{-1}{(x + \Delta)x}$$

$$\approx \frac{-1}{(x + 0)x} = \frac{-1}{x^2} = -x^{-2} \qquad \text{SO} \quad \frac{dy}{dx} = -x^{-2}$$

DIFFERENTIATE $y = \dfrac{1}{x^2}$ WITH RESPECT TO x.

$$\dfrac{\dfrac{1}{(x+\Delta)^2} - \dfrac{1}{x^2}}{\Delta}$$

$$= \dfrac{\left(\dfrac{1}{x+\Delta}\right)^2 - \left(\dfrac{1}{x}\right)^2}{\Delta}$$

$$= \dfrac{\left(\dfrac{1}{x+\Delta} + \dfrac{1}{x}\right)\left(\dfrac{1}{x+\Delta} - \dfrac{1}{x}\right)}{\Delta}$$

$$= \dfrac{\dfrac{x+(x+\Delta)}{(x+\Delta)x} \times \dfrac{x-(x+\Delta)}{(x+\Delta)x}}{\Delta}$$

$$= \dfrac{\dfrac{2x+\Delta}{(x+\Delta)x} \times \dfrac{-\Delta}{(x+\Delta)x}}{\Delta}$$

$$= \dfrac{2x+\Delta}{(x+\Delta)x} \times \dfrac{-\Delta}{(x+\Delta)x} \times \dfrac{1}{\Delta}$$

$$= \dfrac{-(2x+\Delta)}{\left[(x+\Delta)x\right]^2}$$

$$\approx \dfrac{-(2x+0)}{\left[(x+0)x\right]^2}$$

$$= \dfrac{-2x}{x^4}$$

$$= \dfrac{-2}{x^3}$$

$$= -2x^{-3}$$

SO $\dfrac{dy}{dx} = -2x^{-3}$

BASED ON THESE EXAMPLES, YOU CAN SEE THAT WHEN YOU DIFFERENTIATE $y = x^n$ WITH RESPECT TO x, THE RESULT IS $\dfrac{dy}{dx} = nx^{n-1}$.

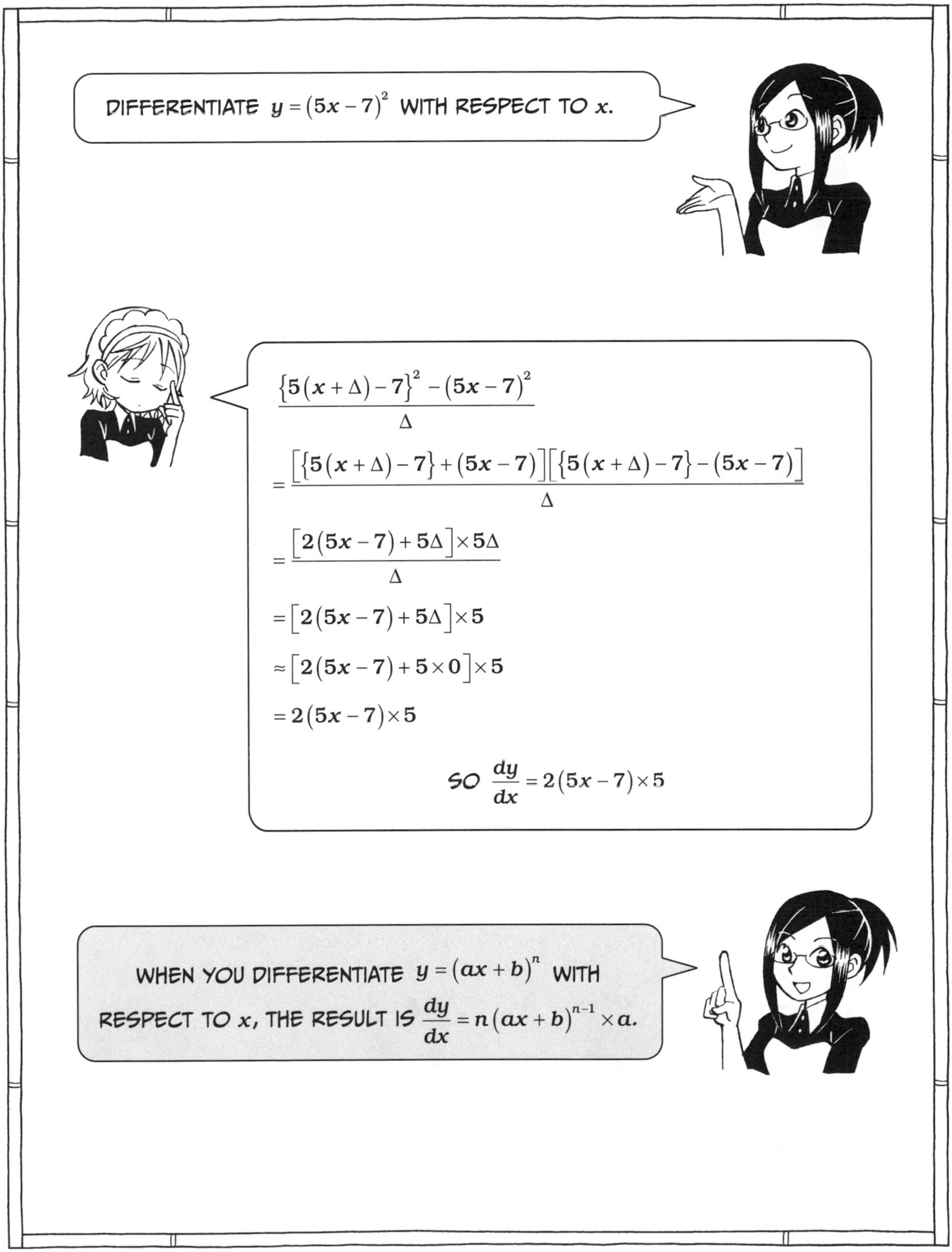
DIFFERENTIATE $y = (5x - 7)^2$ WITH RESPECT TO x.
$$\frac{\{5(x+\Delta)-7\}^2-(5x-7)^2}{\Delta}$$
$$= \frac{\left[\{5(x+\Delta)-7\}+(5x-7)\right]\left[\{5(x+\Delta)-7\}-(5x-7)\right]}{\Delta}$$
$$= \frac{\left[2(5x-7)+5\Delta\right]\times 5\Delta}{\Delta}$$
$$= \left[2(5x-7)+5\Delta\right]\times 5$$
$$\approx \left[2(5x-7)+5\times 0\right]\times 5$$
$$= 2(5x-7)\times 5$$
SO $\frac{dy}{dx} = 2(5x-7)\times 5$
WHEN YOU DIFFERENTIATE $y = (ax+b)^n$ WITH RESPECT TO x, THE RESULT IS $\frac{dy}{dx} = n(ax+b)^{n-1}\times a$.

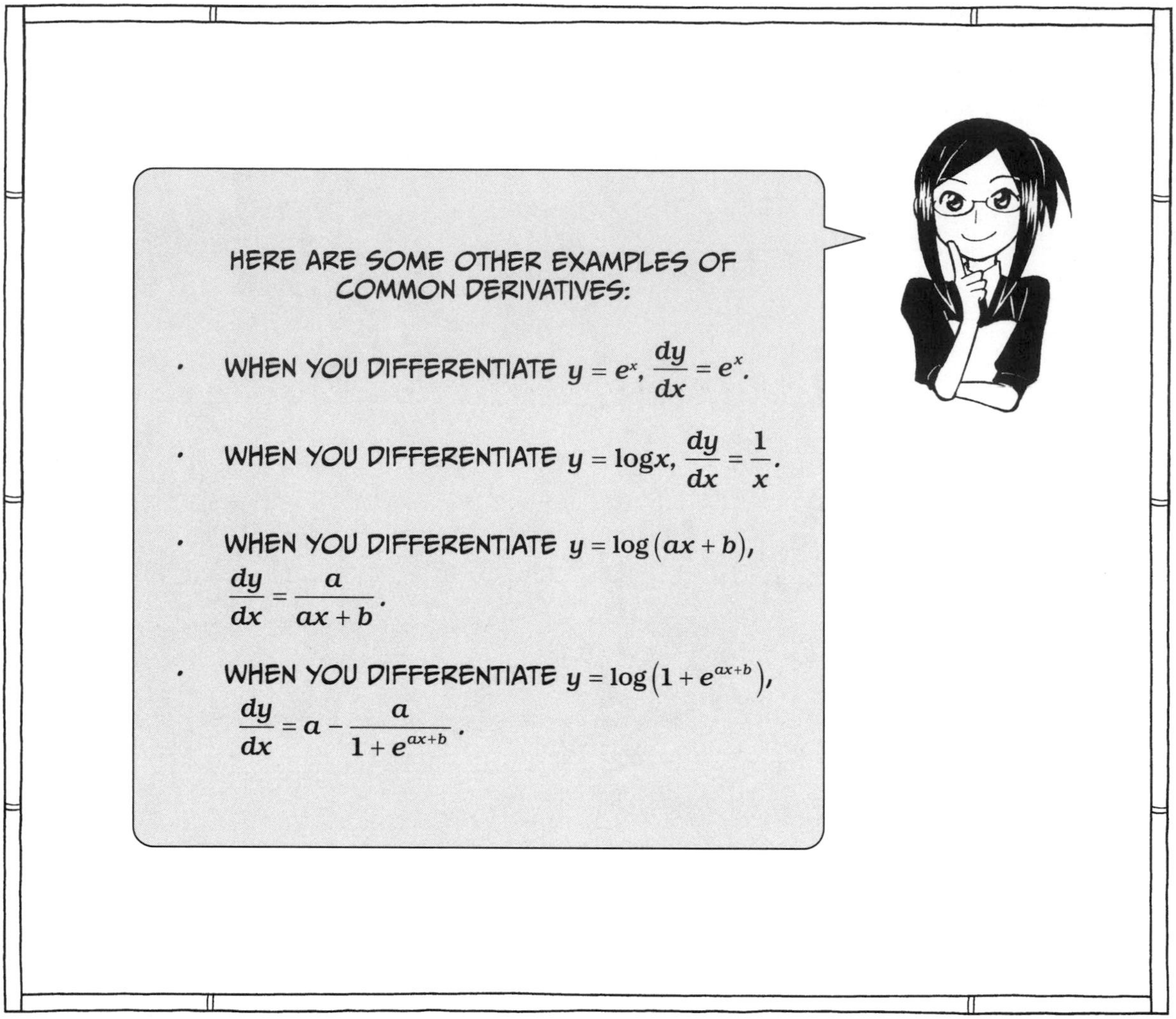
HERE ARE SOME OTHER EXAMPLES OF COMMON DERIVATIVES:
• WHEN YOU DIFFERENTIATE $y = e^x$, $\frac{dy}{dx} = e^x$.
• WHEN YOU DIFFERENTIATE $y = \log x$, $\frac{dy}{dx} = \frac{1}{x}$.
• WHEN YOU DIFFERENTIATE $y = \log(ax + b)$, $\frac{dy}{dx} = \frac{a}{ax + b}$.
• WHEN YOU DIFFERENTIATE $y = \log\left(1 + e^{ax+b}\right)$, $\frac{dy}{dx} = a - \frac{a}{1 + e^{ax+b}}$.

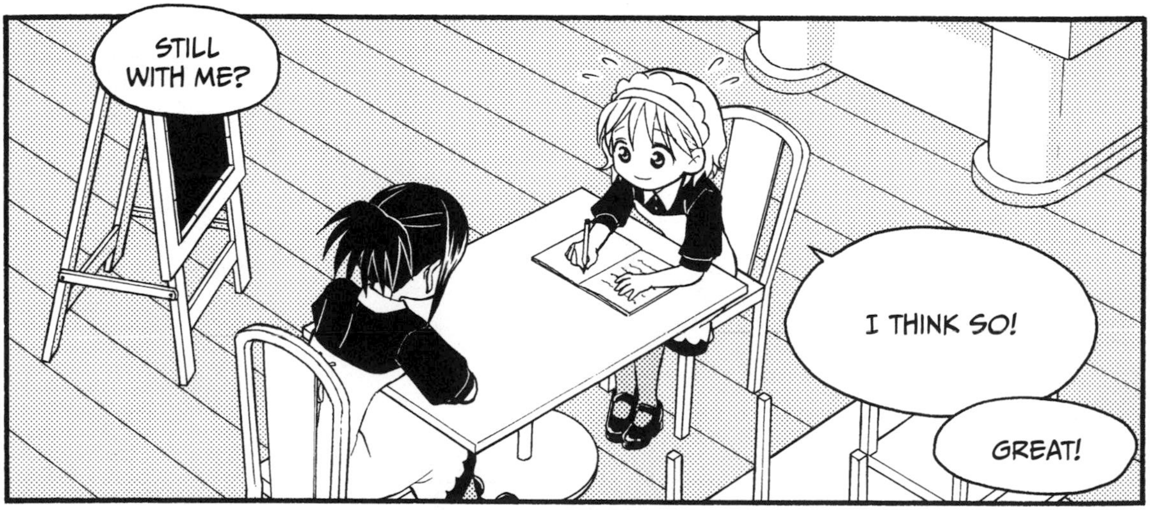
STILL WITH ME?
I THINK SO!
GREAT!

MATRICES

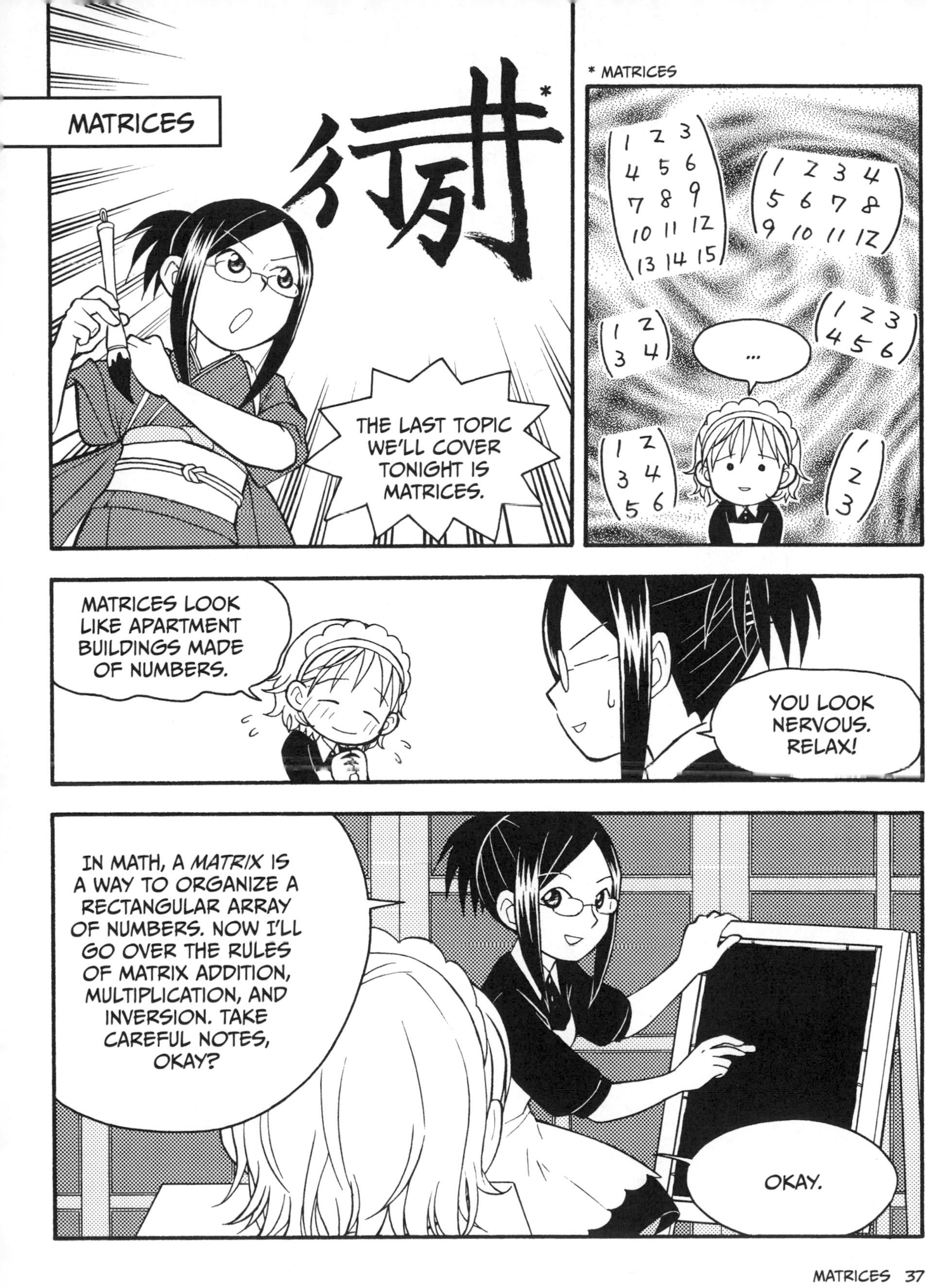
行列*
THE LAST TOPIC WE'LL COVER TONIGHT IS MATRICES.
* MATRICES
...
MATRICES LOOK LIKE APARTMENT BUILDINGS MADE OF NUMBERS.
YOU LOOK NERVOUS. RELAX!
IN MATH, A *MATRIX* IS A WAY TO ORGANIZE A RECTANGULAR ARRAY OF NUMBERS. NOW I'LL GO OVER THE RULES OF MATRIX ADDITION, MULTIPLICATION, AND INVERSION. TAKE CAREFUL NOTES, OKAY?
OKAY.

A MATRIX CAN BE USED TO WRITE EQUATIONS QUICKLY. JUST AS WITH EXPONENTS, MATHEMATICIANS HAVE RULES FOR WRITING THEM.

$$\begin{cases} x_1 + 2x_2 = -1 \\ 3x_1 + 4x_2 = 5 \end{cases}$$ CAN BE WRITTEN AS $\begin{pmatrix} 1 & 2 \\ 3 & 4 \end{pmatrix}\begin{pmatrix} x_1 \\ x_2 \end{pmatrix} = \begin{pmatrix} -1 \\ 5 \end{pmatrix}$

AND $\begin{cases} x_1 + 2x_2 \\ 3x_1 + 4x_2 \end{cases}$ CAN BE WRITTEN AS $\begin{pmatrix} 1 & 2 \\ 3 & 4 \end{pmatrix}\begin{pmatrix} x_1 \\ x_2 \end{pmatrix}$

EXAMPLE

$$\begin{cases} k_1 + 2k_2 + 3k_3 = -3 \\ 4k_1 + 5k_2 + 6k_3 = 8 \\ 10k_1 + 11k_2 + 12k_3 = 2 \\ 13k_1 + 14k_2 + 15k_3 = 7 \end{cases}$$ can be written as $\begin{pmatrix} 1 & 2 & 3 \\ 4 & 5 & 6 \\ 10 & 11 & 12 \\ 13 & 14 & 15 \end{pmatrix}\begin{pmatrix} k_1 \\ k_2 \\ k_3 \end{pmatrix} = \begin{pmatrix} -3 \\ 8 \\ 2 \\ 7 \end{pmatrix}$

If you don't know the values of the expressions, you write the expressions and the matrix like this:

$$\begin{cases} k_1 + 2k_2 + 3k_3 \\ 4k_1 + 5k_2 + 6k_3 \\ 7k_1 + 8k_2 + 9k_3 \\ 10k_1 + 11k_2 + 12k_3 \\ 13k_1 + 14k_2 + 15k_3 \end{cases}$$ $\begin{pmatrix} 1 & 2 & 3 \\ 4 & 5 & 6 \\ 7 & 8 & 9 \\ 10 & 11 & 12 \\ 13 & 14 & 15 \end{pmatrix}\begin{pmatrix} k_1 \\ k_2 \\ k_3 \end{pmatrix}$

Just like an ordinary table, we say matrices have *columns* and *rows*. Each number inside of the matrix is called an *element*.

SUMMARY

$$\begin{cases} a_{11}x_1 + a_{12}x_2 + \cdots + a_{1q}x_q = b_1 \\ a_{21}x_2 + a_{22}x_2 + \cdots + a_{2q}x_q = b_2 \\ \cdots\cdots\cdots\cdots\cdots\cdots\cdots\cdots\cdots\cdots\cdots\cdots \\ a_{p1}x_1 + a_{p2}x_2 + \cdots + a_{pq}x_q = b_p \end{cases}$$ can be written as $\begin{pmatrix} a_{11} & a_{12} & \cdots & a_{1q} \\ a_{21} & a_{22} & \cdots & a_{2q} \\ \vdots & \vdots & \ddots & \vdots \\ a_{p1} & a_{p2} & \cdots & a_{pq} \end{pmatrix}\begin{pmatrix} x_1 \\ x_2 \\ \vdots \\ x_q \end{pmatrix} = \begin{pmatrix} b_1 \\ b_2 \\ \vdots \\ b_p \end{pmatrix}$

$$\begin{cases} a_{11}x_1 + a_{12}x_2 + \cdots + a_{1q}x_q \\ a_{21}x_2 + a_{22}x_2 + \cdots + a_{2q}x_q \\ \cdots\cdots\cdots\cdots\cdots\cdots\cdots\cdots\cdots\cdots \\ a_{p1}x_1 + a_{p2}x_2 + \cdots + a_{pq}x_q \end{cases}$$ can be written as $\begin{pmatrix} a_{11} & a_{12} & \cdots & a_{1q} \\ a_{21} & a_{22} & \cdots & a_{2q} \\ \vdots & \vdots & \ddots & \vdots \\ a_{p1} & a_{p2} & \cdots & a_{pq} \end{pmatrix}\begin{pmatrix} x_1 \\ x_2 \\ \vdots \\ x_q \end{pmatrix}$

ADDING MATRICES

NEXT, I'LL EXPLAIN THE ADDITION OF MATRICES.

CONSIDER THIS: $\begin{pmatrix} 1 & 2 \\ 3 & 4 \end{pmatrix} + \begin{pmatrix} 4 & 5 \\ -2 & 4 \end{pmatrix}$

NOW JUST ADD THE NUMBERS IN THE SAME POSITION: TOP LEFT PLUS TOP LEFT, AND SO ON.

$$\begin{pmatrix} 1+4 & 2+5 \\ 3+(-2) & 4+4 \end{pmatrix} = \begin{pmatrix} 5 & 7 \\ 1 & 8 \end{pmatrix}$$

YOU CAN ONLY ADD MATRICES THAT HAVE THE SAME DIMENSIONS, THAT IS, THE SAME NUMBER OF ROWS AND COLUMNS.

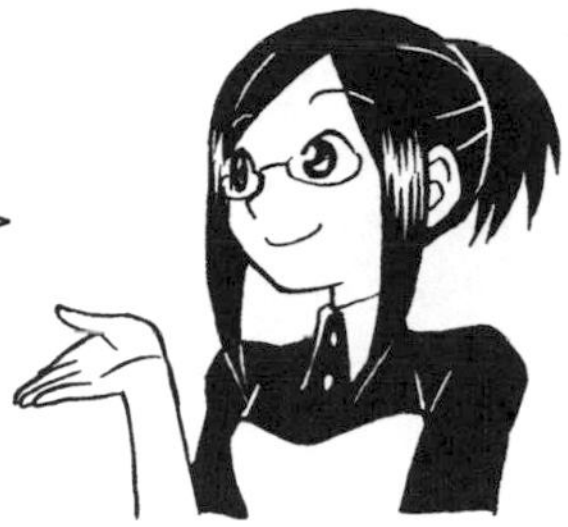

EXAMPLE PROBLEM 1

What is $\begin{pmatrix} 5 & 1 \\ 6 & -9 \end{pmatrix} + \begin{pmatrix} -1 & 3 \\ -3 & 10 \end{pmatrix}$?

ANSWER

$$\begin{pmatrix} 5 & 1 \\ 6 & -9 \end{pmatrix} + \begin{pmatrix} -1 & 3 \\ -3 & 10 \end{pmatrix} = \begin{pmatrix} 5+(-1) & 1+3 \\ 6+(-3) & (-9)+10 \end{pmatrix} = \begin{pmatrix} 4 & 4 \\ 3 & 1 \end{pmatrix}$$

EXAMPLE PROBLEM 2

What is $\begin{pmatrix} 1 & 2 & 3 \\ 4 & 5 & 6 \\ 7 & 8 & 9 \\ 10 & 11 & 12 \\ 13 & 14 & 15 \end{pmatrix} + \begin{pmatrix} 7 & 2 & 3 \\ -1 & 7 & -4 \\ -7 & -3 & 10 \\ 8 & 2 & -1 \\ 7 & 1 & -9 \end{pmatrix}$?

ANSWER

$$\begin{pmatrix} 1 & 2 & 3 \\ 4 & 5 & 6 \\ 7 & 8 & 9 \\ 10 & 11 & 12 \\ 13 & 14 & 15 \end{pmatrix} + \begin{pmatrix} 7 & 2 & 3 \\ -1 & 7 & -4 \\ -7 & -3 & 10 \\ 8 & 2 & -1 \\ 7 & 1 & -9 \end{pmatrix} = \begin{pmatrix} 1+7 & 2+2 & 3+3 \\ 4+(-1) & 5+7 & 6+(-4) \\ 7+(-7) & 8+(-3) & 9+10 \\ 10+8 & 11+2 & 12+(-1) \\ 13+7 & 14+1 & 15+(-9) \end{pmatrix} = \begin{pmatrix} 8 & 4 & 6 \\ 3 & 12 & 2 \\ 0 & 5 & 19 \\ 18 & 13 & 11 \\ 20 & 15 & 6 \end{pmatrix}$$

SUMMARY

Here are two generic matrices.

$$\begin{pmatrix} a_{11} & a_{12} & \cdots & a_{1q} \\ a_{21} & a_{22} & \cdots & a_{2q} \\ \vdots & \vdots & \ddots & \vdots \\ a_{p1} & a_{p2} & \cdots & a_{pq} \end{pmatrix} \begin{pmatrix} b_{11} & b_{12} & \cdots & b_{1q} \\ b_{21} & b_{22} & \cdots & b_{2q} \\ \vdots & \vdots & \ddots & \vdots \\ b_{p1} & b_{p2} & \cdots & b_{pq} \end{pmatrix}$$

You can add them together,

$$\begin{pmatrix} a_{11} & a_{12} & \cdots & a_{1q} \\ a_{21} & a_{22} & \cdots & a_{2q} \\ \vdots & \vdots & \ddots & \vdots \\ a_{p1} & a_{p2} & \cdots & a_{pq} \end{pmatrix} + \begin{pmatrix} b_{11} & b_{12} & \cdots & b_{1q} \\ b_{21} & b_{22} & \cdots & b_{2q} \\ \vdots & \vdots & \ddots & \vdots \\ b_{p1} & b_{p2} & \cdots & b_{pq} \end{pmatrix}$$

like this:

$$\begin{pmatrix} a_{11}+b_{11} & a_{12}+b_{12} & \cdots & a_{1q}+b_{1q} \\ a_{21}+b_{21} & a_{22}+b_{22} & \cdots & a_{2q}+b_{2q} \\ \vdots & \vdots & \ddots & \vdots \\ a_{p1}+b_{p1} & a_{p2}+b_{p2} & \cdots & a_{pq}+b_{pq} \end{pmatrix}$$

And of course, matrix subtraction works the same way. Just subtract the corresponding elements!

MULTIPLYING MATRICES

ON TO MATRIX MULTIPLICATION! WE DON'T MULTIPLY MATRICES IN THE SAME WAY AS WE ADD AND SUBTRACT THEM. IT'S EASIEST TO EXPLAIN BY EXAMPLE, SO LET'S MULTIPLY THE FOLLOWING:

$$\begin{pmatrix} 1 & 2 \\ 3 & 4 \end{pmatrix} \begin{pmatrix} x_1 & y_1 \\ x_2 & y_2 \end{pmatrix}$$

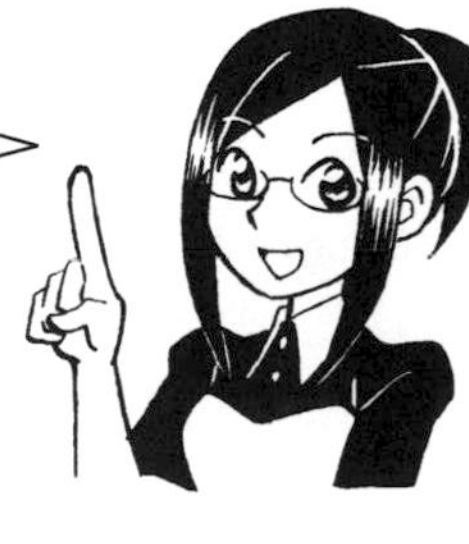

WE MULTIPLY EACH ELEMENT IN THE FIRST COLUMN OF THE LEFT MATRIX BY THE TOP ELEMENT OF THE FIRST COLUMN IN THE RIGHT MATRIX, THEN THE SECOND COLUMN OF THE LEFT MATRIX BY THE SECOND ELEMENT IN THE FIRST COLUMN OF THE RIGHT MATRIX. THEN WE ADD THE PRODUCTS, LIKE THIS:

$$1x_1 + 2x_2$$
$$3x_1 + 4x_2$$

AND THEN WE DO THE SAME WITH THE SECOND COLUMN OF THE RIGHT MATRIX TO GET:

$$1y_1 + 2y_2$$
$$3y_1 + 4y_2$$

SO THE FINAL RESULT IS:

$$\begin{pmatrix} 1x_1 + 2x_2 & 1y_1 + 2y_2 \\ 3x_1 + 4x_2 & 3y_1 + 4y_2 \end{pmatrix}$$

IN MATRIX MULTIPLICATION, FIRST YOU MULTIPLY AND THEN YOU ADD TO GET THE FINAL RESULT. LET'S TRY THIS OUT.

EXAMPLE PROBLEM 1

What is $\begin{pmatrix} 1 & 2 \\ 3 & 4 \end{pmatrix}\begin{pmatrix} 4 & 5 \\ -2 & 4 \end{pmatrix}$?

We know to multiply the elements and then add the terms to simplify. When multiplying, we take the right matrix, column by column, and multiply it by the left matrix.*

ANSWER

$$\begin{pmatrix} 1 & 2 \\ 3 & 4 \end{pmatrix}\begin{pmatrix} 4 \\ -2 \end{pmatrix} = \begin{pmatrix} 1 \times 4 + 2 \times (-2) \\ 3 \times 4 + 4 \times (-2) \end{pmatrix} = \begin{pmatrix} 0 \\ 4 \end{pmatrix} \qquad \text{First column}$$

$$\begin{pmatrix} 1 & 2 \\ 3 & 4 \end{pmatrix}\begin{pmatrix} 5 \\ 4 \end{pmatrix} = \begin{pmatrix} 1 \times 5 + 2 \times 4 \\ 3 \times 5 + 4 \times 4 \end{pmatrix} = \begin{pmatrix} 13 \\ 31 \end{pmatrix} \qquad \text{Second column}$$

So the answer is $\begin{pmatrix} 1 & 2 \\ 3 & 4 \end{pmatrix}\begin{pmatrix} 4 & 5 \\ -2 & 4 \end{pmatrix} = \begin{pmatrix} 0 & 13 \\ 4 & 31 \end{pmatrix}$.

* NOTE THAT THE RESULTING MATRIX WILL HAVE THE SAME NUMBER OF ROWS AS THE FIRST MATRIX AND THE SAME NUMBER OF COLUMNS AS THE SECOND MATRIX.

What is $\begin{pmatrix} 1 & 2 \\ 4 & 5 \\ 7 & 8 \\ 10 & 11 \end{pmatrix} \begin{pmatrix} k_1 & l_1 & m_1 \\ k_2 & l_2 & m_2 \end{pmatrix}$?

ANSWER

$$\begin{pmatrix} 1 & 2 \\ 4 & 5 \\ 7 & 8 \\ 10 & 11 \end{pmatrix} \begin{pmatrix} k_1 \\ k_2 \end{pmatrix} = \begin{pmatrix} k_1 + 2k_2 \\ 4k_1 + 5k_2 \\ 7k_1 + 8k_2 \\ 10k_1 + 11k_2 \end{pmatrix}$$

Multiply the first column of the second matrix by the respective rows of the first matrix.

$$\begin{pmatrix} 1 & 2 \\ 4 & 5 \\ 7 & 8 \\ 10 & 11 \end{pmatrix} \begin{pmatrix} l_1 \\ l_2 \end{pmatrix} = \begin{pmatrix} l_1 + 2l_2 \\ 4l_1 + 5l_2 \\ 7l_1 + 8l_2 \\ 10l_1 + 11l_2 \end{pmatrix}$$

Do the same with the second column.

$$\begin{pmatrix} 1 & 2 \\ 4 & 5 \\ 7 & 8 \\ 10 & 11 \end{pmatrix} \begin{pmatrix} m_1 \\ m_2 \end{pmatrix} = \begin{pmatrix} m_1 + 2m_2 \\ 4m_1 + 5m_2 \\ 7m_1 + 8m_2 \\ 10m_1 + 11m_2 \end{pmatrix}$$

And the third column.

The final answer is just a concatenation of the three answers above.

$$\begin{pmatrix} k_1 + 2k_2 & l_1 + 2l_2 & m_1 + 2m_2 \\ 4k_1 + 5k_2 & 4l_1 + 5l_2 & 4m_1 + 5m_2 \\ 7k_1 + 8k_2 & 7l_1 + 8l_2 & 7m_1 + 8m_2 \\ 10k_1 + 11k_2 & 10l_1 + 11l_2 & 10m_1 + 11m_2 \end{pmatrix}$$

THE RULES OF MATRIX MULTIPLICATION

WHEN MULTIPLYING MATRICES, THERE ARE THREE THINGS TO REMEMBER:

- THE NUMBER OF COLUMNS IN THE FIRST MATRIX MUST EQUAL THE NUMBER OF ROWS IN THE SECOND MATRIX.
- THE RESULT MATRIX WILL HAVE A NUMBER OF ROWS EQUAL TO THE FIRST MATRIX.
- THE RESULT MATRIX WILL HAVE A NUMBER OF COLUMNS EQUAL TO THE SECOND MATRIX.

Can the following pairs of matrices can be multiplied?
If so, how many rows and columns will the resulting matrix have?

EXAMPLE PROBLEM 1

$$\begin{pmatrix} 2 & 3 & 4 \\ -5 & 3 & 6 \end{pmatrix} \begin{pmatrix} 2 \\ -7 \\ 0 \end{pmatrix}$$

ANSWER

Yes! The resulting matrix will have 2 rows and 1 column:

$$\begin{pmatrix} 2 & 3 & 4 \\ -5 & 3 & 6 \end{pmatrix} \begin{pmatrix} 2 \\ -7 \\ 0 \end{pmatrix} = \begin{pmatrix} 2 \times 2 + 3 \times (-7) + 4 \times 0 \\ (-5) \times 2 + 3 \times (-7) + 6 \times 0 \end{pmatrix} = \begin{pmatrix} -17 \\ -31 \end{pmatrix}$$

EXAMPLE PROBLEM 2

$$\begin{pmatrix} 9 & 4 & -1 \\ 7 & -6 & 0 \\ -5 & 3 & 8 \end{pmatrix} \begin{pmatrix} 2 & -2 & 1 \\ 4 & 9 & -7 \end{pmatrix}$$

ANSWER

No. The number of columns in the first matrix is 3, but the number of rows in the second matrix is 2. These matrices cannot be multiplied.

IDENTITY AND INVERSE MATRICES

THE LAST THINGS I'M GOING TO EXPLAIN TONIGHT ARE *IDENTITY MATRICES* AND *INVERSE MATRICES*.

AN IDENTITY MATRIX IS A SQUARE MATRIX WITH ONES ACROSS THE DIAGONAL, FROM TOP LEFT TO BOTTOM RIGHT, AND ZEROS EVERYWHERE ELSE.

HERE IS A 2×2 IDENTITY MATRIX: $\begin{pmatrix} 1 & 0 \\ 0 & 1 \end{pmatrix}$

AND HERE IS A 3×3 IDENTITY MATRIX: $\begin{pmatrix} 1 & 0 & 0 \\ 0 & 1 & 0 \\ 0 & 0 & 1 \end{pmatrix}$

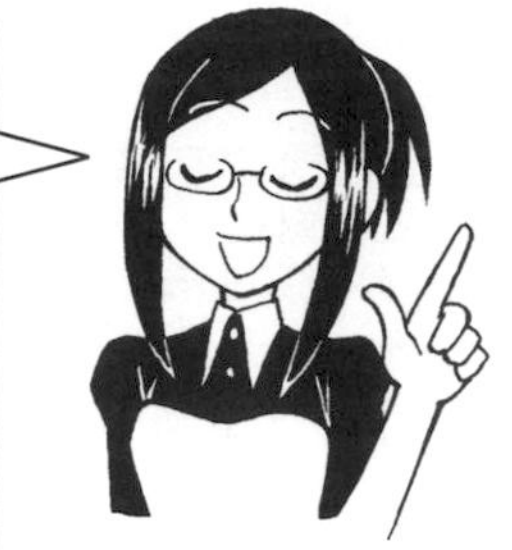

Some square matrices (a matrix that has the same number of rows as columns) are *invertible*. A square matrix multiplied by its inverse will equal an identity matrix of the same size and shape, so it's easy to demonstrate that one matrix is the inverse of another.

For example:

$$\begin{pmatrix} 1 & 2 \\ 3 & 4 \end{pmatrix}\begin{pmatrix} -2 & 1 \\ 1.5 & -0.5 \end{pmatrix} = \begin{pmatrix} 1\times(-2)+2\times1.5 & 1\times1+2\times(-0.5) \\ 3\times(-2)+4\times1.5 & 3\times1+4\times(-0.5) \end{pmatrix} = \begin{pmatrix} 1 & 0 \\ 0 & 1 \end{pmatrix}$$

So $\begin{pmatrix} -2 & 1 \\ 1.5 & -0.5 \end{pmatrix}$ is the inverse of $\begin{pmatrix} 1 & 2 \\ 3 & 4 \end{pmatrix}$.

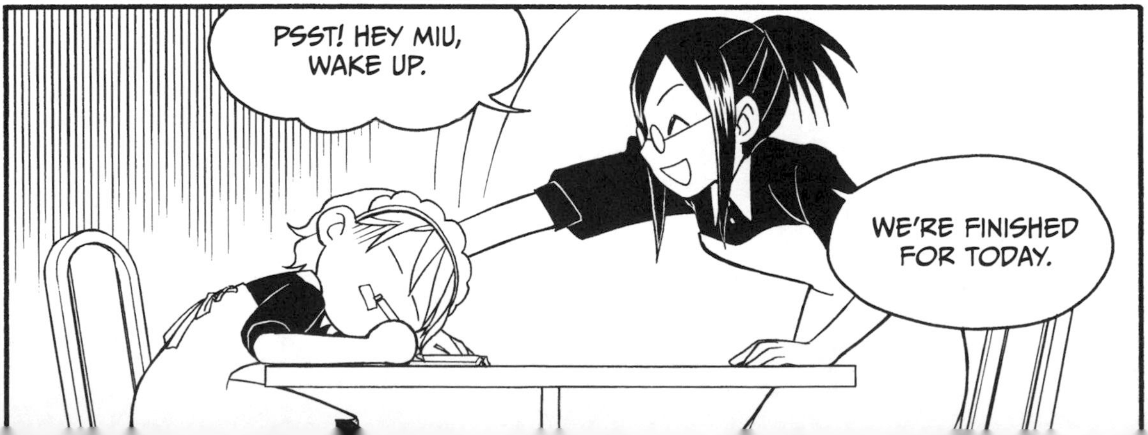
PSST! HEY MIU, WAKE UP.
WE'RE FINISHED FOR TODAY.

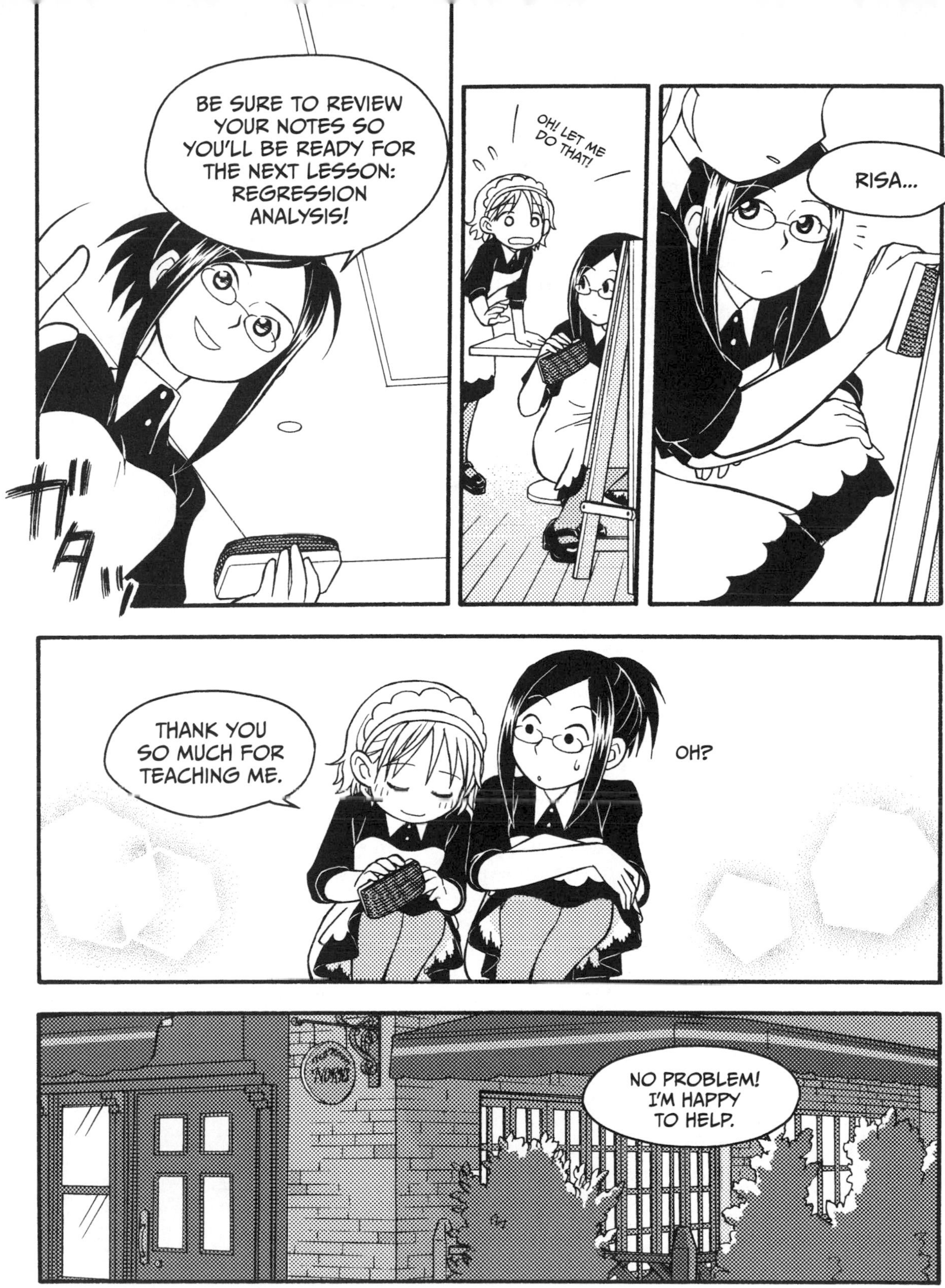
BE SURE TO REVIEW YOUR NOTES SO YOU'LL BE READY FOR THE NEXT LESSON: REGRESSION ANALYSIS!
ガタッ
OH! LET ME DO THAT!
RISA...
THANK YOU SO MUCH FOR TEACHING ME.
OH?
NO PROBLEM! I'M HAPPY TO HELP.

STATISTICAL DATA TYPES

Now that you've had a little general math refresher, it's time for a refreshing chaser of *statistics*, a branch of mathematics that deals with the interpretation and analysis of data. Let's dive right in.

We can categorize data into two types. Data that can be measured with numbers is called *numerical data*, and data that cannot be measured is called *categorical data*. Numerical data is sometimes called *quantitative data*, and categorical data is sometimes called *qualitative data*. These names are subjective and vary based on the field and the analyst. Table 1-1 shows examples of numerical and categorical data.

TABLE 1-1: NUMERICAL VS. CATEGORICAL DATA

	Number of books read per month	Age (in years)	Place where person most often reads	Gender
Person A	4	20	Train	Female
Person B	2	19	Home	Male
Person C	10	18	Café	Male
Person D	14	22	Library	Female

Numerical Data (Number of books read per month, Age) — **Categorical Data** (Place where person most often reads, Gender)

Number of books read per month and *Age* are both examples of numerical data, while *Place where person most often reads* and *Gender* are not typically represented by numbers. However, categorical data can be converted into numerical data, and vice versa. Table 1-2 gives an example of how numerical data can be converted to categorical.

TABLE 1-2: CONVERTING NUMERICAL DATA TO CATEGORICAL DATA

	Number of books read per month	→	Number of books read per month
Person A	4		Few
Person B	2		Few
Person C	10		Many
Person D	14		Many

In this conversion, the analyst has converted the values 1 to 5 into the category *Few*, values 6 to 9 into the category *Average*, and values 10 and higher into the category *Many*. The ranges are up to the discretion of the researcher. Note that these three categories (Few, Average, Many) are *ordinal*, meaning that they can be ranked in order: Many is more than Average is more than Few. Some categories cannot be easily ordered. For instance, how would one easily order the categories Brown, Purple, Green?

Table 1-3 provides an example of how categorical data can be converted to numerical data.

TABLE 1-3: CONVERTING CATEGORICAL DATA TO NUMERICAL DATA

	Favorite season		Spring	Summer	Autumn	Winter
Person A	Spring		1	0	0	0
Person B	Summer		0	1	0	0
Person C	Autumn		0	0	1	0
Person D	Winter		0	0	0	1

In this case, we have converted the categorical data *Favorite season*, which has four categories (Spring, Summer, Autumn, Winter), into binary data in four columns. The data is described as binary because it takes on one of two values: *Favorite* is represented by 1 and *Not Favorite* is represented by 0.

It is also possible to represent this data with three columns. Why can we omit one column? Because we know each respondent's favorite season even if a column is omitted. For example, if the first three columns (Spring, Summer, Autumn) are 0, you know Winter must be 1, even if it isn't shown.

In multiple regression analysis, we need to ensure that our data is *linearly independent*; that is, no set of J columns shown can be used to exactly infer the content of another column within that set. Ensuring linear independence is often done by deleting the last column of data. Because the following statement is true, we can delete the Winter column from Table 1-3:

$$(Winter) = 1 - (Spring) - (Summer) - (Autumn)$$

In regression analysis, we must be careful to recognize which variables are numerical, ordinal, and categorical so we use the variables correctly.

HYPOTHESIS TESTING

Statistical methods are often used to test scientific hypotheses. A *hypothesis* is a proposed statement about the relationship between variables or the properties of a single variable, describing a phenomenon or concept. We collect data and use hypothesis testing to decide whether our hypothesis is supported by the data.

We set up a hypothesis test by stating not one but two hypotheses, called the *null hypothesis* (H_0) and the *alternative hypothesis* (H_a). The null hypothesis is the default hypothesis we wish to disprove, usually stating that there is a specific relationship (or none at all) between variables or the properties of a single variable. The alternative hypothesis is the hypothesis we are trying to prove. If our data differs enough from what we would expect if the null hypothesis were true, we can reject the null and accept the alternative hypothesis. Let's consider a very simple example, with the following hypotheses:

H_0: Children order on average 10 cups of hot chocolate per month.

H_a: Children do not order on average 10 cups of hot chocolate per month.

We're proposing statements about a single variable—the number of hot chocolates ordered per month—and checking if it has a certain property: having an average of 10. Suppose we observed five children for a month and found that they ordered 7, 9, 10, 11, and 13 cups of hot chocolate, respectively. We assume these five children are a representative *sample* of the total *population* of all hot chocolate–drinking children. The average of these five children's orders is 10. In this case, we cannot prove that the null hypothesis is false, since the value proposed in our null hypothesis (10) is indeed the average of this sample.

However, suppose we observed a sample of five different children for a month and they ordered 29, 30, 31, 32, and 35 cups of hot chocolate, respectively. The average of these five children's orders is 31.4; in fact, not a single child came anywhere close to drinking only 10 cups of hot chocolate. On the basis of this data, we would assert that we should reject the null hypothesis.

In this example, we've stated hypotheses about a single variable: the number of cups each child orders per month. But when we're looking at the relationship between two or more variables, as we do in regression analysis, our null hypothesis usually states that there is no relationship between the variables being tested, and the alternative hypothesis states that there is a relationship.

MEASURING VARIATION

Suppose Miu and Risa had a karaoke competition with some friends from school. They competed in two teams of five. Table 1-4 shows how they scored.

TABLE 1-4: KARAOKE SCORES FOR TEAM MIU AND TEAM RISA

Team member	Score
Miu	48
Yuko	32
Aiko	88
Maya	61
Marie	71
Average	60

Team member	Score
Risa	67
Asuka	55
Nana	61
Yuki	63
Rika	54
Average	60

There are multiple statistics we can use to describe the "center" of a data set. Table 1-4 shows the average of the data for each team, also known as the *mean*. This is calculated by adding the scores of each member of the group and dividing by the number of members in the group. Each of the karaoke groups has a mean score of 60.

We could also define the center of these data sets as being the middle number of each group when the scores are put in order. This is the *median* of the data. To find the median, write the scores in increasing order (for Team Miu, this is 32, 48, 61, 71, 88) and the median is the number in the middle of this list. For Team Miu, the median is Maya's score of 61. The median happens to be 61 for Team Risa as well, with Nana having the median score on this team. If there were an even number of members on each team, we would usually take the mean of the two middle scores.

So far, the statistics we've calculated seem to indicate that the two sets of scores are the same. But what do you notice when we put the scores on a number line (see Figure 1-1)?

Team Miu
Yuko
Miu
Maya
Marie
Aiko
0 10 20 30 40 50 60 70 80 90 100
Score
Team Risa
Nana
Asuka
Yuki
Rika
Risa
0 10 20 30 40 50 60 70 80 90 100
Score

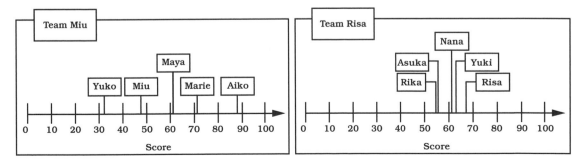

FIGURE 1-1: KARAOKE SCORES FOR TEAM MIU AND TEAM RISA ON NUMBER LINES

Team Miu's scores are much more spread out than Team Risa's. Thus, we say that the data sets have different *variation*.

There are several ways to measure variation, including the sum of squared deviations, variance, and standard deviation. Each of these measures share the following characteristics:

- All of them measure the spread of the data from the mean.
- The greater the variation in the data, the greater the value of the measure.
- The minimum value of the measures is zero—that happens only if your data doesn't vary at all!

SUM OF SQUARED DEVIATIONS

The *sum of squared deviations* is a measure often used during regression analysis. It is calculated as follows:

sum of (individual score – mean score)^2,

which is written mathematically as

$$\sum (x - \bar{x})^2.$$

The sum of squared deviations is not often used on its own to describe variation because it has a fatal shortcoming—its value increases as the number of data points increases. As you have more and more numbers, the sum of their differences from the mean gets bigger and bigger.

VARIANCE

This shortcoming is alleviated by calculating the *variance*:

$$\frac{\sum (x - \bar{x})^2}{n - 1}$$, where n = the number of data points.

This calculation is also called the *unbiased sample variance*, because the denominator is the number of data points minus 1 rather than simply the number of data points. In research studies that use data from samples, we usually subtract 1 from the number of data points to adjust for the fact that we are using a sample of the population, rather than the entire population. This increases the variance.

This reduced denominator is called the *degrees of freedom*, because it represents the number of values that are free to vary. For practical purposes, it is the number of cases (for example, observations or groups) minus 1. So if we were looking at Team Miu and

Team Risa as samples of the entire karaoke-singing population, we'd say there were 4 degrees of freedom when calculating their statistics, since there are five members on each team. We subtract 1 from the number of singers because they are just a sample of all possible singers in the world and we want to overestimate the variance among them.

The units of the variance are not the same as the units of the observed data. Instead, variance is expressed in units squared, in this case "points squared."

STANDARD DEVIATION

Like variance, the *standard deviation* shows whether all the data points are clustered together or spread out. The standard deviation is actually just the square root of the variance:

$$\sqrt{\text{variance}}$$

Researchers usually use standard deviation as the measure of variation because the units of the standard deviation are the same as those of the original data. For our karaoke singers, the standard deviation is reported in "points."

Let's calculate the sum of squared deviations, variance, and standard deviation for Team Miu (see Table 1-5).

TABLE 1-5: MEASURING VARIATION OF SCORES FOR TEAM MIU

Measure of variation	Calculation
Sum of squared deviations	$(48-60)^2 + (32-60)^2 + (88-60)^2 + (61-60)^2 + (71-60)^2$ $= (-12)^2 + (-28)^2 + 28^2 + 1^2 + 11^2$ $= 1834$
Variance	$\dfrac{1834}{5-1} = 458.8$
Standard deviation	$\sqrt{458.5} = 21.4$

Now let's do the same for Team Risa (see Table 1-6).

TABLE 1-6: MEASURING VARIATION OF SCORES FOR TEAM RISA

Measure of variation	Calculation
Sum of squared deviations	$(67-60)^2 + (55-60)^2 + (61-60)^2 + (63-60)^2 + (54-60)^2$ $= 7^2 + (-5)^2 + 1^2 + 3^2 + (-6)^2$ $= 120$
Variance	$\dfrac{120}{5-1} = 30$
Standard deviation	$\sqrt{30} = 5.5$

We see that Team Risa's standard deviation is 5.5 points, whereas Team Miu's is 21.4 points. Team Risa's karaoke scores vary less than Team Miu's, so Team Risa has more consistent karaoke performers.

PROBABILITY DENSITY FUNCTIONS

We use probability to model events that we cannot predict with certainty. Although we can accurately predict many future events— such as whether running out of gas will cause a car to stop running or how much rocket fuel it would take to get to Mars—many physical, chemical, biological, social, and strategic problems are so complex that we cannot hope to know all of the variables and forces that affect the outcome.

A simple example is the flipping of a coin. We do not know all of the physical forces involved in a single coin flip—temperature, torque, spin, landing surface, and so on. However, we expect that over the course of many flips, the variance in all these factors will cancel out, and we will observe an equal number of heads and tails. Table 1-5 shows the results of flipping a billion quarters in number of flips and percentage of flips.

TABLE 1-5: TALLY OF A BILLION COIN FLIPS

	Number of flips	Percentage of flips
Heads	499,993,945	49.99939%
Tails	500,006,054	50.00061%
Stands on its edge	1	0.0000001%

As we might have guessed, the percentages of heads and tails are both very close to 50%. We can summarize what we know about coin flips in a probability density function, $P(x)$, which we can apply to any given coin flip, as shown here:

$$P(\text{Heads}) = .5, \ P(\text{Tails}) = .5, \ P(\text{Stands on its edge}) < 1 \times 10^{-9}$$

But what if we are playing with a cheater? Perhaps someone has weighted the coin so that $P(x)$ is now this:

$$P(\text{Heads}) = .3, \ P(\text{Tails}) = .7, \ P(\text{Stands on its edge}) = 0$$

What do we expect to happen on a single flip? Will it always be tails? What will the average be after a billion flips?

Not all events have so few possibilities as these coin examples. We often wish to model data that can be continuously measured. For example, height is a continuous measurement. We could measure your height down to the nearest meter, centimeter, millimeter, or . . . nanometer. As we begin dealing with data where the possibilities lie on a continuous space, we need to use continuous functions to represent the probability of events.

A *probability density function* allows us to to compute the probability that the data lies within a given range of values. We can plot a probability density function as a curve, where the x-axis represents the *event space*, or the possible values the result can take, and the y-axis is $f(x)$, or the probability density function value of x. The area under the curve between two possible values represents the probability of getting a result between those two values.

NORMAL DISTRIBUTIONS

One important probability density function is the *normal distribution* (see Figure 1-2), also called the *bell curve* because of its symmetrical shape, which researchers use to model many events.

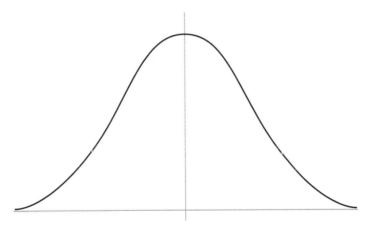

FIGURE 1-2: A NORMAL DISTRIBUTION

The standard normal distribution probability density function can be expressed as follows:

$$f(x) = \frac{1}{\sqrt{2\pi}} e^{\frac{-x^2}{2}}$$

The mean of the standard normal distribution function is zero. When we plot the function, its peak or *maximum* is at the mean and thus at zero. The tails of the distribution fall symmetrically on either side of the mean in a bell shape and extend to infinity, approaching, but never quite touching, the x-axis. The standard normal distribution has a standard deviation of 1. Because the mean is zero and the standard deviation is 1, this distribution is also written as N(0,1).

The area under the curve is equal to 1 (100%), since the value will definitely fall somewhere beneath the curve. The further from the mean a value is, the less probable that value is, as represented by the diminishing height of the curve. You may have seen a curve like this describing the distribution of test scores. Most test takers have a score that is close to the mean. A few people score exceptionally high, and a few people score very low.

CHI-SQUARED DISTRIBUTIONS

Not all data is best modeled by a normal distribution. The *chi-squared* (χ^2) *distribution* is a probability density function that fits the distribution of the sum of squares. That means chi-squared distributions can be used to estimate variation. The chi-squared probability density function is shown here:

$$f(x) = \begin{cases} \dfrac{1}{2^{\frac{k}{2}} \int_0^{\infty} x^{\frac{k}{2}-1} e^{-x} dx} \times x^{\frac{k}{2}-1} \times e^{-\frac{x}{2}} & , x > 0 \\ 0, & x \leq 0 \end{cases}$$

The sum of squares can never be negative, and we see that $f(x)$ is exactly zero for negative numbers. When the probability density function of x is the one shown above, we say, "x follows a chi-squared distribution with k degree(s) of freedom."

The chi-squared distribution is related to the standard normal distribution. In fact, if you take $Z_1, Z_2, \ldots, Z_k$, as a set of independent, identically distributed standard normal random variables and then take the sum of squares of these variables like this,

$$X = Z_1^2 + Z_2^2 + \cdots + Z_k^2,$$

then X is a chi-squared random variable with k degrees of freedom. Thus, we will use the chi-squared distribution of k to represent sums of squares of a set of k normal random variables.

In Figure 1-3, we plot two chi-squared density curves, one for $k = 2$ degrees of freedom and another for $k = 10$ degrees of freedom.

When $k = 2$

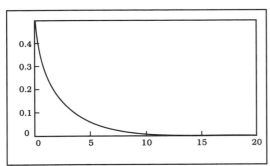

When $k = 10$

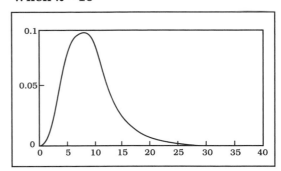

FIGURE 1-3: CHI-SQUARED DENSITY CURVES FOR 2 DEGREES OF FREEDOM (LEFT) AND 10 DEGREES OF FREEDOM (RIGHT)

Notice the differences. What is the limit of the density functions as x goes to infinity? Where is the peak of the functions?

PROBABILITY DENSITY DISTRIBUTION TABLES

Let's say we have a data set with a variable X that follows a chi-squared distribution, with 5 degrees of freedom. If we wanted to know for some point x whether the probability P of $X > x$ is less than a target probability—also known as the *critical value* of the statistic—we must integrate a density curve to calculate that probability. By *integrate*, we mean find the area under the relevant portion of the curve, illustrated in Figure 1-4.

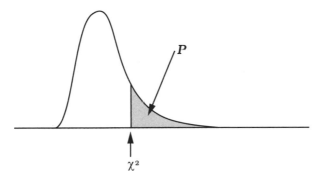

FIGURE 1-4: THE PROBABILITY P THAT A VALUE X EXCEEDS THE CRITICAL CHI-SQUARED VALUE x

Since that is cumbersome to do by hand, we use a computer or, if one is unavailable, a distribution table we find in a book. Distribution tables summarize features of a density curve in many ways. In the case of the chi-squared distribution, the distribution table gives us the point x such that the probability that $X > x$ is equal to a probability P. Statisticians often choose $P = .05$, meaning there is only a 5% chance that a randomly selected value of X will be greater than x. The value of P is known as a p-value.

We use a chi-squared probability distribution table (Table 1-6) to see where our degrees of freedom and our p-value intersect. This number gives us the value of χ^2 (our test statistic). The probability of a chi-squared of this magnitude is equal to or less than the p at the top of the column.

TABLE 1-6: CHI-SQUARED PROBABILITY DISTRIBUTION TABLE

degrees of freedom	.995	.99	.975	.95	.05	.025	.01	.005
1	0.000039	0.0002	0.0010	0.0039	3.8415	5.0239	6.6349	7.8794
2	0.0100	0.0201	0.0506	0.1026	5.9915	7.3778	9.2104	10.5965
3	0.0717	0.1148	0.2158	0.3518	7.8147	9.3484	11.3449	12.8381
4	0.2070	0.2971	0.4844	0.7107	9.4877	11.1433	13.2767	14.8602
5	0.4118	0.5543	0.8312	1.1455	11.0705	12.8325	15.0863	16.7496
6	0.6757	0.8721	1.2373	1.6354	12.5916	14.4494	16.8119	18.5475
7	0.9893	1.2390	1.6899	2.1673	14.0671	16.0128	18.4753	20.2777
8	1.3444	1.6465	2.1797	2.7326	15.5073	17.5345	20.0902	21.9549
9	1.7349	2.0879	2.7004	3.3251	16.9190	19.0228	21.6660	23.5893
10	2.1558	2.5582	3.2470	3.9403	18.3070	20.4832	23.2093	25.1881

To read this table, identify the k degrees of freedom in the first column to determine which row to use. Then select a value for p. For instance, if we selected $p = .05$ and had degrees of freedom $k = 5$, then we would find where the the fifth column and the fifth row intersect (highlighted in Table 1-6). We see that $x = 11.0705$. This means that for a chi-squared random variable and 5 degrees of freedom, the probability of getting a draw $X = 11.0705$ or greater is .05. In other words, the area under the curve corresponding to chi-squared values of 11.0705 or greater is equal to 11% of the total area under the curve.

If we observed a chi-squared random variable with 5 degrees of freedom to have a value of 6.1, is the probability more or less than .05?

F DISTRIBUTIONS

The *F* distribution is just a ratio of two separate chi-squared distributions, and it is used to compare the variance of two samples. As a result, it has two different degrees of freedom, one for each sample.

This is the probability density function of an *F* distribution:

$$f(x) = \begin{cases} \dfrac{\left(\int_0^\infty x^{\frac{v_1+v_2}{2}-1}e^{-x}dx\right) \times (v_1)^{\frac{v_1}{2}} \times (v_2)^{\frac{v_2}{2}}}{\left(\int_0^\infty x^{\frac{v_1}{2}-1}e^{-x}dx\right) \times \left(\int_0^\infty x^{\frac{v_2}{2}-1}e^{-x}dx\right)} \times \dfrac{x^{\frac{v_1}{2}-1}}{\left(v_1 \times x + v_2\right)^{\frac{v_1+v_2}{2}}}, & x > 0 \\[6ex] 0, & x \le 0 \end{cases}$$

If the probability density function of *X* is the one shown above, in statistics, we say, "*X* follows an *F* distribution with degrees of freedom v_1 and v_2."

When $v_1 = 5$ and $v_2 = 10$ and when $v_1 = 10$ and $v_2 = 5$, we get slightly different curves, as shown in Figure 1-5.

When $v_1 = 5$ and $v_2 = 10$

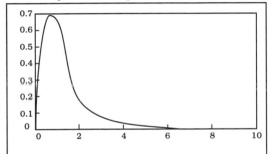

When $v_1 = 10$ and $v_2 = 5$

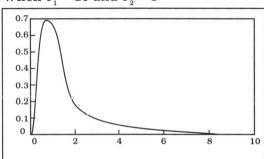

FIGURE 1-5: *F* DISTRIBUTION DENSITY CURVES FOR 5 AND 10 RESPECTIVE DEGREES OF FREEDOM (LEFT) AND 10 AND 5 RESPECTIVE DEGREES OF FREEDOM (RIGHT)

Figure 1-6 shows a graph of an *F* distribution with degrees of freedom v_1 and v_2. This shows the *F* value as a point on the horizontal axis, and the total area of the shaded part to the right is the probability *P* that a variable with an *F* distribution has a value greater than the selected *F* value.

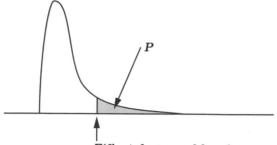

F(first degree of freedom, second degree of freedom; P)

FIGURE 1-6: THE PROBABILITY P THAT A VALUE x EXCEEDS THE CRITICAL F VALUE

Table 1-7 shows the F distribution table when $p = .05$.

TABLE 1-7: F PROBABILITY DISTRIBUTION TABLE FOR $p = .05$

v_2 \ v_1	1	2	3	4	5	6	7	8	9	10
1	161.4	199.5	215.7	224.6	230.2	264.0	236.8	238.9	240.5	241.9
2	18.5	19.0	19.2	19.2	19.3	19.3	19.4	19.4	19.4	19.4
3	10.1	9.6	9.3	9.1	9.0	8.9	8.9	8.8	8.8	8.8
4	7.7	6.9	6.6	6.4	6.3	6.2	6.1	6.0	6.0	6.0
5	6.6	5.8	5.4	5.2	5.1	5.0	4.9	4.8	4.8	4.7
6	6.0	5.1	4.8	4.5	4.4	4.3	4.2	4.1	4.1	4.1
7	5.6	4.7	4.3	4.1	4.0	3.9	3.8	3.7	3.7	3.6
8	5.3	4.5	4.1	3.8	3.7	3.6	3.5	3.4	3.4	3.3
9	5.1	4.3	3.9	3.6	3.5	3.4	3.3	3.2	3.2	3.1
10	5.0	4.1	3.7	3.5	3.3	3.2	3.1	3.1	3.0	3.0
11	4.8	4.0	3.6	3.4	3.2	3.1	3.1	2.9	2.9	2.9
12	4.7	3.9	3.5	3.3	3.1	3.0	2.9	2.8	2.8	2.8

Using an F distribution table is similar to using a chi-squared distribution table, only this time the column headings across the top give the degrees of freedom for one sample and the row labels give the degrees of freedom for the other sample. A separate table is used for each common p-value.

In Table 1-7, when $v_1 = 1$ and $v_2 = 12$, the critical value is 4.7. This means that when we perform a statistical test, we calculate our test statistic and compare it to the critical value of 4.7 from this table; if our calculated test statistic is greater than 4.7, our result is considered *statistically significant*. In this table, for any test statistic greater than the number in the table, the p-value is less than .05. This means that when $v_1 = 1$ and $v_2 = 12$, the probability of an F statistic of 4.7 or higher occurring when your null hypothesis is true is 5%, so there's only a 5% chance of rejecting the null hypothesis when it is actually true.

Let's look at another example. Table 1-8 shows the F distribution table when $p = .01$.

TABLE 1-8: F PROBABILITY DISTRIBUTION TABLE FOR $p = .01$

v_2 \ v_1	1	2	3	4	5	6	7	8	9	10
1	4052.2	4999.3	5403.5	5624.3	5764.0	5859.0	5928.3	5981.0	6022.4	6055.9
2	98.5	99.0	99.2	99.3	99.3	99.3	99.4	99.4	99.4	99.4
3	34.1	30.8	29.5	28.7	28.2	27.9	27.7	27.5	27.3	27.2
4	21.2	18.8	16.7	16.0	15.5	15.2	15.0	14.8	14.7	14.5
5	16.3	13.3	12.1	11.4	11.0	10.7	10.5	10.3	10.2	10.1
6	13.7	10.9	9.8	9.1	8.7	8.5	8.3	8.1	8.0	7.9
7	12.2	9.5	8.5	7.8	7.5	7.2	7.0	6.8	6.7	6.6
8	11.3	8.6	7.6	7.0	6.6	6.4	6.2	6.0	5.9	5.8
9	10.6	8.0	7.0	6.4	6.1	5.8	5.6	5.5	5.4	5.6
10	10.0	7.6	6.6	6.0	5.6	5.4	5.2	5.1	4.9	4.8
11	9.6	7.2	6.2	5.7	5.3	5.1	4.9	4.7	4.6	4.5
12	9.3	6.9	6.0	5.4	5.1	4.8	4.6	4.5	4.4	4.3

Now when $v_1 = 1$ and $v_2 = 12$, the critical value is 9.3. The probability that a sample statistic as large or larger than 9.3 would occur if your null hypothesis is true is only .01. Thus, there is a very small probability that you would incorrectly reject the null hypothesis. Notice that when $p = .01$, the critical value is larger than when $p = .05$. For constant v_1 and v_2, as the p-value goes down, the critical value goes up.

2
SIMPLE
REGRESSION
ANALYSIS

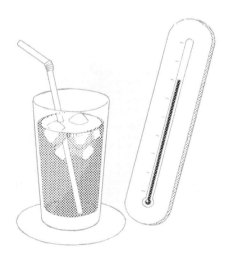

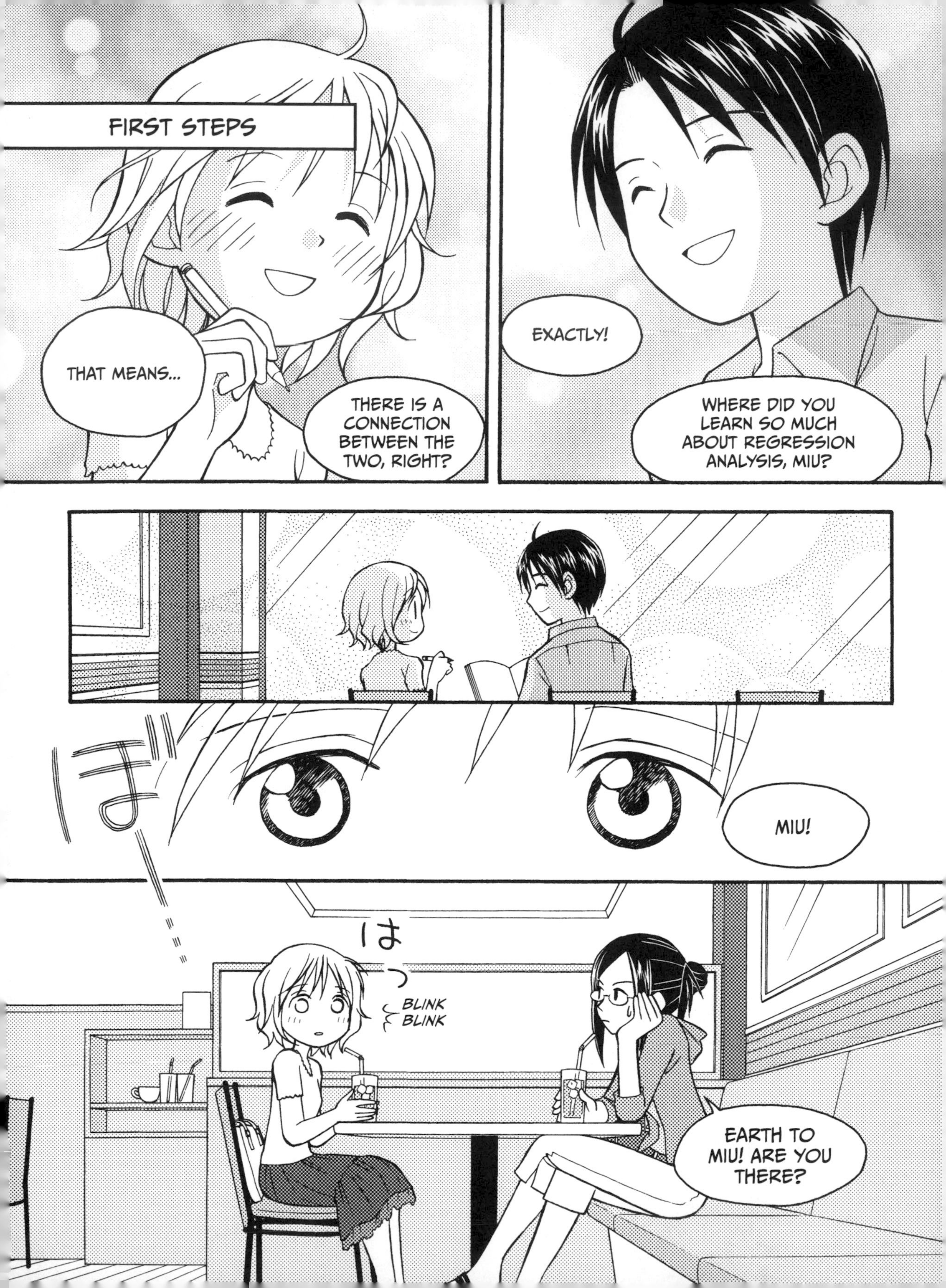
FIRST STEPS
THAT MEANS...
THERE IS A CONNECTION BETWEEN THE TWO, RIGHT?
EXACTLY!
WHERE DID YOU LEARN SO MUCH ABOUT REGRESSION ANALYSIS, MIU?
ぼ
MIU!
はっ
BLINK BLINK
EARTH TO MIU! ARE YOU THERE?

YOU WERE STARING AT THAT COUPLE.

I GOT IT!

ACK, YOU CAUGHT ME!

IT'S JUST... THEY'RE STUDYING TOGETHER.

I WISH I COULD STUDY WITH HIM LIKE THAT.

THAT'S WHY I AM TEACHING YOU! AND THERE'S NO CRYING IN STATISTICS!

I'M SORRY!

PAT

PAT

THERE, THERE.

WE'RE FINALLY DOING REGRESSION ANALYSIS TODAY. DOESN'T THAT CHEER YOU UP?

YES. I WANT TO LEARN.

SIGH

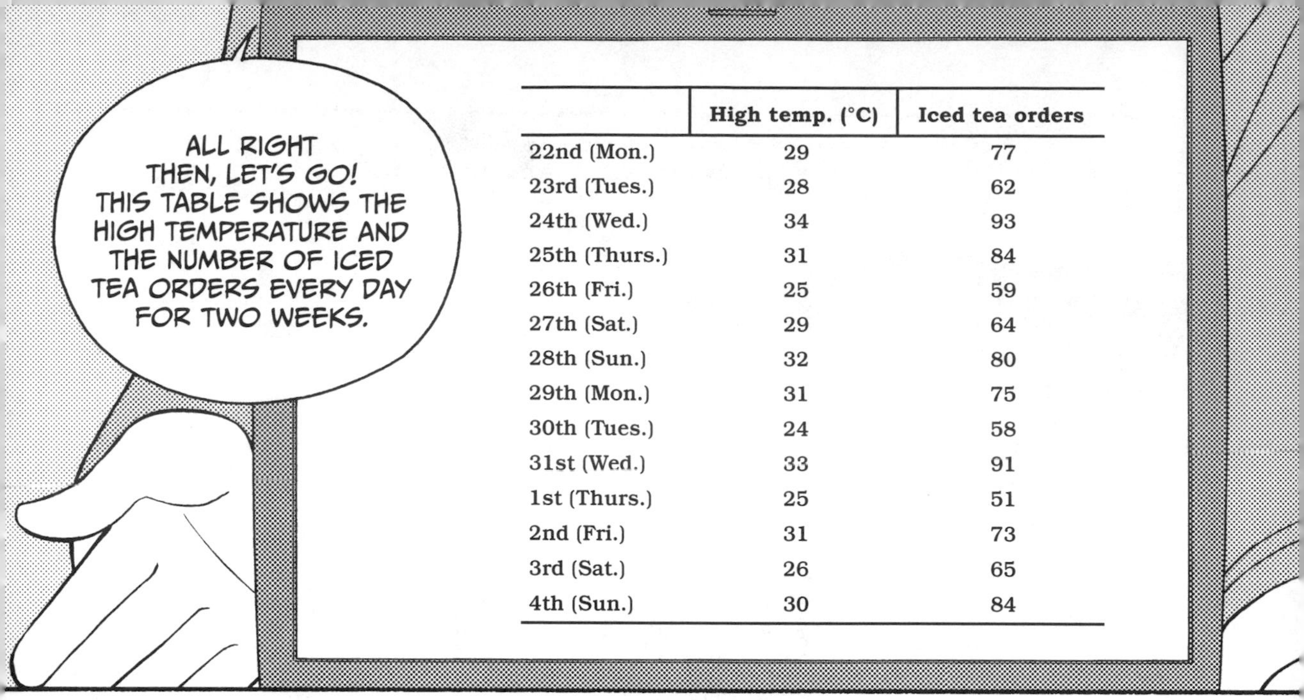
ALL RIGHT THEN, LET'S GO! THIS TABLE SHOWS THE HIGH TEMPERATURE AND THE NUMBER OF ICED TEA ORDERS EVERY DAY FOR TWO WEEKS.

	High temp. (°C)	Iced tea orders
22nd (Mon.)	29	77
23rd (Tues.)	28	62
24th (Wed.)	34	93
25th (Thurs.)	31	84
26th (Fri.)	25	59
27th (Sat.)	29	64
28th (Sun.)	32	80
29th (Mon.)	31	75
30th (Tues.)	24	58
31st (Wed.)	33	91
1st (Thurs.)	25	51
2nd (Fri.)	31	73
3rd (Sat.)	26	65
4th (Sun.)	30	84

PLOTTING THE DATA

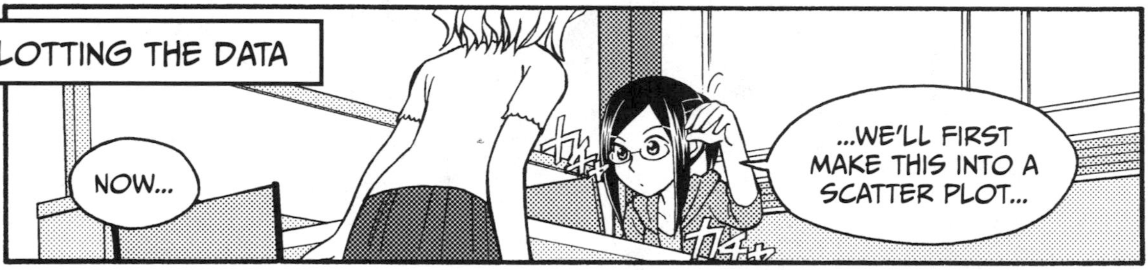
NOW...
...WE'LL FIRST MAKE THIS INTO A SCATTER PLOT...

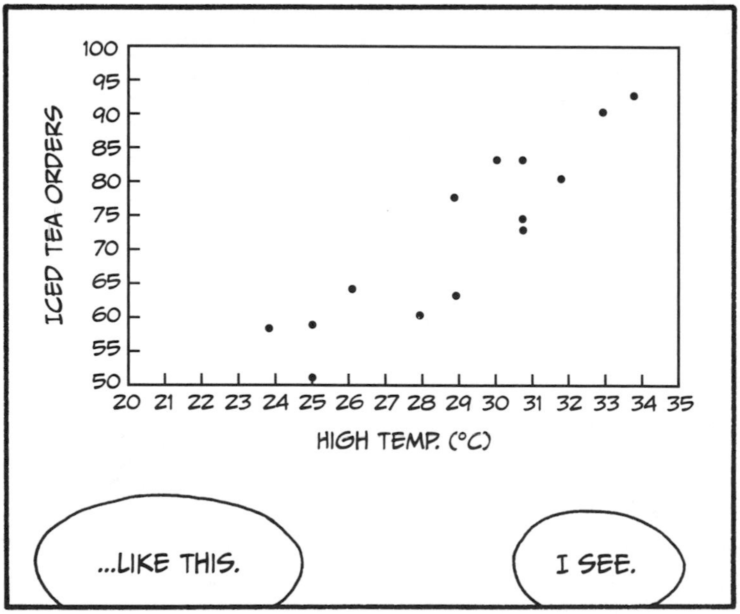
ICED TEA ORDERS
100
95
90
85
80
75
70
65
60
55
50
20 21 22 23 24 25 26 27 28 29 30 31 32 33 34 35
HIGH TEMP. (°C)
...LIKE THIS.
I SEE.

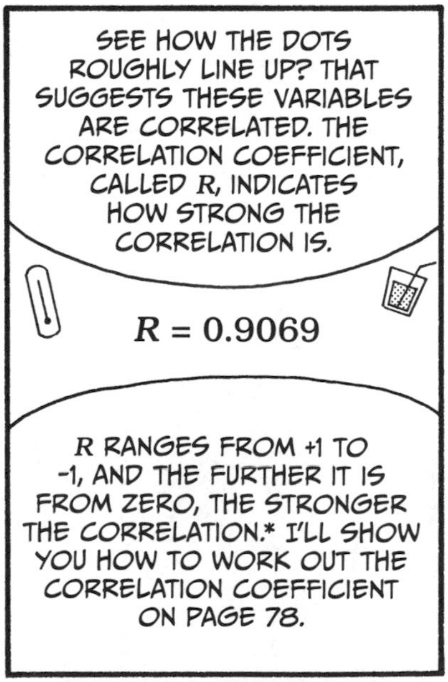
SEE HOW THE DOTS ROUGHLY LINE UP? THAT SUGGESTS THESE VARIABLES ARE CORRELATED. THE CORRELATION COEFFICIENT, CALLED *R*, INDICATES HOW STRONG THE CORRELATION IS.
$R = 0.9069$
R RANGES FROM +1 TO −1, AND THE FURTHER IT IS FROM ZERO, THE STRONGER THE CORRELATION.* I'LL SHOW YOU HOW TO WORK OUT THE CORRELATION COEFFICIENT ON PAGE 78.

* A POSITIVE *R* VALUE INDICATES A POSITIVE RELATIONSHIP, MEANING AS *x* INCREASES, SO DOES *y*. A NEGATIVE *R* VALUE MEANS AS THE *x* VALUE INCREASES, THE *y* VALUE DECREASES.

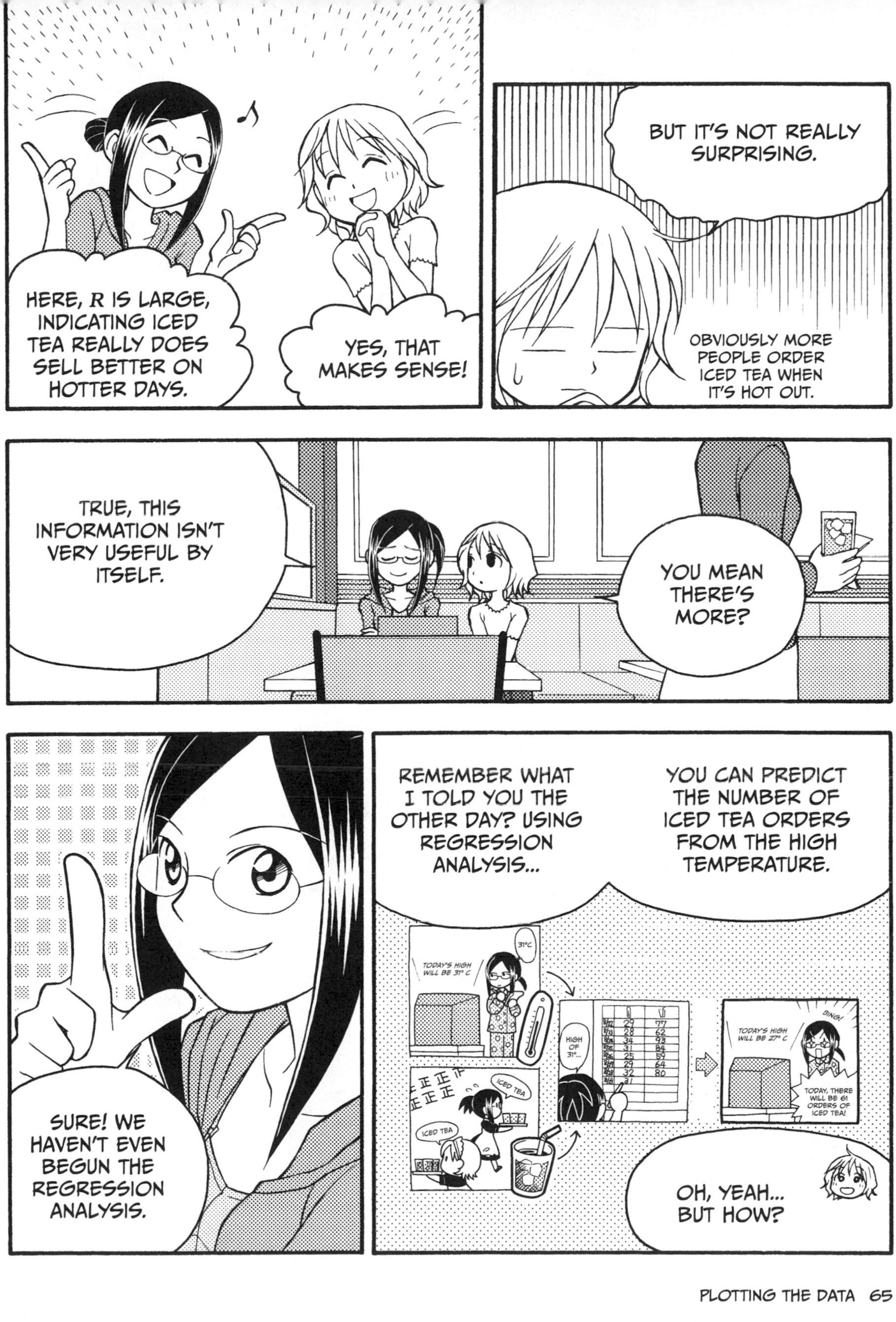
HERE, R IS LARGE, INDICATING ICED TEA REALLY DOES SELL BETTER ON HOTTER DAYS.
YES, THAT MAKES SENSE!
BUT IT'S NOT REALLY SURPRISING.
OBVIOUSLY MORE PEOPLE ORDER ICED TEA WHEN IT'S HOT OUT.
TRUE, THIS INFORMATION ISN'T VERY USEFUL BY ITSELF.
YOU MEAN THERE'S MORE?
SURE! WE HAVEN'T EVEN BEGUN THE REGRESSION ANALYSIS.
REMEMBER WHAT I TOLD YOU THE OTHER DAY? USING REGRESSION ANALYSIS...
YOU CAN PREDICT THE NUMBER OF ICED TEA ORDERS FROM THE HIGH TEMPERATURE.
TODAY'S HIGH WILL BE 31° C
31°C
ICED TEA
ICED TEA
HIGH OF 31°...
TODAY'S HIGH WILL BE 27° C
BING!
TODAY, THERE WILL BE 61 ORDERS OF ICED TEA!
OH, YEAH... BUT HOW?

THE REGRESSION EQUATION

BASICALLY, THE GOAL OF REGRESSION ANALYSIS IS...
ARE YOU READY?
HOLD ON! LET ME GRAB A PENCIL.
...TO OBTAIN THE REGRESSION EQUATION...
...IN THE FORM OF $y = ax + b$.
ICED TEA ORDERS
HIGH TEMP. (°C)
ICED TEA ORDERS
HIGH TEMP. (°C)
$y=ax+b$
WHAT CAN THAT TELL US?
IF YOU INPUT A HIGH TEMPERATURE FOR x...
$y=ax+b$
SCRITCH
SCRATCH
...YOU CAN PREDICT HOW MANY ORDERS OF ICED TEA THERE WILL BE (y).

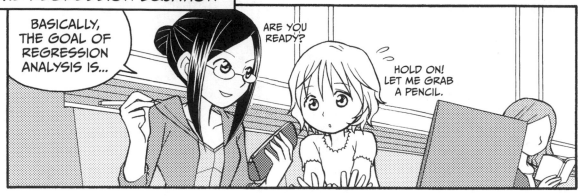

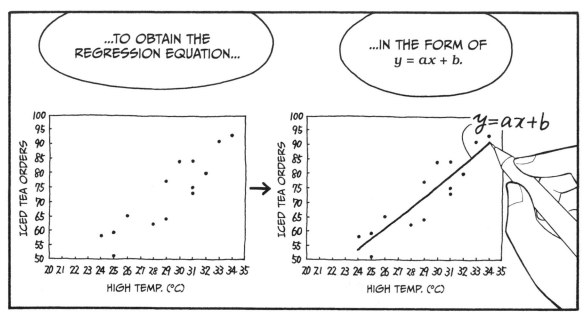

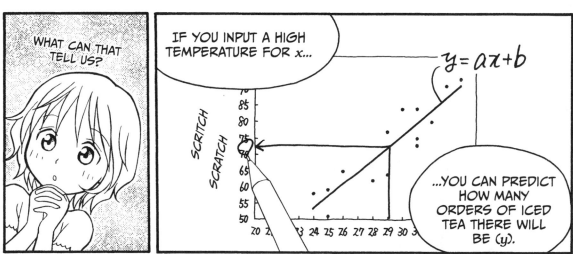

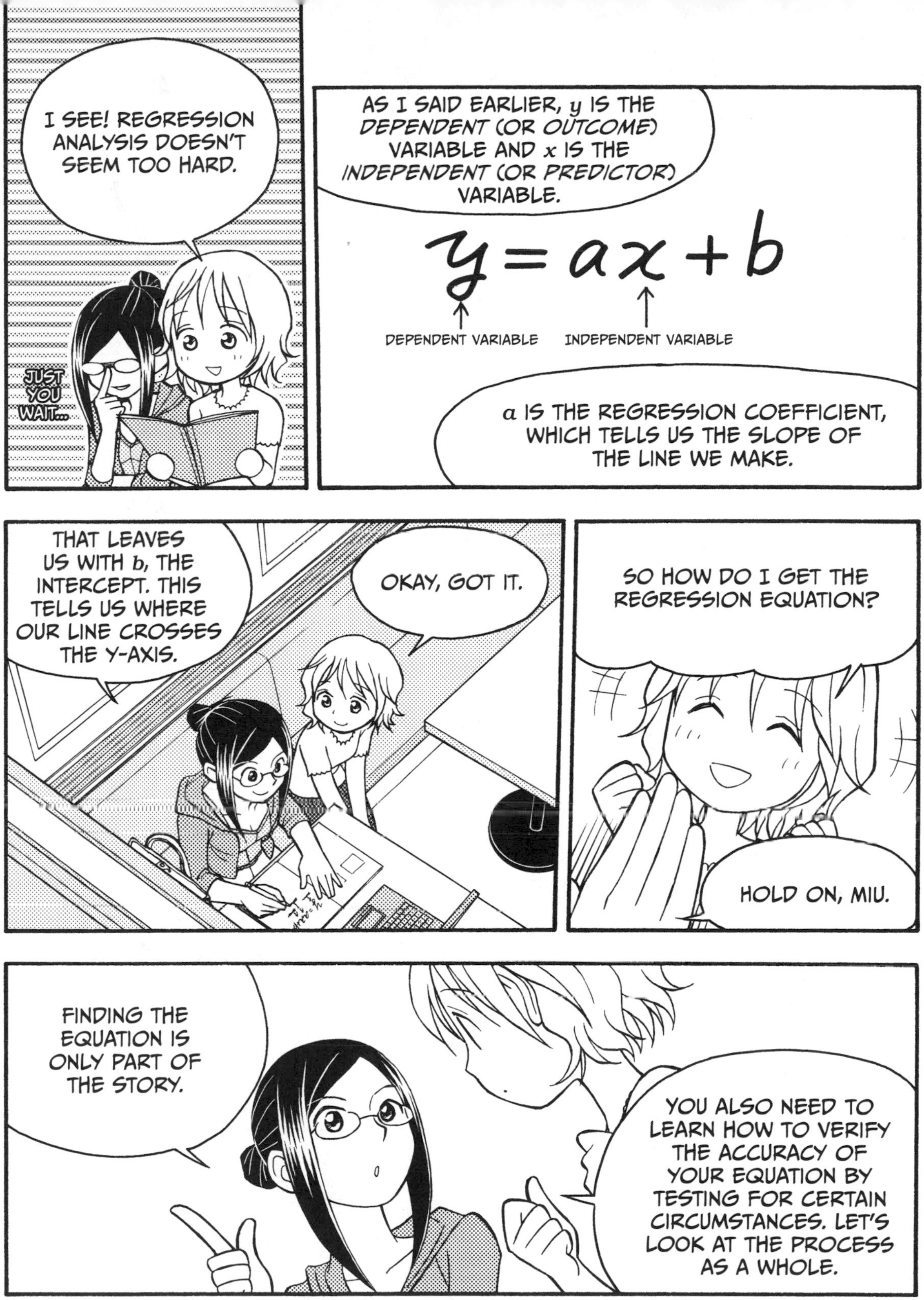
I SEE! REGRESSION ANALYSIS DOESN'T SEEM TOO HARD.
JUST YOU WAIT...
AS I SAID EARLIER, *y* IS THE *DEPENDENT* (OR *OUTCOME*) VARIABLE AND *x* IS THE *INDEPENDENT* (OR *PREDICTOR*) VARIABLE.
$y = ax + b$
DEPENDENT VARIABLE
INDEPENDENT VARIABLE
a IS THE REGRESSION COEFFICIENT, WHICH TELLS US THE SLOPE OF THE LINE WE MAKE.
THAT LEAVES US WITH *b*, THE INTERCEPT. THIS TELLS US WHERE OUR LINE CROSSES THE Y-AXIS.
OKAY, GOT IT.
SO HOW DO I GET THE REGRESSION EQUATION?
HOLD ON, MIU.
FINDING THE EQUATION IS ONLY PART OF THE STORY.
YOU ALSO NEED TO LEARN HOW TO VERIFY THE ACCURACY OF YOUR EQUATION BY TESTING FOR CERTAIN CIRCUMSTANCES. LET'S LOOK AT THE PROCESS AS A WHOLE.

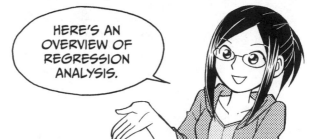

GENERAL REGRESSION ANALYSIS PROCEDURE

HERE'S AN OVERVIEW OF REGRESSION ANALYSIS.

STEP 1

DRAW A SCATTER PLOT OF THE INDEPENDENT VARIABLE VERSUS THE DEPENDENT VARIABLE. IF THE DOTS LINE UP, THE VARIABLES MAY BE CORRELATED.

y
x

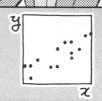

STEP 2

CALCULATE THE REGRESSION EQUATION.

y
y=ax+b
x

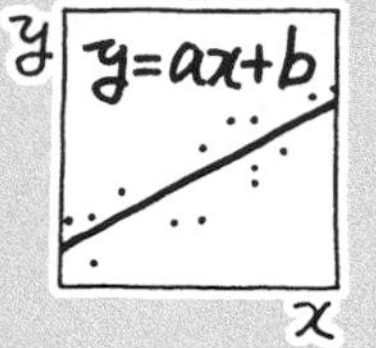

STEP 3

CALCULATE THE CORRELATION COEFFICIENT (*R*) AND ASSESS OUR POPULATION AND ASSUMPTIONS.

y
WHAT'S R?
?
x

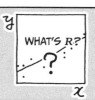

STEP 4

CONDUCT THE ANALYSIS OF VARIANCE.

y
Ax+B
x

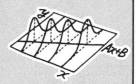

STEP 5

CALCULATE THE CONFIDENCE INTERVALS.

y
x

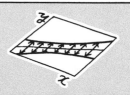

REGRESSION DIAGNOSTICS (STEPS 4–5)

STEP 6

MAKE A PREDICTION!

y
x

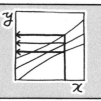

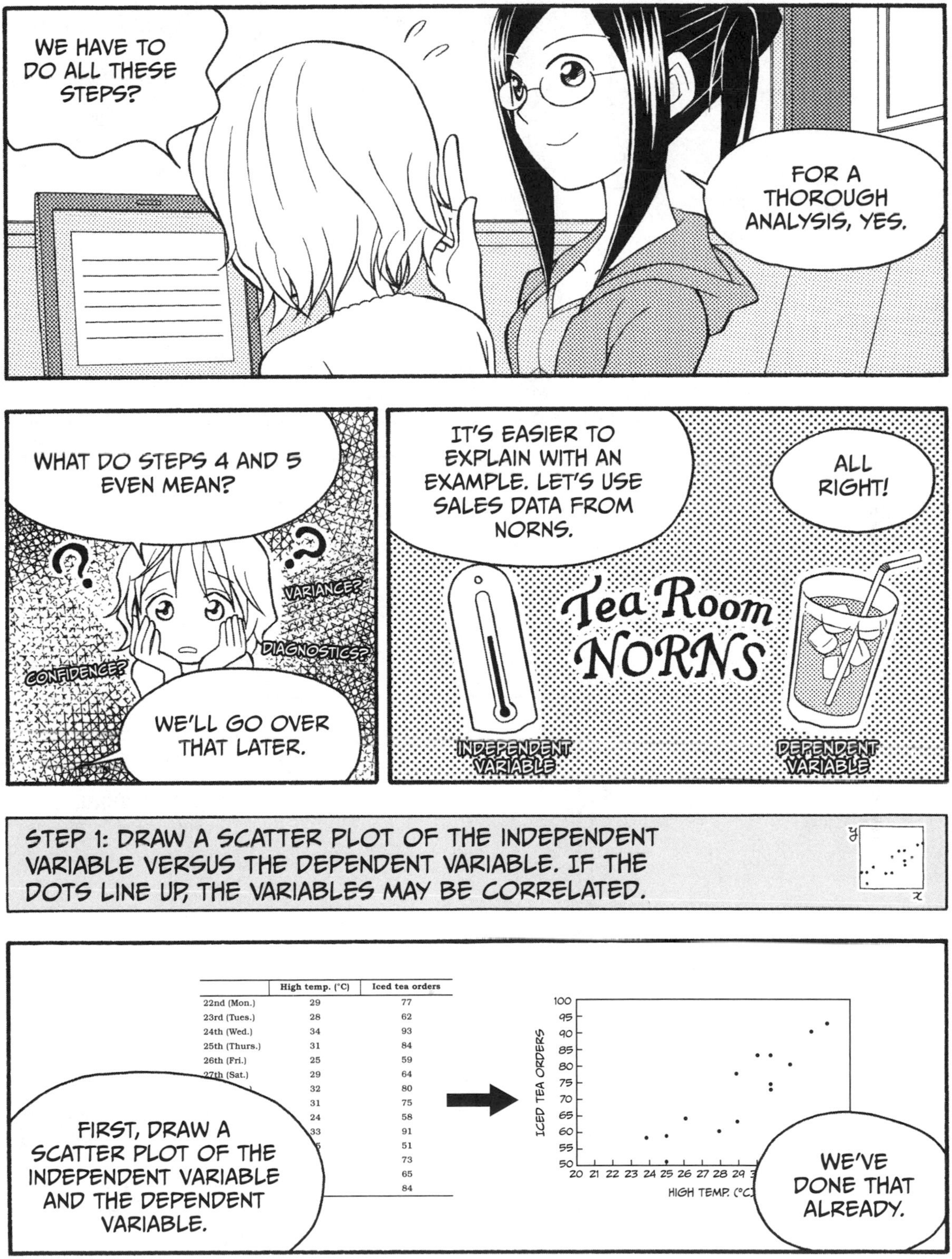
WE HAVE TO DO ALL THESE STEPS?
FOR A THOROUGH ANALYSIS, YES.
WHAT DO STEPS 4 AND 5 EVEN MEAN?
VARIANCE?
DIAGNOSTICS?
CONFIDENCE?
WE'LL GO OVER THAT LATER.
IT'S EASIER TO EXPLAIN WITH AN EXAMPLE. LET'S USE SALES DATA FROM NORNS.
ALL RIGHT!
Tea Room NORNS
INDEPENDENT VARIABLE
DEPENDENT VARIABLE
STEP 1: DRAW A SCATTER PLOT OF THE INDEPENDENT VARIABLE VERSUS THE DEPENDENT VARIABLE. IF THE DOTS LINE UP, THE VARIABLES MAY BE CORRELATED.
y
x
High temp. (°C) | Iced tea orders
22nd (Mon.) 29 77
23rd (Tues.) 28 62
24th (Wed.) 34 93
25th (Thurs.) 31 84
26th (Fri.) 25 59
27th (Sat.) 29 64
32 80
31 75
24 58
33 91
51
73
65
84
FIRST, DRAW A SCATTER PLOT OF THE INDEPENDENT VARIABLE AND THE DEPENDENT VARIABLE.
ICED TEA ORDERS
100 95 90 85 80 75 70 65 60 55 50
20 21 22 23 24 25 26 27 28 29
HIGH TEMP. (°C)
WE'VE DONE THAT ALREADY.

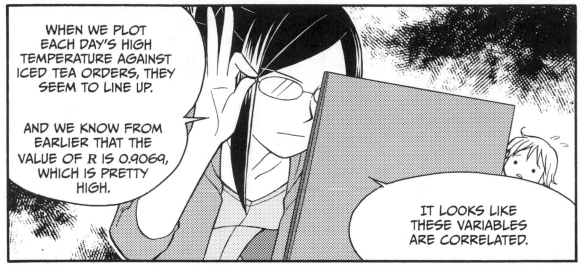

WHEN WE PLOT EACH DAY'S HIGH TEMPERATURE AGAINST ICED TEA ORDERS, THEY SEEM TO LINE UP.
AND WE KNOW FROM EARLIER THAT THE VALUE OF R IS 0.9069, WHICH IS PRETTY HIGH.
IT LOOKS LIKE THESE VARIABLES ARE CORRELATED.
DO YOU REALLY LEARN ANYTHING FROM ALL THOSE DOTS? WHY NOT JUST CALCULATE R?
ICED TEA ORDERS
HIGH TEMP. (°C)
THE SHAPE OF OUR DATA IS IMPORTANT!
ICED TEA ORDERS
HIGH TEMP. (°C)
LOOK AT THIS CHART. RATHER THAN FLOWING IN A LINE, THE DOTS ARE SCATTERED RANDOMLY.
ICED TEA ORDERS
y = 0.2x + 69.5
HIGH TEMP. (°C)
YOU CAN STILL FIND A REGRESSION EQUATION, BUT IT'S MEANINGLESS. THE LOW R VALUE CONFIRMS IT, BUT THE SCATTER PLOT LETS YOU SEE IT WITH YOUR OWN EYES.
ALWAYS DRAW A PLOT FIRST TO GET A SENSE OF THE DATA'S SHAPE.
OH, I SEE. PLOTS...ARE... IMPORTANT!

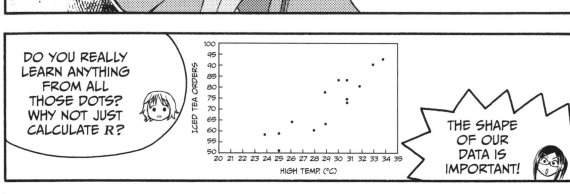

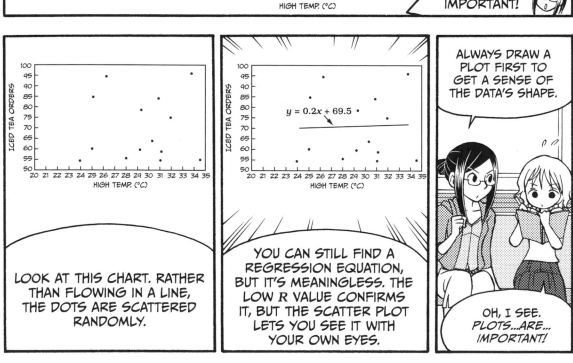

STEP 2: CALCULATE THE REGRESSION EQUATION.

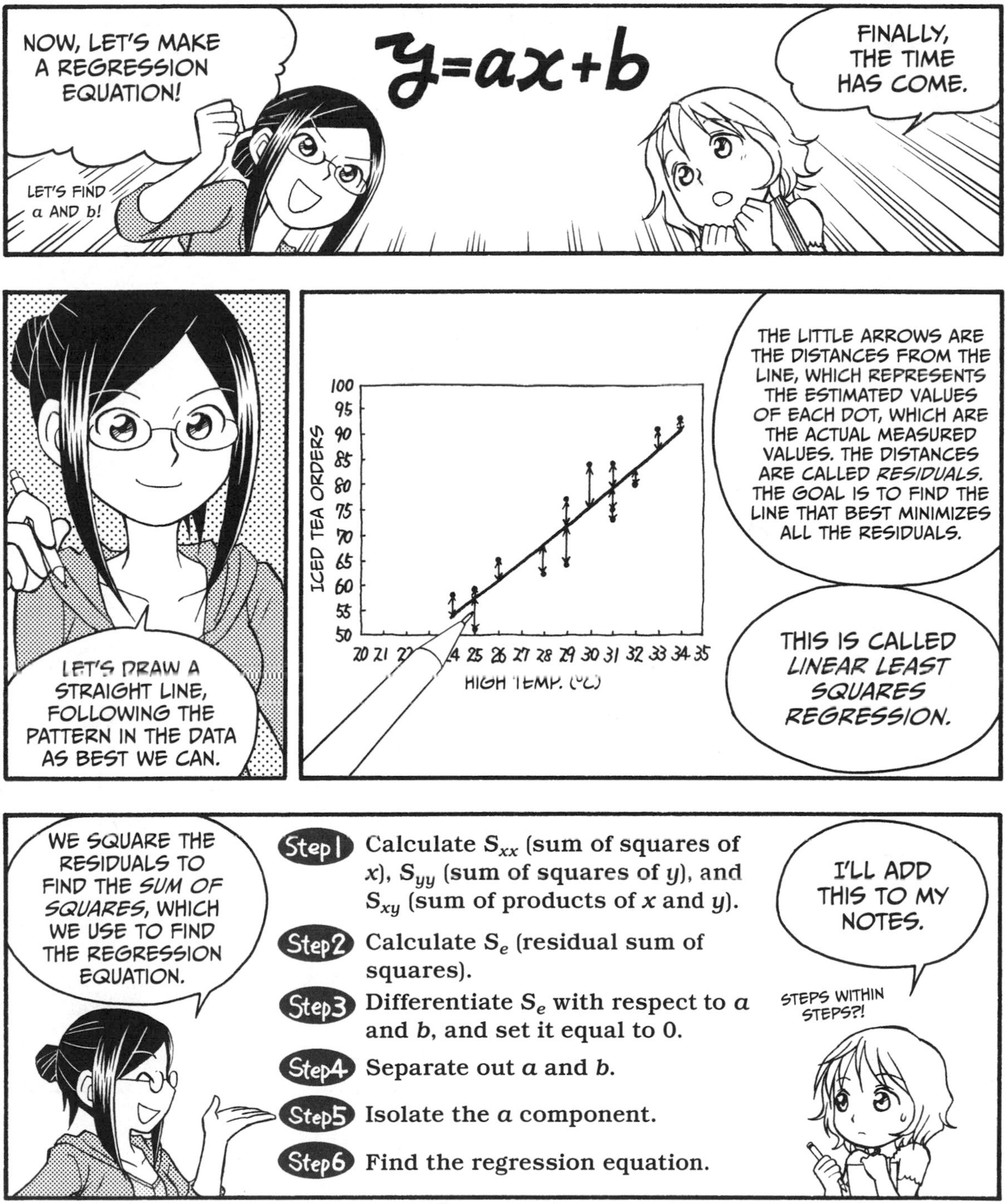

NOW, LET'S MAKE A REGRESSION EQUATION!
y=ax+b
LET'S FIND a AND b!
FINALLY, THE TIME HAS COME.
LET'S DRAW A STRAIGHT LINE, FOLLOWING THE PATTERN IN THE DATA AS BEST WE CAN.
ICED TEA ORDERS
100 95 90 85 80 75 70 65 60 55 50
20 21 22 23 24 25 26 27 28 29 30 31 32 33 34 35
HIGH TEMP. (°C)
THE LITTLE ARROWS ARE THE DISTANCES FROM THE LINE, WHICH REPRESENTS THE ESTIMATED VALUES OF EACH DOT, WHICH ARE THE ACTUAL MEASURED VALUES. THE DISTANCES ARE CALLED RESIDUALS. THE GOAL IS TO FIND THE LINE THAT BEST MINIMIZES ALL THE RESIDUALS.
THIS IS CALLED LINEAR LEAST SQUARES REGRESSION.
WE SQUARE THE RESIDUALS TO FIND THE SUM OF SQUARES, WHICH WE USE TO FIND THE REGRESSION EQUATION.
Step1 Calculate S_{xx} (sum of squares of x), S_{yy} (sum of squares of y), and S_{xy} (sum of products of x and y).
Step2 Calculate S_e (residual sum of squares).
Step3 Differentiate S_e with respect to a and b, and set it equal to 0.
Step4 Separate out a and b.
Step5 Isolate the a component.
Step6 Find the regression equation.
I'LL ADD THIS TO MY NOTES.
STEPS WITHIN STEPS?!

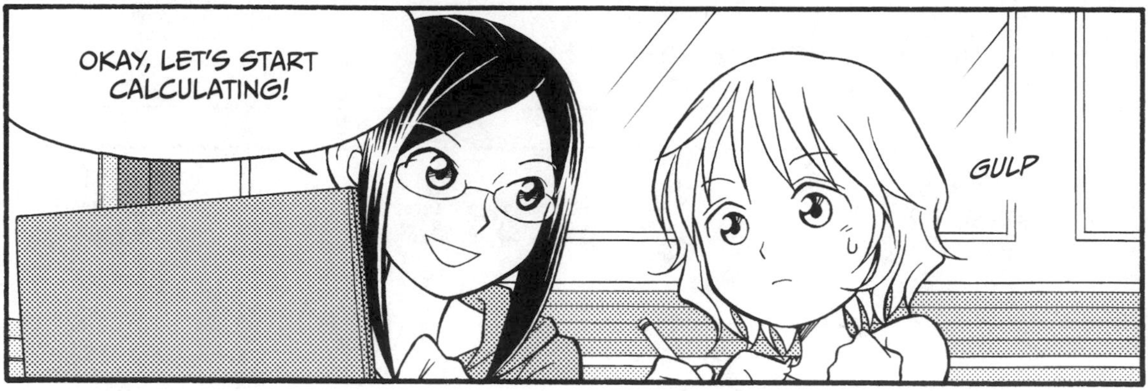
OKAY, LET'S START CALCULATING!
GULP

Step 1 Find

- The sum of squares of x, S_{xx}: $(x - \bar{x})^2$
- The sum of squares of y, S_{yy}: $(y - \bar{y})^2$
- The sum of products of x and y, S_{xy}: $(x - \bar{x})(y - \bar{y})$

Note: The bar over a variable (like $\bar{x}$) is a notation that means *average*. We can call this variable x-bar.

	High temp. in °C x	Iced tea orders y	$x - \bar{x}$	$y - \bar{y}$	$(x - \bar{x})^2$	$(y - \bar{y})^2$	$(x - \bar{x})(y - \bar{y})$
22nd (Mon.)	29	77	−0.1	4.4	0.0	19.6	−0.6
23rd (Tues.)	28	62	−1.1	−10.6	1.3	111.8	12.1
24th (Wed.)	34	93	4.9	20.4	23.6	417.3	99.2
25th (Thurs.)	31	84	1.9	11.4	3.4	130.6	21.2
26th (Fri.)	25	59	−4.1	−13.6	17.2	184.2	56.2
27th (Sat.)	29	64	−0.1	−8.6	0.0	73.5	1.2
28th (Sun.)	32	80	2.9	7.4	8.2	55.2	21.2
29th (Mon.)	31	75	1.9	2.4	3.4	5.9	4.5
30th (Tues.)	24	58	−5.1	−14.6	26.4	212.3	74.9
31st (Wed.)	33	91	3.9	18.4	14.9	339.6	71.1
1st (Thurs.)	25	51	−4.1	−21.6	17.2	465.3	89.4
2nd (Fri.)	31	73	1.9	0.4	3.4	0.2	0.8
3rd (Sat.)	26	65	−3.1	−7.6	9.9	57.8	23.8
4th (Sun.)	30	84	0.9	11.4	0.7	130.6	9.8
Sum	408	1016	0	0	129.7	2203.4	484.9
Average	29.1	72.6					
	$\bar{x}$	$\bar{y}$			S_{xx}	S_{yy}	S_{xy}

* SOME OF THE FIGURES IN THIS CHAPTER ARE ROUNDED FOR THE SAKE OF PRINTING, BUT CALCULATIONS ARE DONE USING THE FULL, UNROUNDED VALUES RESULTING FROM THE RAW DATA UNLESS OTHERWISE STATED.

Step 2 Find the residual sum of squares, S_e.

- y is the observed value.
- $\hat{y}$ is the the estimated value based on our regression equation.
- $y - \hat{y}$ is called the residual and is written as e.

Note: The caret in $\hat{y}$ is affectionately called a *hat*, so we call this parameter estimate y-hat.

	High temp. in °C x	Actual iced tea orders y	Predicted iced tea orders $\hat{y} = ax + b$	Residuals (e) $y - \hat{y}$	Squared residuals $(y - \hat{y})^2$
22nd (Mon.)	29	77	$a \times 29 + b$	$77 - (a \times 29 + b)$	$[77 - (a \times 29 + b)]^2$
23rd (Tues.)	28	62	$a \times 28 + b$	$62 - (a \times 28 + b)$	$[62 - (a \times 28 + b)]^2$
24th (Wed.)	34	93	$a \times 34 + b$	$93 - (a \times 34 + b)$	$[93 - (a \times 34 + b)]^2$
25th (Thurs.)	31	84	$a \times 31 + b$	$84 - (a \times 31 + b)$	$[84 - (a \times 31 + b)]^2$
26th (Fri.)	25	59	$a \times 25 + b$	$59 - (a \times 25 + b)$	$[59 - (a \times 25 + b)]^2$
27th (Sat.)	29	64	$a \times 29 + b$	$64 - (a \times 29 + b)$	$[64 - (a \times 29 + b)]^2$
28th (Sun.)	32	80	$a \times 32 + b$	$80 - (a \times 32 + b)$	$[80 - (a \times 32 + b)]^2$
29th (Mon.)	31	75	$a \times 31 + b$	$75 - (a \times 31 + b)$	$[75 - (a \times 31 + b)]^2$
30th (Tues.)	24	58	$a \times 24 + b$	$58 - (a \times 24 + b)$	$[58 - (a \times 24 + b)]^2$
31st (Wed.)	33	91	$a \times 33 + b$	$91 - (a \times 33 + b)$	$[91 - (a \times 33 + b)]^2$
1st (Thurs.)	25	51	$a \times 25 + b$	$51 - (a \times 25 + b)$	$[51 - (a \times 25 + b)]^2$
2nd (Fri.)	31	73	$a \times 31 + b$	$73 - (a \times 31 + b)$	$[73 - (a \times 31 + b)]^2$
3rd (Sat.)	26	65	$a \times 26 + b$	$65 - (a \times 26 + b)$	$[65 - (a \times 26 + b)]^2$
4th (Sun.)	30	84	$a \times 30 + b$	$84 - (a \times 30 + b)$	$[84 - (a \times 30 + b)]^2$
Sum	408	1016	$408a + 14b$	$1016 - (408a + 14b)$	S_e ←
Average	29.1	72.6	$29.1a + b$ $= \bar{x}a + b$	$72.6 - (29.1a + b)$ $= \bar{y} - (\bar{x}a + b)$	$= \dfrac{S_e}{14}$

$$\downarrow \quad \downarrow$$
$$\bar{x} \quad \bar{y}$$

$$S_e = \left[77 - (a \times 29 + b)\right]^2 + \cdots + \left[84 - (a \times 30 + b)\right]^2$$

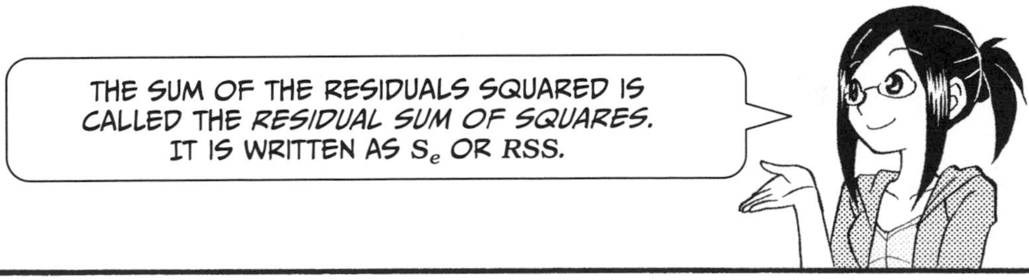

THE SUM OF THE RESIDUALS SQUARED IS CALLED THE *RESIDUAL SUM OF SQUARES*. IT IS WRITTEN AS S_e OR RSS.

Step 3 Differentiate S_e with respect to a and b, and set it equal to 0.
When differentiating $y = (ax + b)^n$ with respect to x, the result is
$$\frac{dy}{dx} = n(ax + b)^{n-1} \times a.$$

· Differentiate with respect to a.

$$\frac{dS_e}{da} = 2\big[77 - (29a + b)\big] \times (-29) + \cdots + 2\big[84 - (30a + b)\big] \times (-30) = 0 \quad \textbf{❶}$$

· Differentiate with respect to b.

$$\frac{dS_e}{db} = 2\big[77 - (29a + b)\big] \times (-1) + \cdots + 2\big[84 - (30a + b)\big] \times (-1) = 0 \quad \textbf{❷}$$

Step 4 Rearrange ❶ and ❷ from the previous step.

Rearrange ❶.

$$2\big[77 - (29a + b)\big] \times (-29) + \cdots + 2\big[84 - (30a + b)\big] \times (-30) = 0$$

$$\big[77 - (29a + b)\big] \times (-29) + \cdots + \big[84 - (30a + b)\big] \times (-30) = 0 \quad \text{DIVIDE BOTH SIDES BY 2.}$$

$$29\big[(29a + b) - 77\big] + \cdots + 30\big[(30a + b) - 84\big] = 0 \quad \text{MULTIPLY BY -1.}$$

$$(29 \times 29a + 29 \times b - 29 \times 77) + \cdots + (30 \times 30a + 30 \times b - 30 \times 84) = 0 \quad \text{MULTIPLY.}$$

❸ $(29^2 + \cdots + 30^2)a + (29 + \cdots + 30)b - (29 \times 77 + \cdots + 30 \times 84) = 0 \quad \text{SEPARATE OUT } a \text{ AND } b.$

Rearrange ❷.

$$2\big[77 - (29a + b)\big] \times (-1) + \cdots + 2\big[84 - (30a + b)\big] \times (-1) = 0$$

$$\big[77 - (29a + b)\big] \times (-1) + \cdots + \big[84 - (30a + b)\big] \times (-1) = 0 \quad \text{DIVIDE BOTH SIDES BY 2.}$$

$$\big[(29a + b) - 77\big] + \cdots + \big[(30a + b) - 84\big] = 0 \quad \text{MULTIPLY BY -1.}$$

$$(29 + \cdots + 30)a + \underbrace{b + \cdots + b}_{14} - (77 + \cdots + 84) = 0 \quad \text{SEPARATE OUT } a \text{ AND } b.$$

$$(29 + \cdots + 30)a + 14b - (77 + \cdots + 84) = 0$$

$$14b = (77 + \cdots + 84) - (29 + \cdots + 30)a \quad \text{SUBTRACT 14b FROM BOTH SIDES AND MULTIPLY BY -1.}$$

❹ $b = \dfrac{77 + \cdots + 84}{14} - \dfrac{29 + \cdots + 30}{14}a \quad \text{ISOLATE } b \text{ ON THE LEFT SIDE OF THE EQUATION.}$

❺ $b = \bar{y} - \bar{x}a \quad \text{THE COMPONENTS IN ❹ ARE THE AVERAGES OF } y \text{ AND } x.$

Step 5 Plug the value of b found in ❹ into line ❸ (❸ and ❹ are the results from Step 4).

❸ $\left(29^2 + \cdots + 30^2\right)a + \left(29 + \cdots + 30\right)\overset{❹}{\left(\dfrac{77 + \cdots + 84}{14} - \dfrac{29 + \cdots + 30}{14}a\right)} - \left(29 \times 77 + \cdots + 30 \times 84\right) = 0$ ◂ NOW a IS THE ONLY VARIABLE.

$\left(29^2 + \cdots + 30^2\right)a + \dfrac{\left(29 + \cdots + 30\right)\left(77 + \cdots + 84\right)}{14} - \dfrac{\left(29 + \cdots + 30\right)^2}{14}a - \left(29 \times 77 + \cdots + 30 \times 84\right) = 0$

$\left[\left(29^2 + \cdots + 30^2\right) - \dfrac{\left(29 + \cdots + 30\right)^2}{14}\right]a + \dfrac{\left(29 + \cdots + 30\right)\left(77 + \cdots + 84\right)}{14} - \left(29 \times 77 + \cdots + 30 \times 84\right) = 0$ ◂ COMBINE THE a TERMS.

$\left[\left(29^2 + \cdots + 30^2\right) - \dfrac{\left(29 + \cdots + 30\right)^2}{14}\right]a = \left(29 \times 77 + \cdots + 30 \times 84\right) - \dfrac{\left(29 + \cdots + 30\right)\left(77 + \cdots + 84\right)}{14}$ ◂ TRANSPOSE.

Rearrange the left side of the equation.

$\left(29^2 + \cdots + 30^2\right) - \dfrac{\left(29 + \cdots + 30\right)^2}{14}$

$= \left(29^2 + \cdots + 30^2\right) - 2 \times \dfrac{\left(29 + \cdots + 30\right)^2}{14} + \dfrac{\left(29 + \cdots + 30\right)^2}{14}$ ◂ WE ADD AND SUBTRACT $\dfrac{\left(29 + \cdots + 30\right)^2}{14}$.

$= \left(29^2 + \cdots + 30^2\right) - 2 \times \left(29 + \cdots + 30\right) \times \dfrac{29 + \cdots + 30}{14} + \left(\dfrac{29 + \cdots + 30}{14}\right)^2 \times 14$ ◂ THE LAST TERM IS MULTIPLIED BY $\dfrac{14}{14}$.

$= \left(29^2 + \cdots + 30^2\right) - 2 \times \left(29 + \cdots + 30\right) \times \bar{x} + \left(\bar{x}\right)^2 \times 14$ $\bar{x} = \dfrac{29 + \cdots + 30}{14}$

$= \left(29^2 + \cdots + 30^2\right) - 2 \times \left(29 + \cdots + 30\right) \times \bar{x} + \underbrace{\left(\bar{x}\right)^2 + \cdots + \left(\bar{x}\right)^2}_{14}$

$= \left[29^2 - 2 \times 29 \times \bar{x} + \left(\bar{x}\right)^2\right] + \cdots + \left[30^2 - 2 \times 30 \times \bar{x} + \left(\bar{x}\right)^2\right]$

$= \left(29 - \bar{x}\right)^2 + \cdots + \left(30 - \bar{x}\right)^2$

$= S_{xx}$

Rearrange the right side of the equation.

$\left(29 \times 77 + \cdots + 30 \times 84\right) - \dfrac{\left(29 + \cdots + 30\right)\left(77 + \cdots + 84\right)}{14}$

$= \left(29 \times 77 + \cdots + 30 \times 84\right) - \dfrac{29 + \cdots + 30}{14} \times \dfrac{77 + \cdots + 84}{14} \times 14$

$= \left(29 \times 77 + \cdots + 30 \times 84\right) - \bar{x} \times \bar{y} \times 14$

$= \left(29 \times 77 + \cdots + 30 \times 84\right) - \bar{x} \times \bar{y} \times 14 - \bar{x} \times \bar{y} \times 14 + \bar{x} \times \bar{y} \times 14$ ◂ WE ADD AND SUBTRACT $\bar{x} \times \bar{y} \times 14$.

$= \left(29 \times 77 + \cdots + 30 \times 84\right) - \dfrac{29 + \cdots + 30}{14} \times \bar{y} \times 14 - \bar{x} \times \dfrac{77 + \cdots + 84}{14} \times 14 + \bar{x} \times \bar{y} \times 14$

$= \left(29 \times 77 + \cdots + 30 \times 84\right) - \left(29 + \cdots + 30\right)\bar{y} - \bar{x}\left(77 + \cdots + 84\right) + \bar{x} \times \bar{y} \times 14$

$= \left(29 \times 77 + \cdots + 30 \times 84\right) - \left(29 + \cdots + 30\right)\bar{y} - \left(77 + \cdots + 84\right)\bar{x} + \underbrace{\bar{x} \times \bar{y} + \cdots + \bar{x} \times \bar{y}}_{14}$

$= \left(29 - \bar{x}\right)\left(77 - \bar{y}\right) + \cdots + \left(30 - \bar{x}\right)\left(84 - \bar{y}\right)$

$= S_{xy}$

$S_{xx}a = S_{xy}$

❻ $a = \dfrac{S_{xy}}{S_{xx}}$ ◂ ISOLATE a ON THE LEFT SIDE OF THE EQUATION.

Step 6 **Calculate the regression equation.**

From ❻ in Step 5, $a = \frac{S_{xy}}{S_{xx}}$. From ❺ in Step 4, $b = \bar{y} - \bar{x}a$.

If we plug in the values we calculated in Step 1,

$$\begin{cases} a = \dfrac{S_{xx}}{S_{xy}} = \dfrac{484.9}{129.7} = 3.7 \\ b = \bar{y} - \bar{x}a = 72.6 - 29.1 \times 3.7 = -36.4 \end{cases}$$

then the regression equation is

$$y = 3.7x - 36.4.$$

It's that simple!

Note: The values shown are rounded for the sake of printing, but the result (36.4) was calculated using the full, unrounded values.

y=3.7x−36.4
ICED TEA ORDERS
100 95 90 85 80 75 70 65 60 55 50
20 21 22 23 24 25 26 27 28 29 30 31 32 33 34 35
HIGH TEMP. (°C)
WE DID IT! WE ACTUALLY DID IT!
NICE JOB!
THE RELATIONSHIP BETWEEN THE RESIDUALS AND THE SLOPE a AND INTERCEPT b IS ALWAYS
$a = \frac{\text{sum of products of } x \text{ and } y}{\text{sum of squares of } x} = \frac{S_{xy}}{S_{xx}}$
$b = \bar{y} - \bar{x}a$
THIS IS TRUE FOR ANY LINEAR REGRESSION.

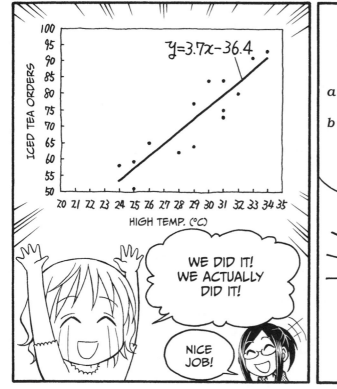

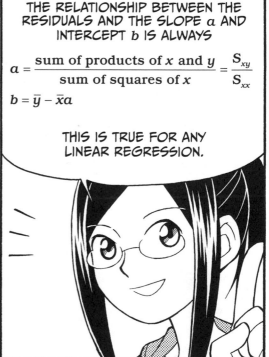

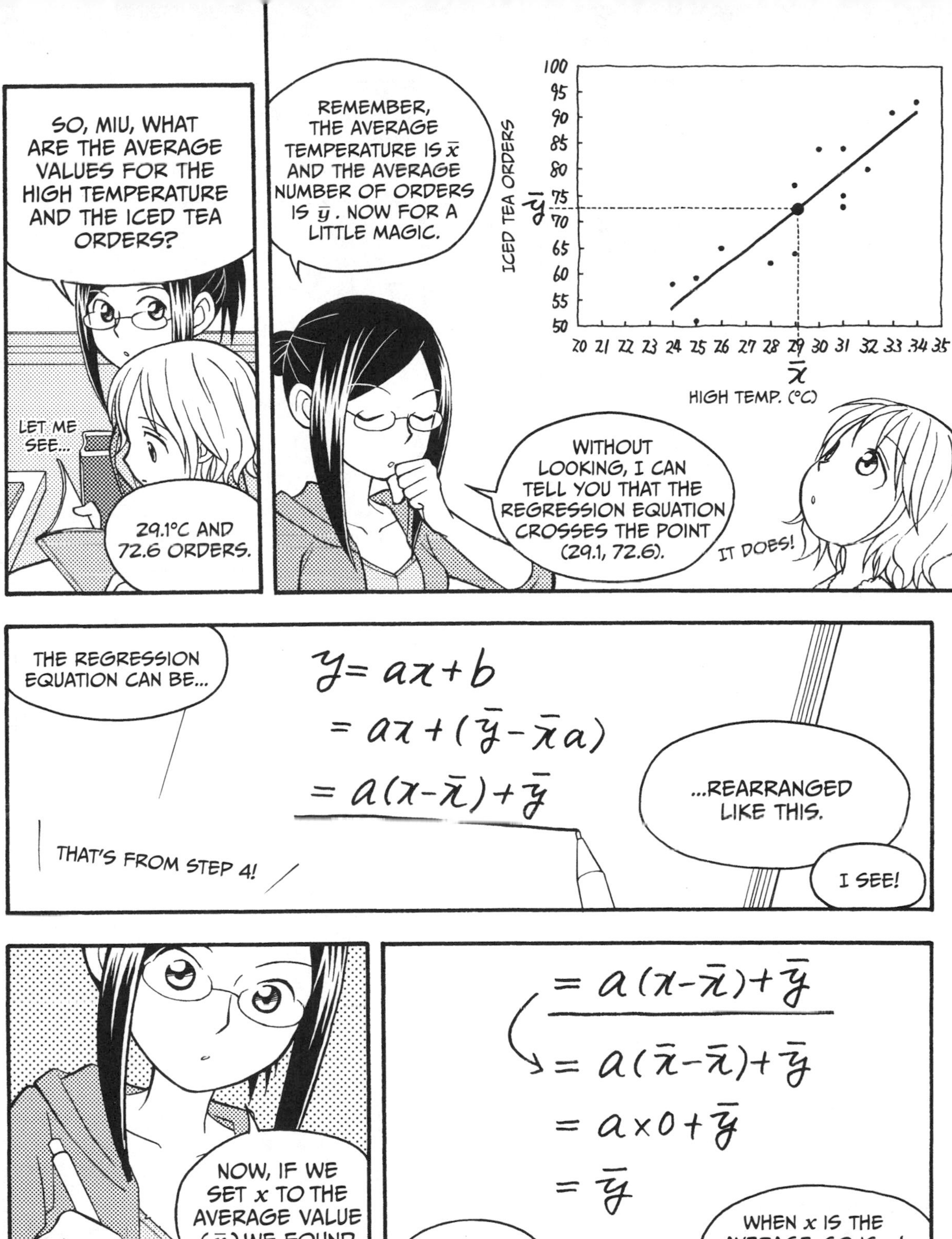
SO, MIU, WHAT ARE THE AVERAGE VALUES FOR THE HIGH TEMPERATURE AND THE ICED TEA ORDERS?
LET ME SEE...
29.1°C AND 72.6 ORDERS.
REMEMBER, THE AVERAGE TEMPERATURE IS $\bar{x}$ AND THE AVERAGE NUMBER OF ORDERS IS $\bar{y}$. NOW FOR A LITTLE MAGIC.
ICED TEA ORDERS
$\bar{y}$
100
95
90
85
80
75
70
65
60
55
50
20 21 22 23 24 25 26 27 28 29 30 31 32 33 34 35
$\bar{x}$
HIGH TEMP. (°C)
WITHOUT LOOKING, I CAN TELL YOU THAT THE REGRESSION EQUATION CROSSES THE POINT (29.1, 72.6).
IT DOES!
THE REGRESSION EQUATION CAN BE...
$y = ax + b$
$= ax + (\bar{y} - \bar{x}a)$
$= a(x - \bar{x}) + \bar{y}$
THAT'S FROM STEP 4!
...REARRANGED LIKE THIS.
I SEE!
NOW, IF WE SET x TO THE AVERAGE VALUE ($\bar{x}$) WE FOUND BEFORE...
$= a(x - \bar{x}) + \bar{y}$
$= a(\bar{x} - \bar{x}) + \bar{y}$
$= a \times 0 + \bar{y}$
$= \bar{y}$
SEE WHAT HAPPENS?
WHEN x IS THE AVERAGE, SO IS y!

STEP 3: CALCULATE THE CORRELATION COEFFICIENT (R) AND ASSESS OUR POPULATION AND ASSUMPTIONS.

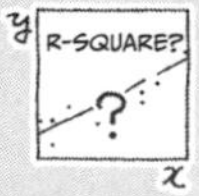

R-SQUARE?

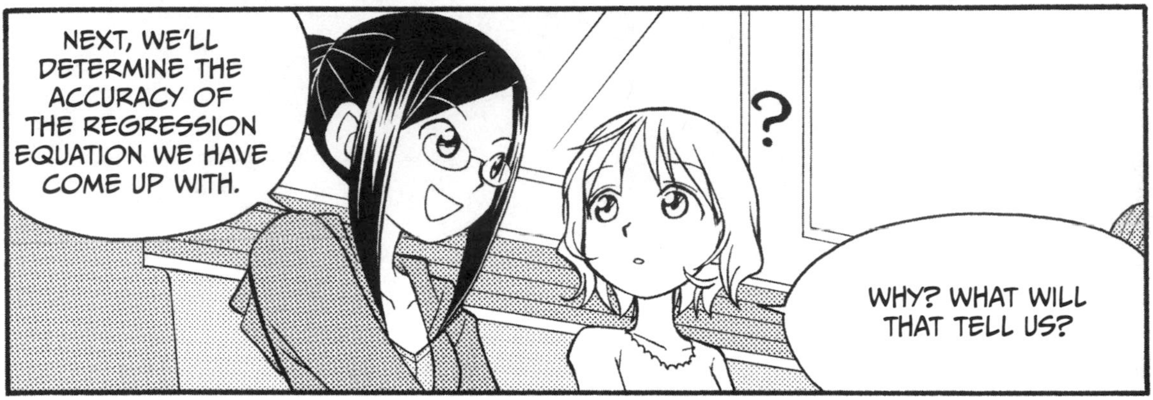

NEXT, WE'LL DETERMINE THE ACCURACY OF THE REGRESSION EQUATION WE HAVE COME UP WITH.
?
WHY? WHAT WILL THAT TELL US?

OUR DATA AND ITS REGRESSION EQUATION

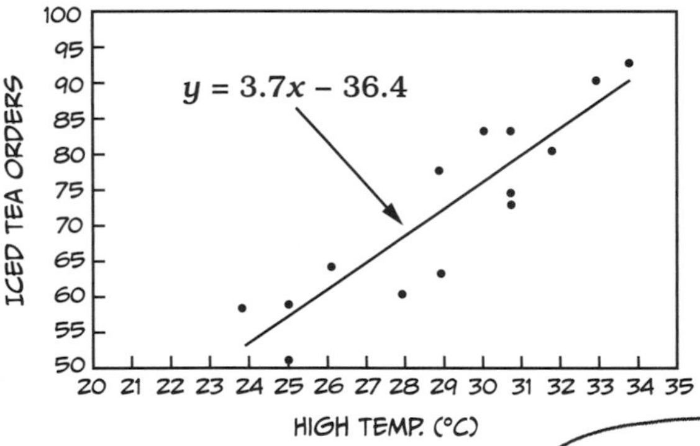

$y = 3.7x - 36.4$
ICED TEA ORDERS
100 95 90 85 80 75 70 65 60 55 50
20 21 22 23 24 25 26 27 28 29 30 31 32 33 34 35
HIGH TEMP. (°C)

EXAMPLE DATA AND ITS REGRESSION EQUATION

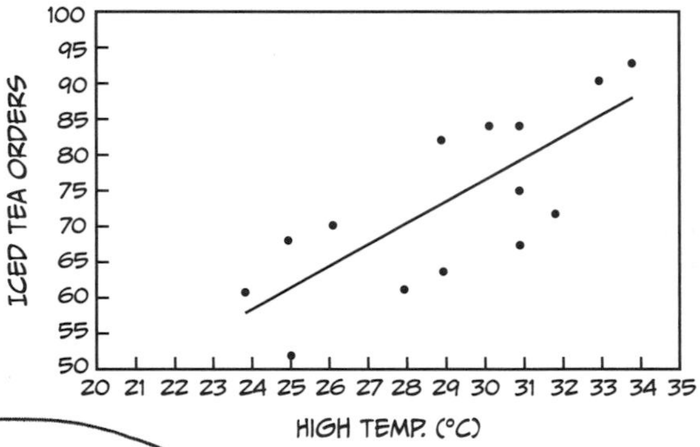

ICED TEA ORDERS
100 95 90 85 80 75 70 65 60 55 50
20 21 22 23 24 25 26 27 28 29 30 31 32 33 34 35
HIGH TEMP. (°C)

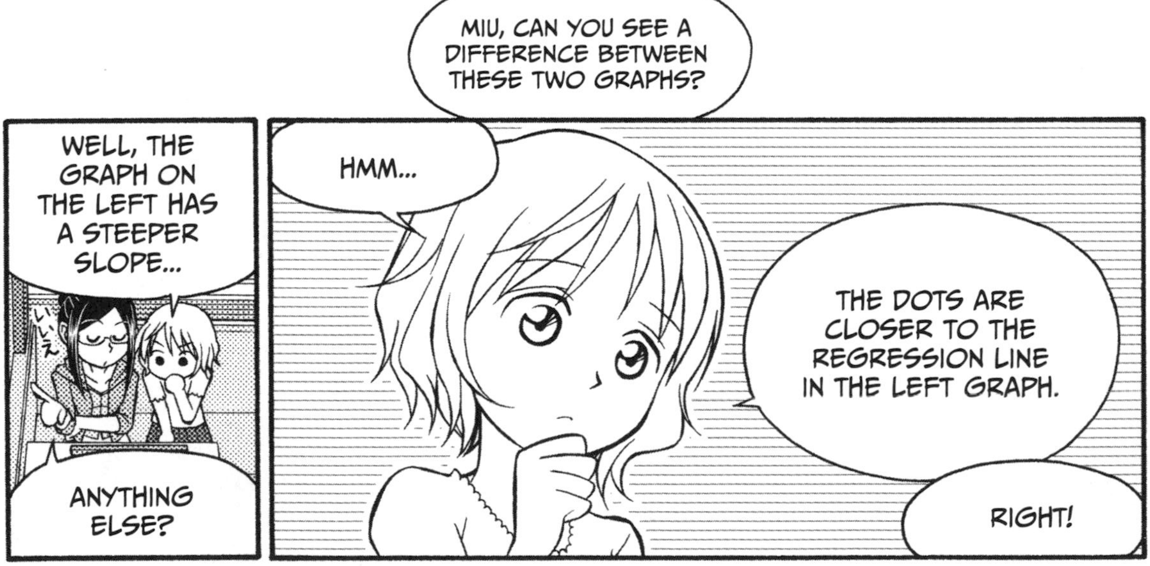

MIU, CAN YOU SEE A DIFFERENCE BETWEEN THESE TWO GRAPHS?
WELL, THE GRAPH ON THE LEFT HAS A STEEPER SLOPE...
いいえ
ANYTHING ELSE?
HMM...
THE DOTS ARE CLOSER TO THE REGRESSION LINE IN THE LEFT GRAPH.
RIGHT!

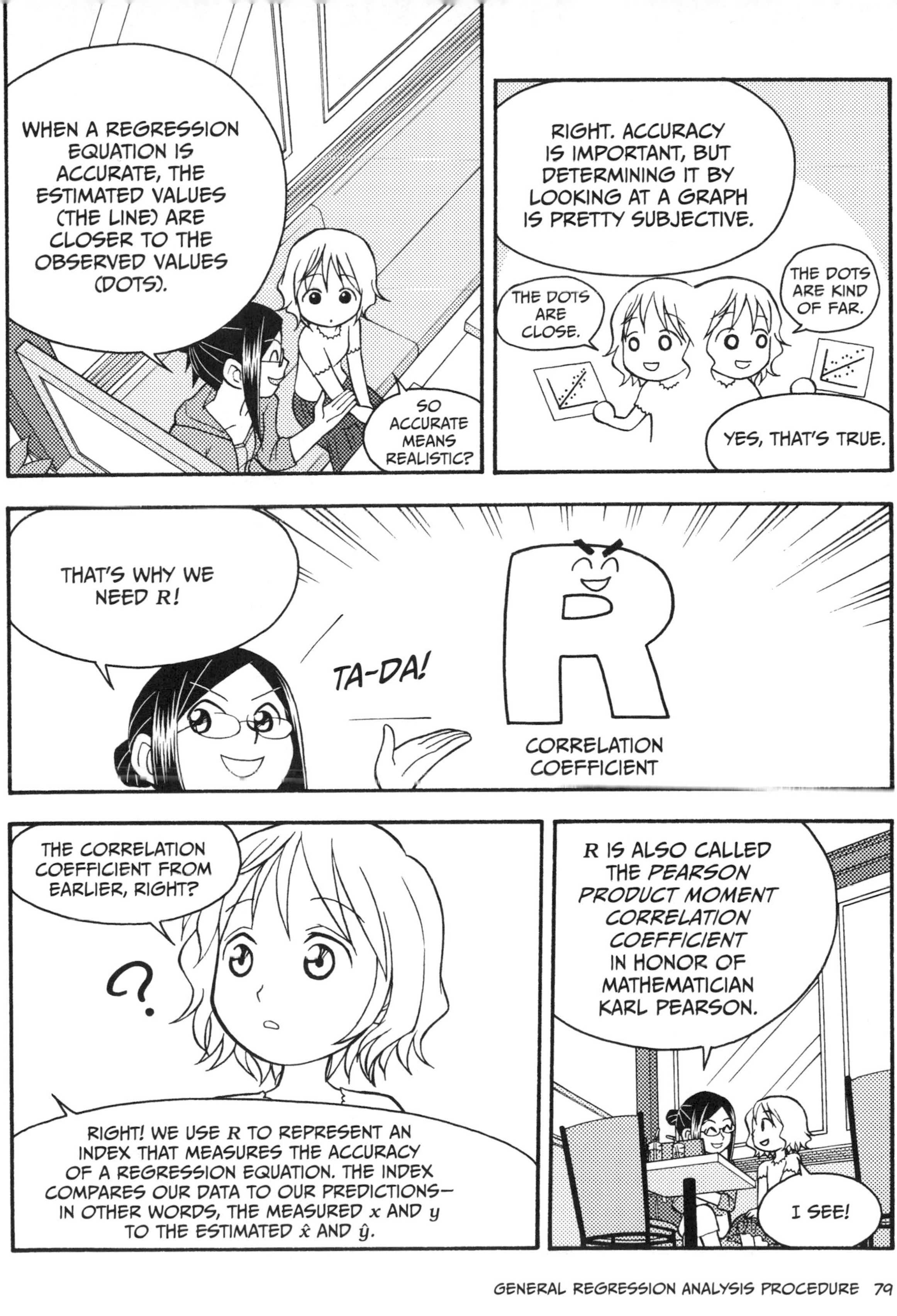
WHEN A REGRESSION EQUATION IS ACCURATE, THE ESTIMATED VALUES (THE LINE) ARE CLOSER TO THE OBSERVED VALUES (DOTS).
SO ACCURATE MEANS REALISTIC?
RIGHT. ACCURACY IS IMPORTANT, BUT DETERMINING IT BY LOOKING AT A GRAPH IS PRETTY SUBJECTIVE.
THE DOTS ARE CLOSE.
THE DOTS ARE KIND OF FAR.
YES, THAT'S TRUE.
THAT'S WHY WE NEED *R*!
TA-DA!
CORRELATION COEFFICIENT
THE CORRELATION COEFFICIENT FROM EARLIER, RIGHT?
?
RIGHT! WE USE *R* TO REPRESENT AN INDEX THAT MEASURES THE ACCURACY OF A REGRESSION EQUATION. THE INDEX COMPARES OUR DATA TO OUR PREDICTIONS— IN OTHER WORDS, THE MEASURED x AND y TO THE ESTIMATED $\hat{x}$ AND $\hat{y}$.
R IS ALSO CALLED THE *PEARSON PRODUCT MOMENT CORRELATION COEFFICIENT* IN HONOR OF MATHEMATICIAN KARL PEARSON.
I SEE!

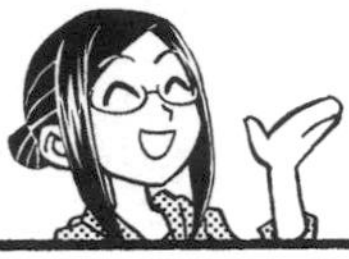
HERE'S THE EQUATION. WE CALCULATE THESE LIKE WE DID S_{xx} AND S_{xy} BEFORE.

$$R = \frac{\text{sum of products } y \text{ and } \hat{y}}{\sqrt{\text{sum of squares of } y \times \text{sum of squares of } \hat{y}}} = \frac{S_{y\hat{y}}}{\sqrt{S_{yy} \times S_{\hat{y}\hat{y}}}}$$

$$= \frac{1812.3}{\sqrt{2203.4 \times 1812.3}} = 0.9069$$

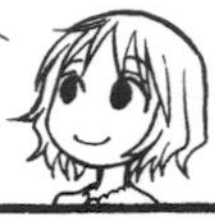
THAT'S NOT TOO BAD!

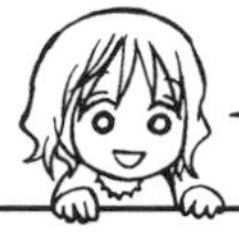
THIS LOOKS FAMILIAR.
REGRESSION FUNCTION!

	Actual values y	Estimated values $\hat{y} = 3.7x - 36.4$	$y - \bar{y}$	$\hat{y} - \bar{\hat{y}}$	$(y - \bar{y})^2$	$(\hat{y} - \bar{\hat{y}})^2$	$(y - \bar{y})(\hat{y} - \bar{\hat{y}})$	$(y - \hat{y})^2$
22nd (Mon.)	77	72.0	4.4	–0.5	19.6	0.3	–2.4	24.6
23rd (Tues.)	62	68.3	–10.6	–4.3	111.8	18.2	45.2	39.7
24th (Wed.)	93	90.7	20.4	18.2	417.3	329.6	370.9	5.2
25th (Thurs.)	84	79.5	11.4	6.9	130.6	48.2	79.3	20.1
26th (Fri.)	59	57.1	–13.6	–15.5	184.2	239.8	210.2	3.7
27th (Sat.)	64	72.0	–8.6	–0.5	73.5	0.3	4.6	64.6
28th (Sun.)	80	83.3	7.4	10.7	55.2	114.1	79.3	10.6
29th (Mon.)	75	79.5	2.4	6.9	5.9	48.2	16.9	20.4
30th (Tues.)	58	53.3	–14.6	–19.2	212.3	369.5	280.1	21.6
31st (Wed.)	91	87.0	18.4	14.4	339.6	207.9	265.7	16.1
1st (Thurs.)	51	57.1	–21.6	–15.5	465.3	239.8	334.0	37.0
2nd (Fri.)	73	79.5	0.4	6.9	0.2	48.2	3.0	42.4
3rd (Sat.)	65	60.8	–7.6	–11.7	57.3	138.0	88.9	17.4
4th (Sun.)	84	75.8	11.4	3.2	130.6	10.3	36.6	67.6
Sum	1016	1016	0	0	2203.4	1812.3	1812.3	391.1
Average	72.6	72.6						
	↓ $\bar{y}$	↓ $\bar{\hat{y}}$			↓ S_{yy}	↓ $S_{\hat{y}\hat{y}}$	↓ $S_{y\hat{y}}$	↓ S_e

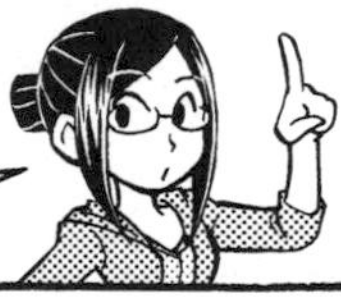
S_e ISN'T NECESSARY FOR CALCULATING R, BUT I INCLUDED IT BECAUSE WE'LL NEED IT LATER.

IF WE SQUARE *R*, IT'S CALLED THE *COEFFICIENT OF DETERMINATION* AND IS WRITTEN AS R^2.

R^2 CAN BE AN INDICATOR OF...

I AM A COEFFICIENT OF DETERMINATION.

I AM A CORRELATION COEFFICIENT.

I AM A CORRELATION COEFFICIENT, TOO.

...HOW MUCH VARIANCE IS EXPLAINED BY OUR REGRESSION EQUATION.

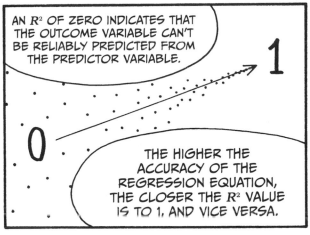

AN R^2 OF ZERO INDICATES THAT THE OUTCOME VARIABLE CAN'T BE RELIABLY PREDICTED FROM THE PREDICTOR VARIABLE.

THE HIGHER THE ACCURACY OF THE REGRESSION EQUATION, THE CLOSER THE R^2 VALUE IS TO 1, AND VICE VERSA.

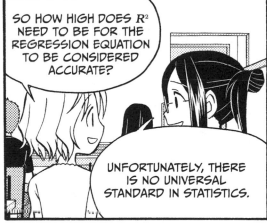

SO HOW HIGH DOES R^2 NEED TO BE FOR THE REGRESSION EQUATION TO BE CONSIDERED ACCURATE?

UNFORTUNATELY, THERE IS NO UNIVERSAL STANDARD IN STATISTICS.

BUT GENERALLY WE WANT A VALUE OF AT LEAST .5.

LOWEST... .5...

NOW TRY FINDING THE VALUE OF R^2.

SURE THING.

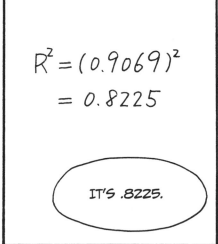

$$R^2 = (0.9069)^2 = 0.8225$$

IT'S .8225.

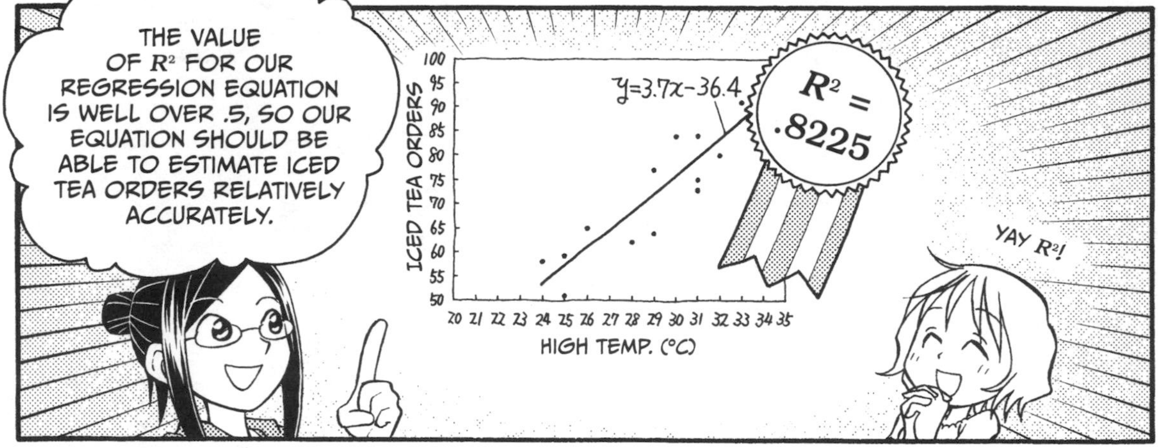
THE VALUE OF R^2 FOR OUR REGRESSION EQUATION IS WELL OVER .5, SO OUR EQUATION SHOULD BE ABLE TO ESTIMATE ICED TEA ORDERS RELATIVELY ACCURATELY.
ICED TEA ORDERS
HIGH TEMP. (°C)
y=3.7x−36.4
R^2 = .8225
YAY R^2!

$$R^2 = \left(\begin{matrix}\text{correlation}\\ \text{coefficient}\end{matrix}\right)^2 = \frac{a \times S_{xy}}{S_{yy}} = 1 - \frac{S_e}{S_{yy}}$$

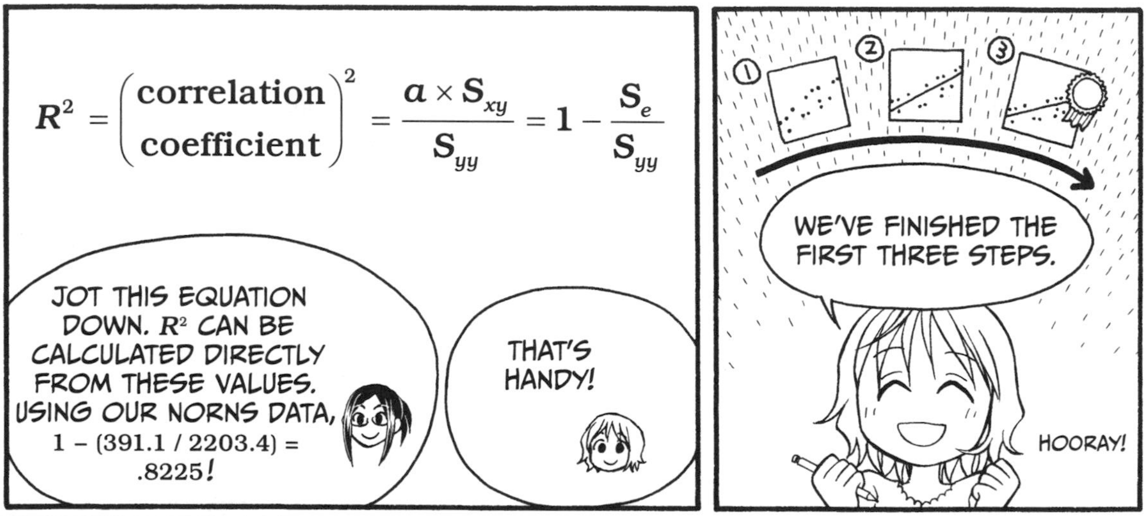
JOT THIS EQUATION DOWN. R^2 CAN BE CALCULATED DIRECTLY FROM THESE VALUES. USING OUR NORNS DATA, 1 − (391.1 / 2203.4) = .8225!
THAT'S HANDY!
WE'VE FINISHED THE FIRST THREE STEPS.
HOORAY!

SAMPLES AND POPULATIONS

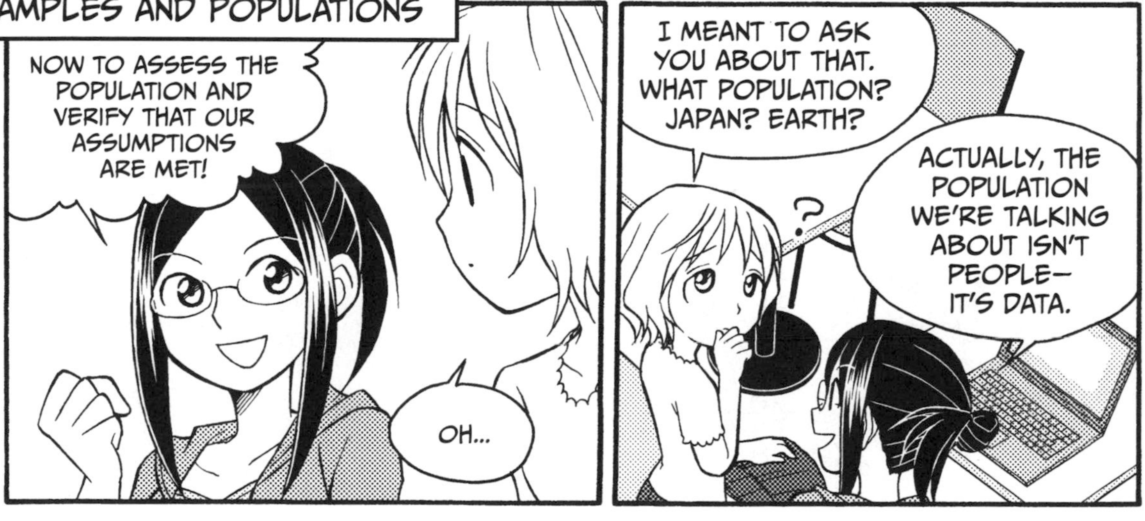
NOW TO ASSESS THE POPULATION AND VERIFY THAT OUR ASSUMPTIONS ARE MET!
OH...
I MEANT TO ASK YOU ABOUT THAT. WHAT POPULATION? JAPAN? EARTH?
?
ACTUALLY, THE POPULATION WE'RE TALKING ABOUT ISN'T PEOPLE— IT'S DATA.

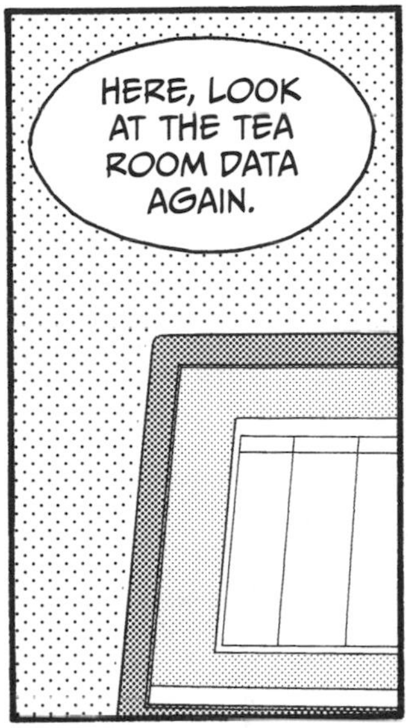
HERE, LOOK AT THE TEA ROOM DATA AGAIN.

	High temp. (°C)	Iced tea orders
22nd (Mon.)	29	77
23rd (Tues.)	28	62
24th (Wed.)	34	93
25th (Thurs.)	31	84
26th (Fri.)	25	59
27th (Sat.)	29	64
28th (Sun.)	32	80
29th (Mon.)	31	75
30th (Tues.)	24	58
31st (Wed.)	33	91
1st (Thurs.)	25	51
2nd (Fri.)	31	73
3rd (Sat.)	26	65
4th (Sun.)	30	84

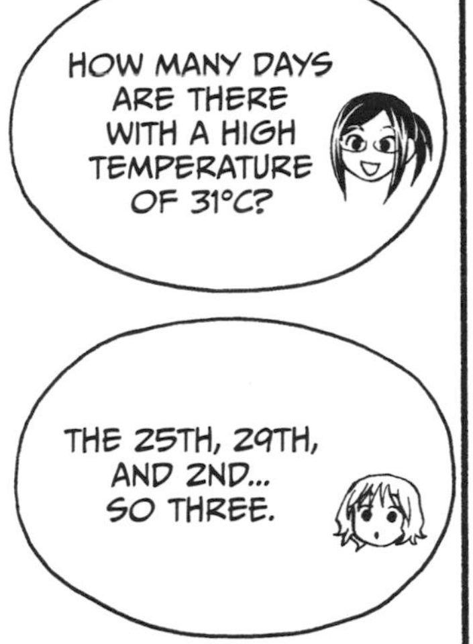
HOW MANY DAYS ARE THERE WITH A HIGH TEMPERATURE OF 31°C?
THE 25TH, 29TH, AND 2ND... SO THREE.

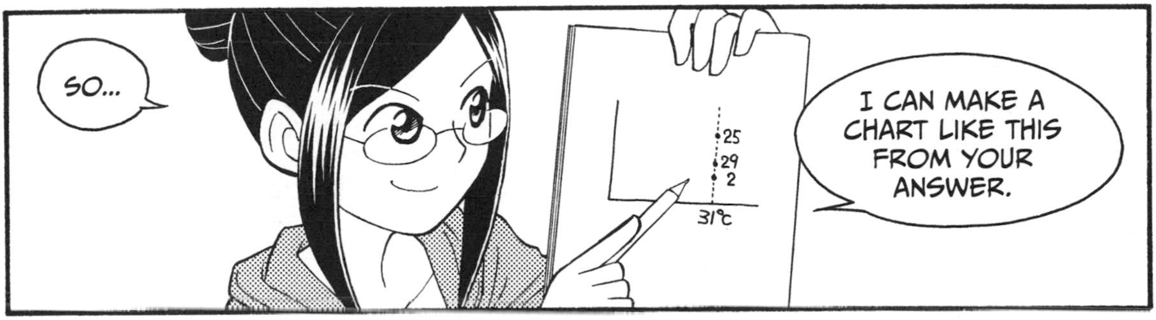
SO...
25
29
2
31°C
I CAN MAKE A CHART LIKE THIS FROM YOUR ANSWER.

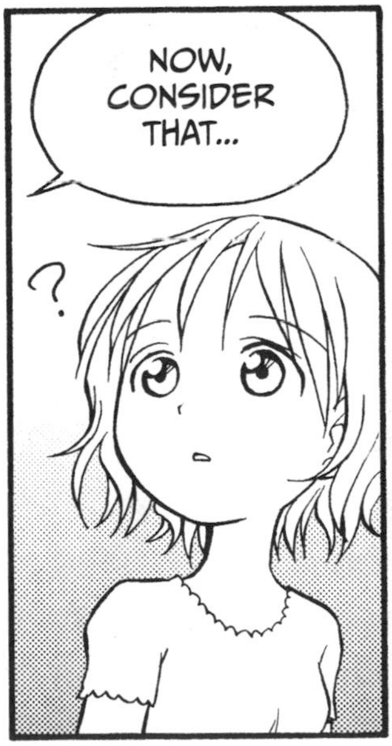
NOW, CONSIDER THAT...
?

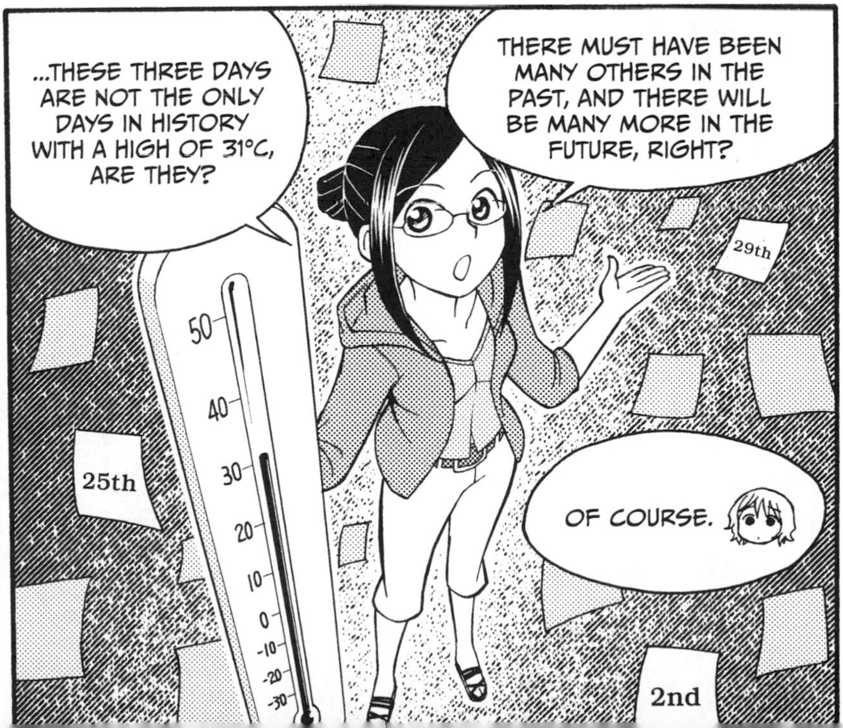
...THESE THREE DAYS ARE NOT THE ONLY DAYS IN HISTORY WITH A HIGH OF 31°C, ARE THEY?
THERE MUST HAVE BEEN MANY OTHERS IN THE PAST, AND THERE WILL BE MANY MORE IN THE FUTURE, RIGHT?
29th
25th
OF COURSE.
2nd

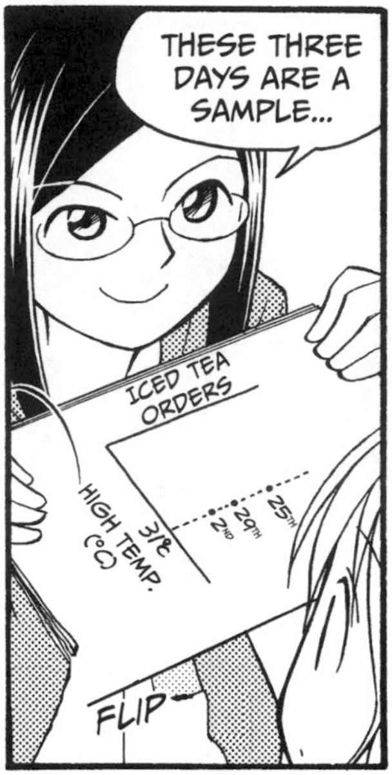
THESE THREE DAYS ARE A SAMPLE...
ICED TEA ORDERS
HIGH TEMP. (°C)
31°C
2ND
29TH
25TH
FLIP

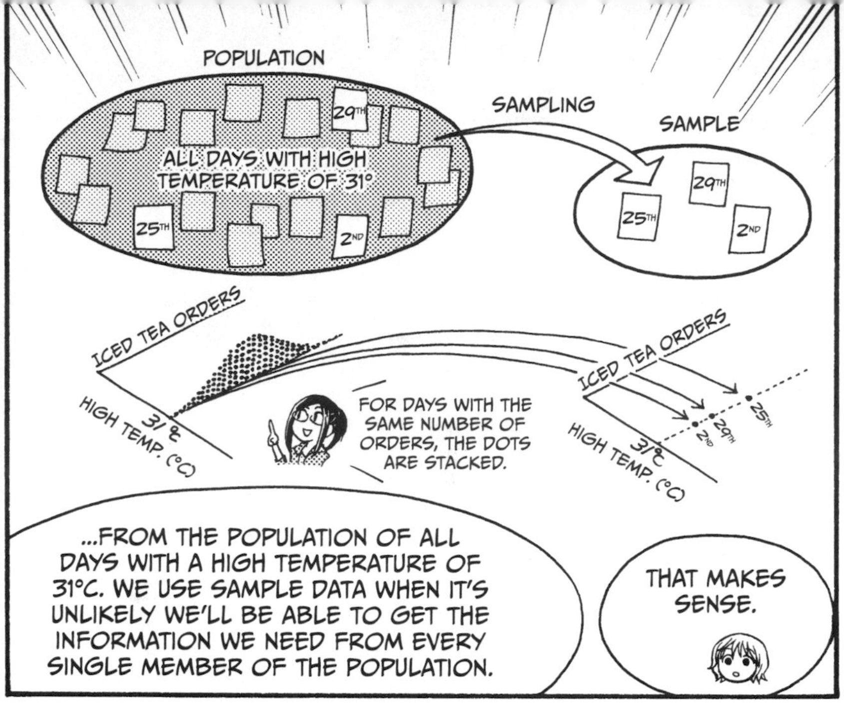
POPULATION
ALL DAYS WITH HIGH TEMPERATURE OF 31°
29TH
25TH
2ND
SAMPLING
SAMPLE
29TH
25TH
2ND
ICED TEA ORDERS
HIGH TEMP. (°C)
31°C
FOR DAYS WITH THE SAME NUMBER OF ORDERS, THE DOTS ARE STACKED.
ICED TEA ORDERS
HIGH TEMP. (°C)
31°C
2ND
29TH
25TH
...FROM THE POPULATION OF ALL DAYS WITH A HIGH TEMPERATURE OF 31°C. WE USE SAMPLE DATA WHEN IT'S UNLIKELY WE'LL BE ABLE TO GET THE INFORMATION WE NEED FROM EVERY SINGLE MEMBER OF THE POPULATION.
THAT MAKES SENSE.

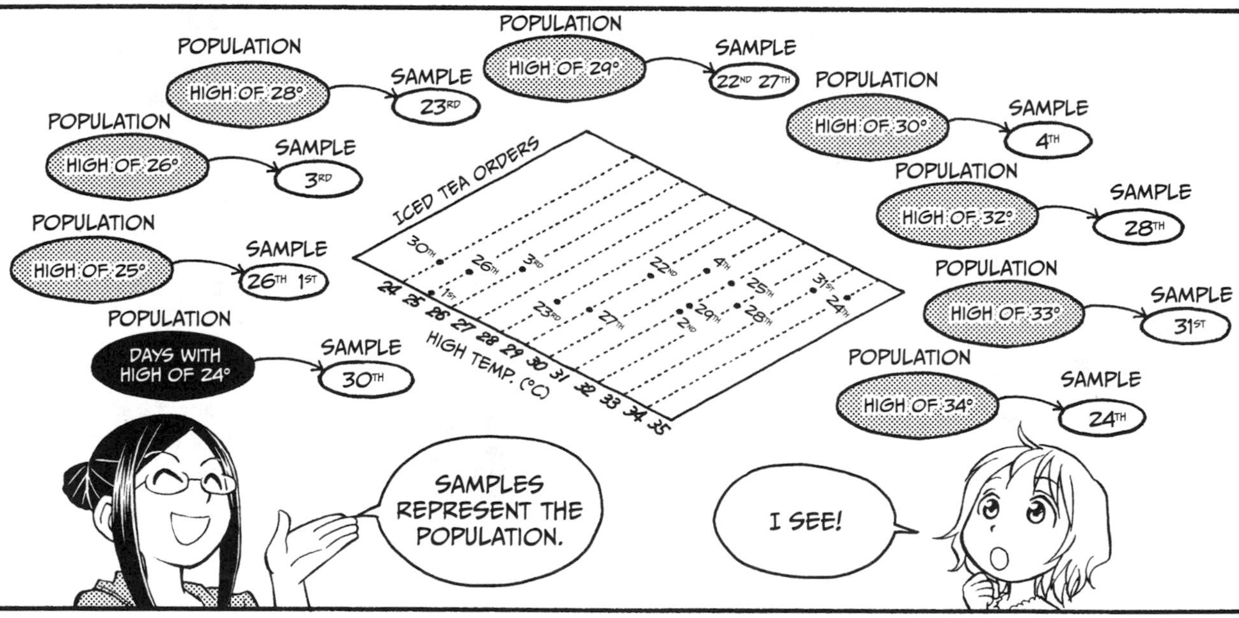
POPULATION
DAYS WITH HIGH OF 24°
SAMPLE
30TH
POPULATION
HIGH OF 25°
SAMPLE
26TH 1ST
POPULATION
HIGH OF 26°
SAMPLE
3RD
POPULATION
HIGH OF 28°
SAMPLE
23RD
POPULATION
HIGH OF 29°
SAMPLE
22ND 27TH
POPULATION
HIGH OF 30°
SAMPLE
4TH
POPULATION
HIGH OF 32°
SAMPLE
28TH
POPULATION
HIGH OF 33°
SAMPLE
31ST
POPULATION
HIGH OF 34°
SAMPLE
24TH
ICED TEA ORDERS
HIGH TEMP. (°C)
24 25 26 27 28 29 30 31 32 33 34 35
SAMPLES REPRESENT THE POPULATION.
I SEE!

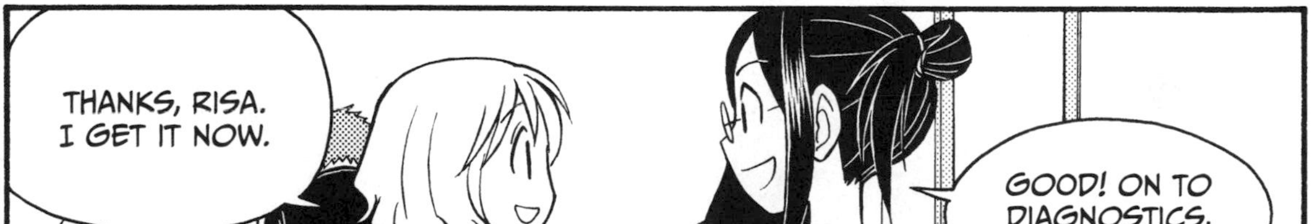
THANKS, RISA. I GET IT NOW.
GOOD! ON TO DIAGNOSTICS, THEN.

ASSUMPTIONS OF NORMALITY

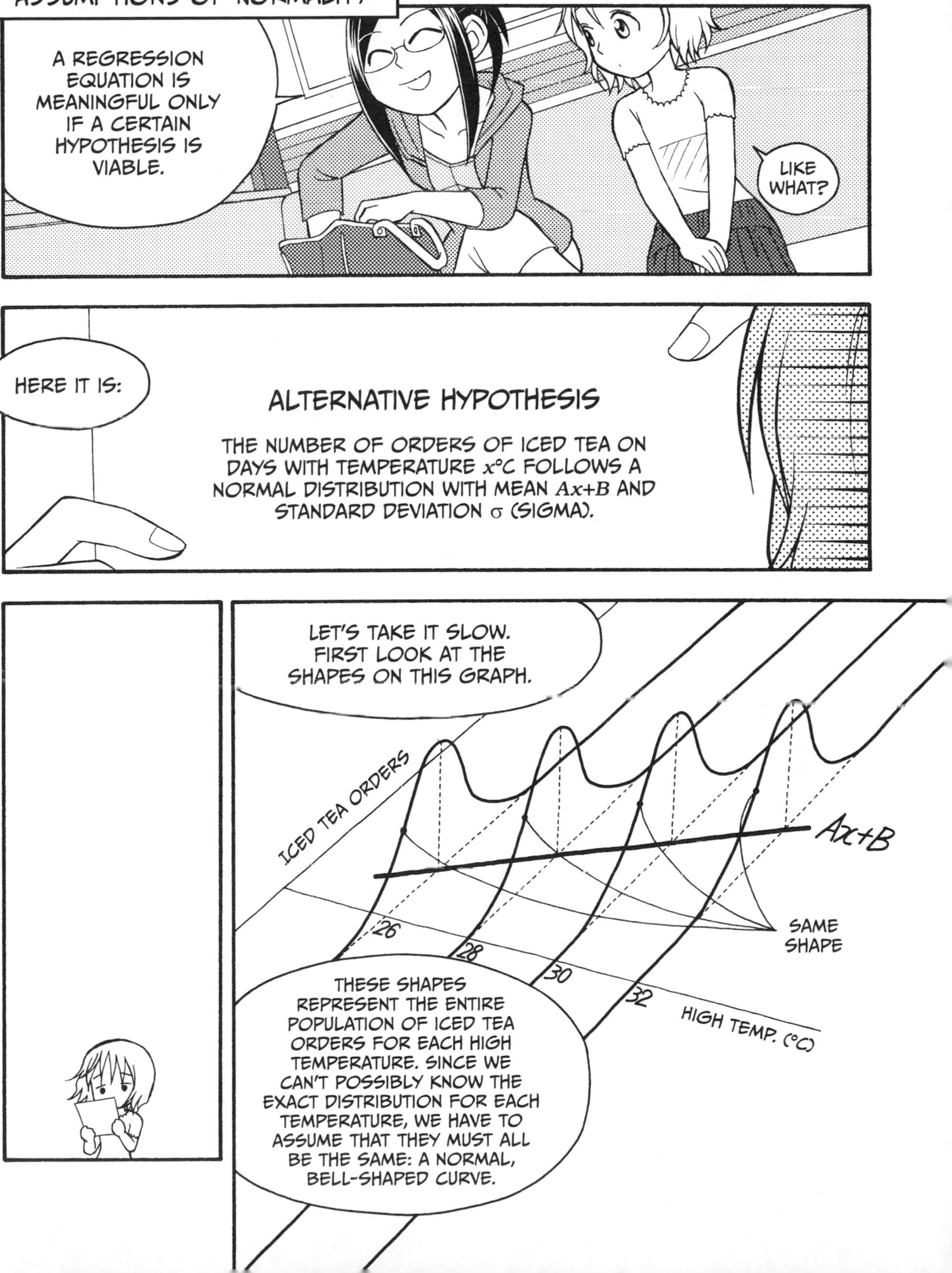
A REGRESSION EQUATION IS MEANINGFUL ONLY IF A CERTAIN HYPOTHESIS IS VIABLE.
LIKE WHAT?
HERE IT IS:
ALTERNATIVE HYPOTHESIS
THE NUMBER OF ORDERS OF ICED TEA ON DAYS WITH TEMPERATURE x°C FOLLOWS A NORMAL DISTRIBUTION WITH MEAN $Ax+B$ AND STANDARD DEVIATION σ (SIGMA).
LET'S TAKE IT SLOW. FIRST LOOK AT THE SHAPES ON THIS GRAPH.
ICED TEA ORDERS
Ax+B
SAME SHAPE
26
28
30
32
HIGH TEMP. (°C)
THESE SHAPES REPRESENT THE ENTIRE POPULATION OF ICED TEA ORDERS FOR EACH HIGH TEMPERATURE. SINCE WE CAN'T POSSIBLY KNOW THE EXACT DISTRIBUTION FOR EACH TEMPERATURE, WE HAVE TO ASSUME THAT THEY MUST ALL BE THE SAME: A NORMAL, BELL-SHAPED CURVE.

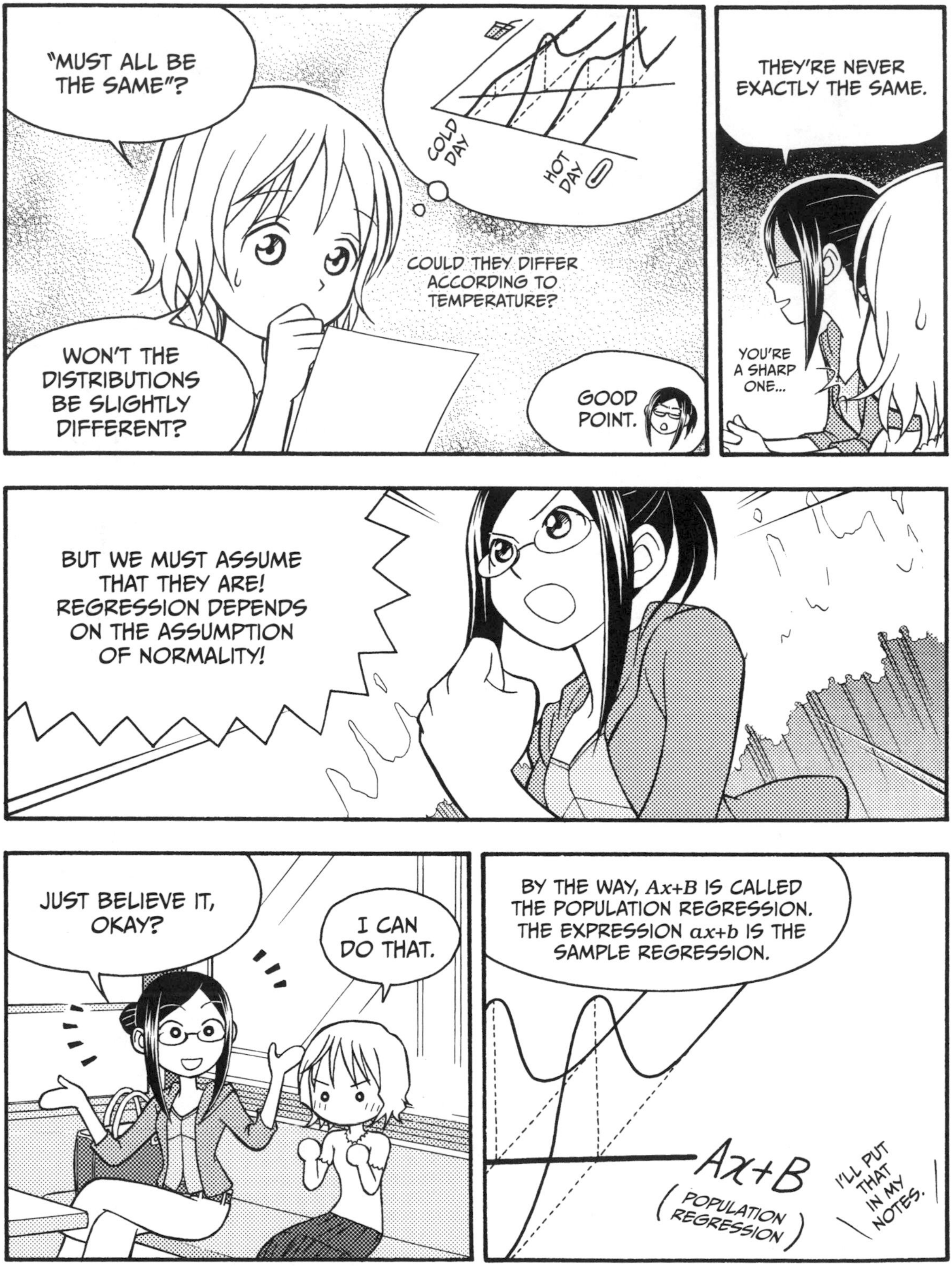
"MUST ALL BE THE SAME"?
COLD DAY
HOT DAY
COULD THEY DIFFER ACCORDING TO TEMPERATURE?
WON'T THE DISTRIBUTIONS BE SLIGHTLY DIFFERENT?
GOOD POINT.
THEY'RE NEVER EXACTLY THE SAME.
YOU'RE A SHARP ONE...
BUT WE MUST ASSUME THAT THEY ARE! REGRESSION DEPENDS ON THE ASSUMPTION OF NORMALITY!
JUST BELIEVE IT, OKAY?
I CAN DO THAT.
BY THE WAY, $Ax+B$ IS CALLED THE POPULATION REGRESSION. THE EXPRESSION $ax+b$ IS THE SAMPLE REGRESSION.
$Ax+B$
(POPULATION REGRESSION)
I'LL PUT THAT IN MY NOTES.

NOW, LET'S RETURN TO THE STORY ABOUT A, B, AND σ.

A, LIKE a, IS A SLOPE. B, LIKE b, IS AN INTERCEPT. AND σ IS THE STANDARD DEVIATION.

A, B, AND σ ARE COEFFICIENTS OF THE ENTIRE POPULATION.

IF THE REGRESSION EQUATION IS

$$y = ax + b$$

- a SHOULD BE CLOSE TO A
- b SHOULD BE CLOSE TO B
- $\sqrt{\dfrac{S_e}{\text{number of individuals} - 2}}$ SHOULD BE CLOSE TO σ

DO YOU RECALL a, b, AND THE STANDARD DEVIATION FOR OUR NORNS DATA?

$y = 3.7x - 36.4$

WELL, THE REGRESSION EQUATION WAS $y = 3.7x - 36.4$, SO...

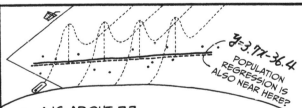

$y = 3.7x - 36.4$

POPULATION REGRESSION IS ALSO NEAR HERE?

- A IS ABOUT 3.7
- B IS ABOUT −36.4
- σ IS ABOUT $\sqrt{\dfrac{391.1}{14 - 2}} = \sqrt{\dfrac{391.1}{12}} = 5.7$

IS THAT RIGHT?

PERFECT!

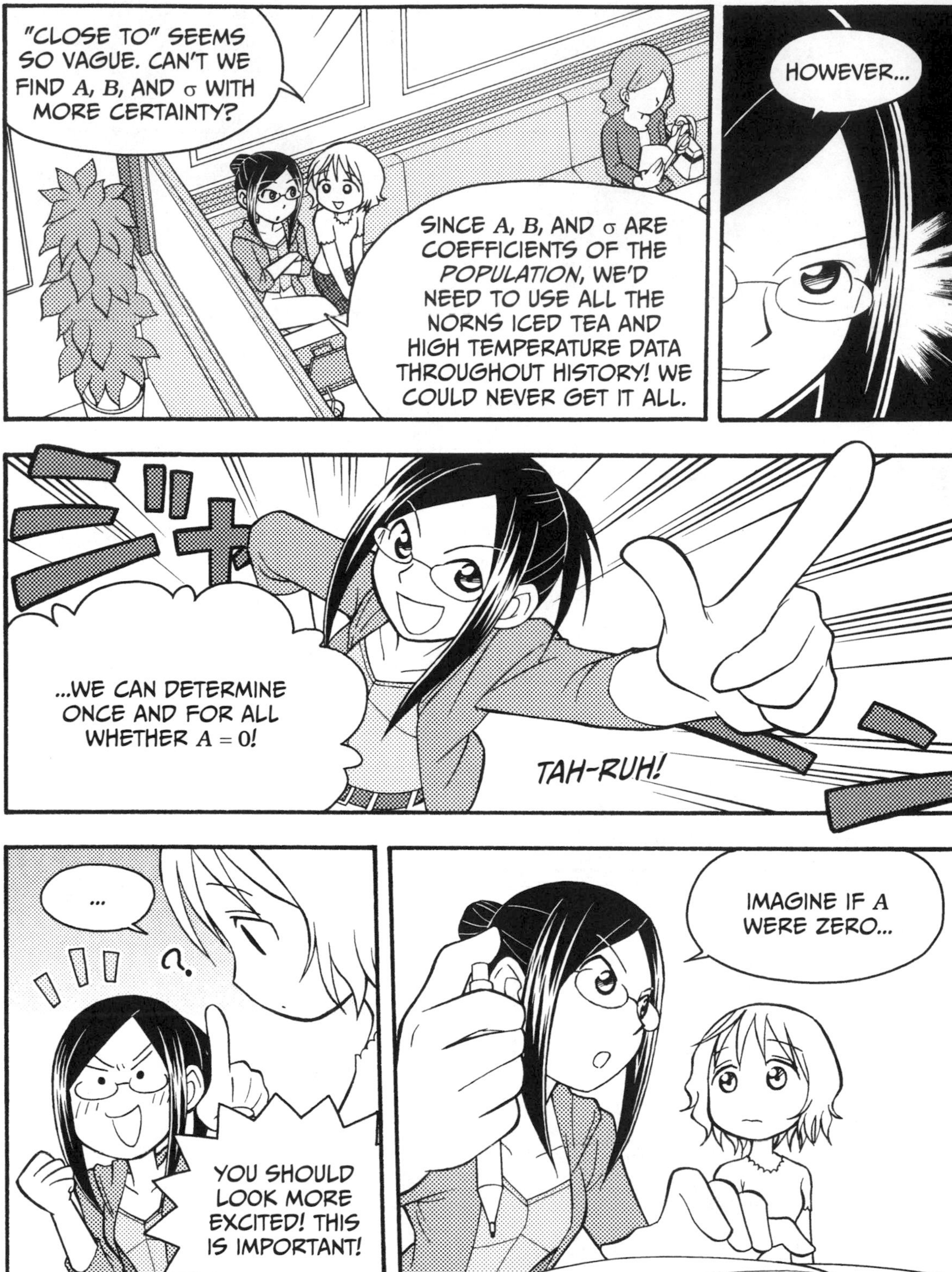
"CLOSE TO" SEEMS SO VAGUE. CAN'T WE FIND A, B, AND σ WITH MORE CERTAINTY?
SINCE A, B, AND σ ARE COEFFICIENTS OF THE *POPULATION*, WE'D NEED TO USE ALL THE NORNS ICED TEA AND HIGH TEMPERATURE DATA THROUGHOUT HISTORY! WE COULD NEVER GET IT ALL.
HOWEVER...
...WE CAN DETERMINE ONCE AND FOR ALL WHETHER $A = 0$!
TAH-RUH!
...
?
YOU SHOULD LOOK MORE EXCITED! THIS IS IMPORTANT!
IMAGINE IF A WERE ZERO...

THAT WOULD MAKE THIS DREADED HYPOTHESIS TRUE!

NULL HYPOTHESIS

THE NUMBER OF ORDERS OF ICED TEA ON DAYS WITH HIGH TEMPERATURE x °C FOLLOWS A NORMAL DISTRIBUTION WITH MEAN B AND STANDARD DEVIATION σ. (A IS ABSENT!)

A IS GONE!

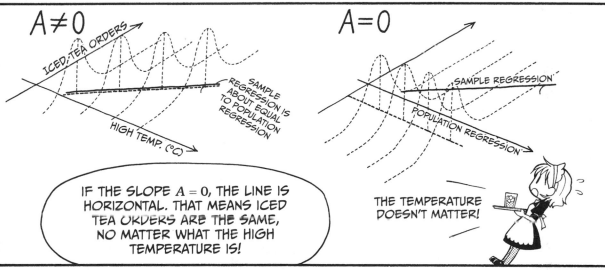

$A \neq 0$

ICED TEA ORDERS

HIGH TEMP. (°C)

SAMPLE REGRESSION IS ABOUT EQUAL TO POPULATION REGRESSION

$A = 0$

SAMPLE REGRESSION

POPULATION REGRESSION

IF THE SLOPE $A = 0$, THE LINE IS HORIZONTAL. THAT MEANS ICED TEA ORDERS ARE THE SAME, NO MATTER WHAT THE HIGH TEMPERATURE IS!

THE TEMPERATURE DOESN'T MATTER!

HOW DO WE FIND OUT ABOUT A?

ANOVA

WE CAN DO AN ANALYSIS OF VARIANCE (ANOVA)!

LET'S DO THE ANALYSIS AND SEE WHAT FATE HAS IN STORE FOR A.

THIS IS GETTING EXCITING.

THE STEPS OF ANOVA

Step 1	Define the population.	The population is "days with a high temperature of x degrees."
Step 2	Set up a null hypothesis and an alternative hypothesis.	Null hypothesis is $A = 0$. Alternative hypothesis is $A \neq 0$.
Step 3	Select which hypothesis test to conduct.	We'll use analysis of one-way variance.
Step 4	Choose the significance level.	We'll use a significance level of .05.
Step 5	Calculate the test statistic from the sample data.	The test statistic is:

$$\frac{a^2}{\left(\dfrac{1}{S_{xx}}\right)} \div \frac{S_e}{\text{number of individuals} - 2}$$

Plug in the values from our sample regression equation:

$$\frac{3.7^2}{\left(\dfrac{1}{129.7}\right)} \div \frac{391.1}{14 - 2} = 55.6$$

The test statistic will follow an F distribution with first degree of freedom 1 and second degree of freedom 12 (number of individuals minus 2), if the null hypothesis is true.

Step 6	Determine whether the p-value for the test statistic obtained in Step 5 is smaller than the significance level.	At significance level .05, with d_1 being 1 and d_2 being 12, the critical value is 4.7472. Our test statistic is 55.6.
Step 7	Decide whether you can reject the null hypothesis.	Since our test statistic is greater than the critical value, we reject the null hypothesis.

THE F STATISTIC LETS US TEST THE SLOPE OF THE LINE BY LOOKING AT VARIANCE. IF THE VARIATION AROUND THE LINE IS MUCH SMALLER THAN THE TOTAL VARIANCE OF Y, THAT'S EVIDENCE THAT THE LINE ACCOUNTS FOR Y'S VARIATION, AND THE STATISTIC WILL BE LARGE. IF THE RATIO IS SMALL, THE LINE DOESN'T ACCOUNT FOR MUCH VARIATION IN Y, AND PROBABLY ISN'T USEFUL!

SO $A \neq 0$, WHAT A RELIEF!

STEP 5: CALCULATE THE CONFIDENCE INTERVALS.

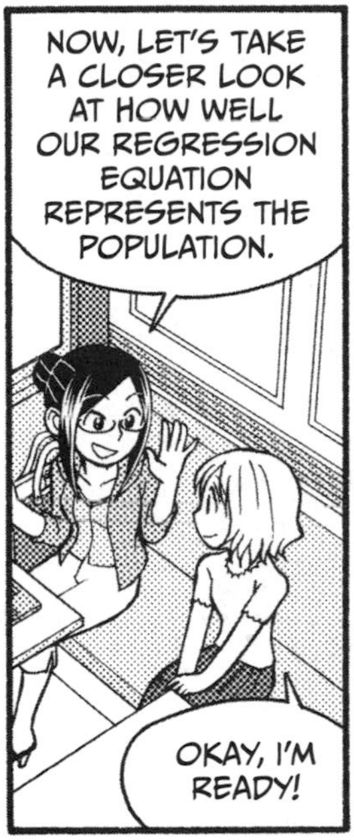
NOW, LET'S TAKE A CLOSER LOOK AT HOW WELL OUR REGRESSION EQUATION REPRESENTS THE POPULATION.
OKAY, I'M READY!

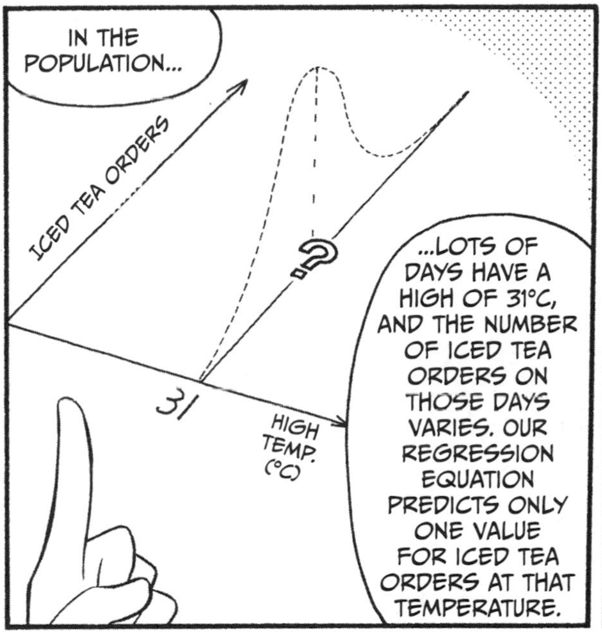
IN THE POPULATION...
ICED TEA ORDERS
31
HIGH TEMP. (°C)
...LOTS OF DAYS HAVE A HIGH OF 31°C, AND THE NUMBER OF ICED TEA ORDERS ON THOSE DAYS VARIES. OUR REGRESSION EQUATION PREDICTS ONLY ONE VALUE FOR ICED TEA ORDERS AT THAT TEMPERATURE.

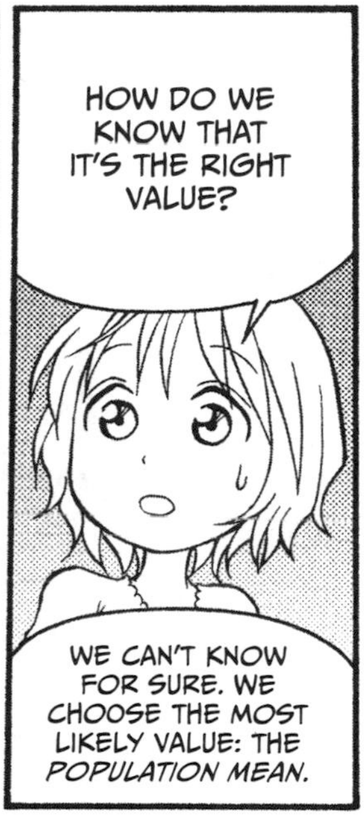
HOW DO WE KNOW THAT IT'S THE RIGHT VALUE?
WE CAN'T KNOW FOR SURE. WE CHOOSE THE MOST LIKELY VALUE: THE *POPULATION MEAN.*

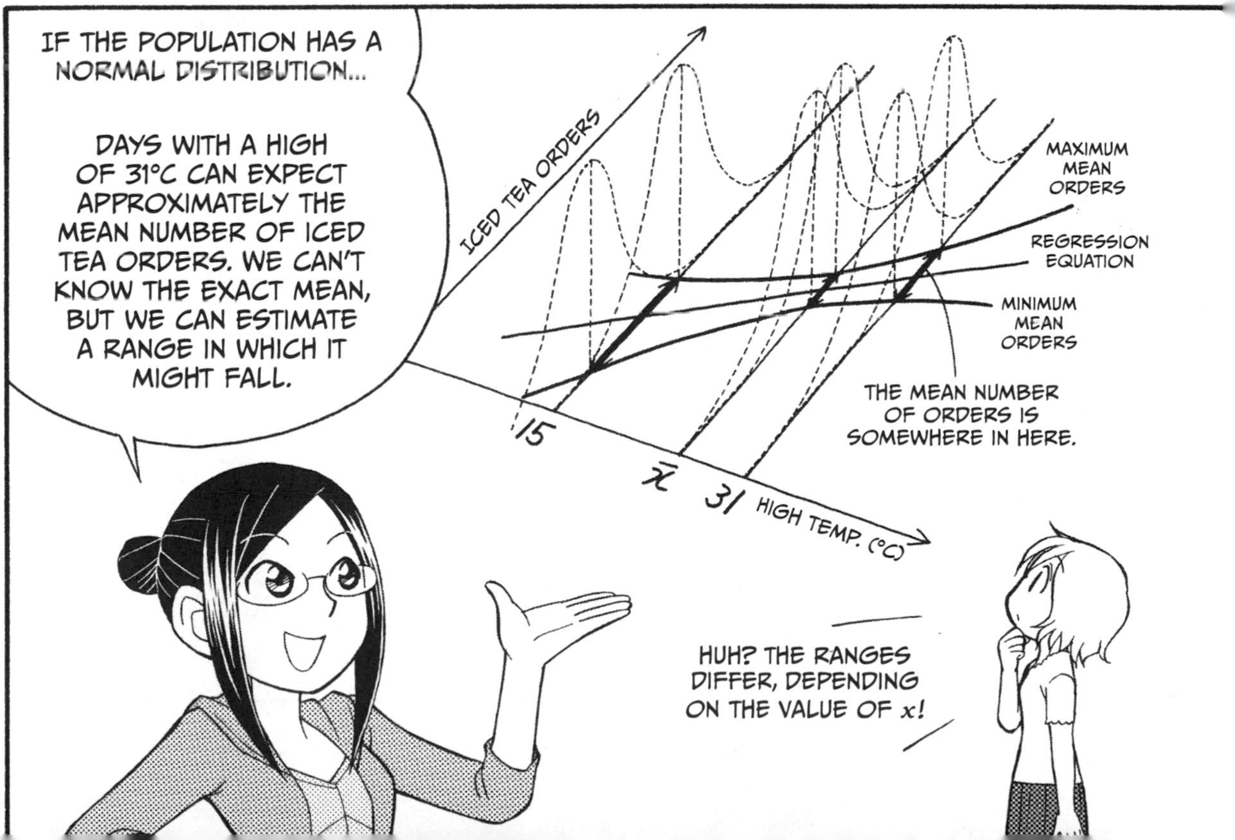
IF THE POPULATION HAS A NORMAL DISTRIBUTION...
DAYS WITH A HIGH OF 31°C CAN EXPECT APPROXIMATELY THE MEAN NUMBER OF ICED TEA ORDERS. WE CAN'T KNOW THE EXACT MEAN, BUT WE CAN ESTIMATE A RANGE IN WHICH IT MIGHT FALL.
ICED TEA ORDERS
MAXIMUM MEAN ORDERS
REGRESSION EQUATION
MINIMUM MEAN ORDERS
THE MEAN NUMBER OF ORDERS IS SOMEWHERE IN HERE.
15
$\bar{x}$
31
HIGH TEMP. (°C)
HUH? THE RANGES DIFFER, DEPENDING ON THE VALUE OF x!

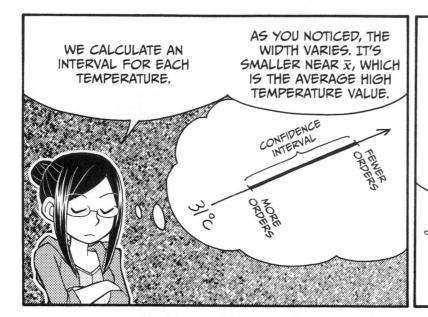

WE CALCULATE AN INTERVAL FOR EACH TEMPERATURE.

AS YOU NOTICED, THE WIDTH VARIES. IT'S SMALLER NEAR $\bar{x}$, WHICH IS THE AVERAGE HIGH TEMPERATURE VALUE.

CONFIDENCE INTERVAL

FEWER ORDERS

MORE ORDERS

31°C

EVEN THIS INTERVAL ISN'T ABSOLUTELY GUARANTEED TO CONTAIN THE TRUE POPULATION MEAN. OUR CONFIDENCE IS DETERMINED BY THE *CONFIDENCE COEFFICIENT.*

SOUNDS FAMILIAR!

NOW, CONFIDENCE...

...IS NO ORDINARY COEFFICIENT.

THERE IS NO EQUATION TO CALCULATE IT, NO SET RULE.

YOU CHOOSE THE CONFIDENCE COEFFICIENT, AND YOU CAN MAKE IT ANY PERCENTAGE YOU WANT.

I WILL MAKE IT 42%.

?

WHEN CALCULATING A CONFIDENCE INTERVAL, YOU CHOOSE THE CONFIDENCE COEFFICIENT FIRST.

YOU WOULD THEN SAY "A 42% CONFIDENCE INTERVAL FOR ICED TEA ORDERS WHEN THE TEMPERATURE IS 31°C IS 30 TO 35 ORDERS," FOR EXAMPLE!

I CHOOSE?

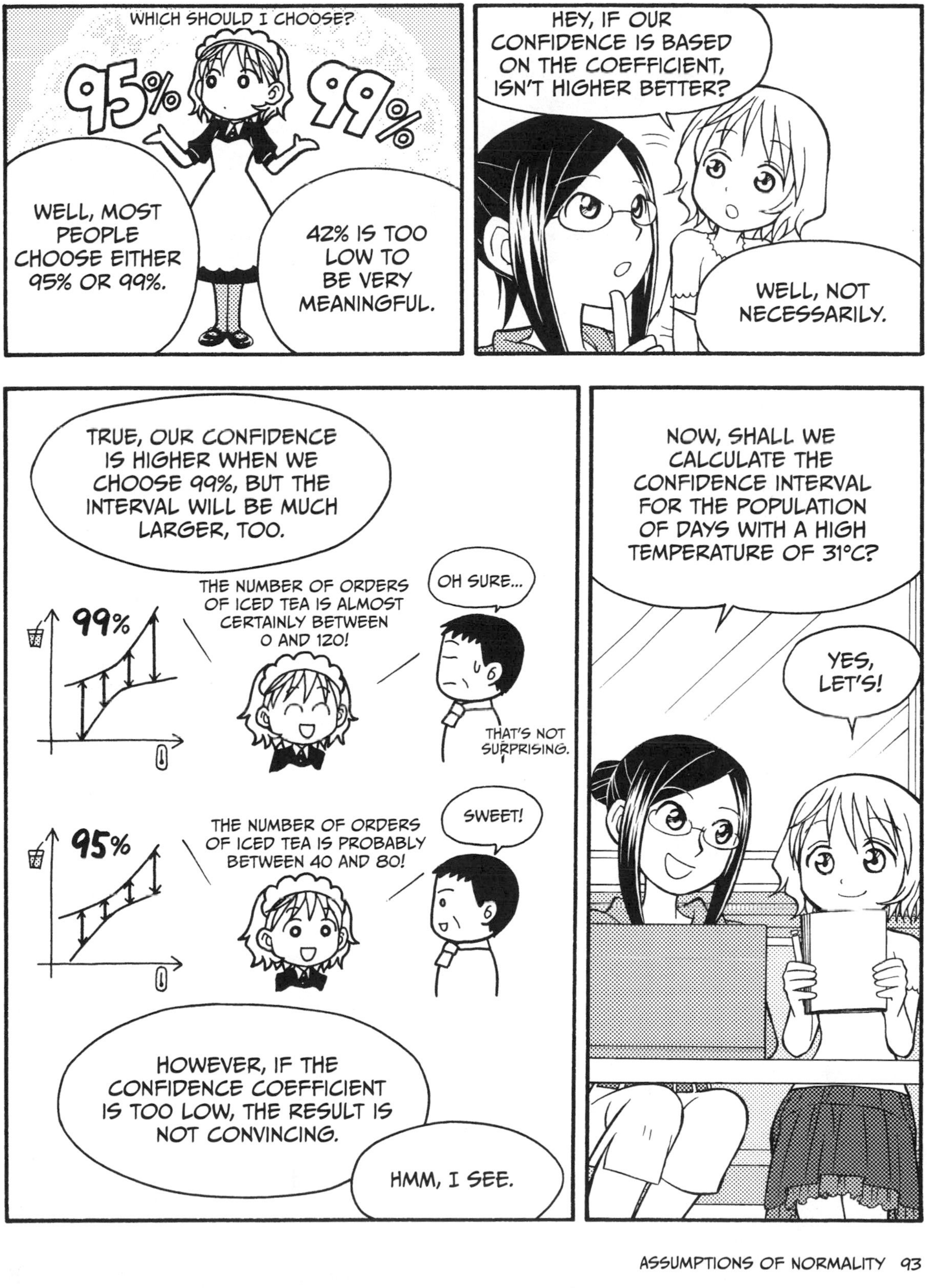
WHICH SHOULD I CHOOSE?
95%
99%
WELL, MOST PEOPLE CHOOSE EITHER 95% OR 99%.
42% IS TOO LOW TO BE VERY MEANINGFUL.
HEY, IF OUR CONFIDENCE IS BASED ON THE COEFFICIENT, ISN'T HIGHER BETTER?
WELL, NOT NECESSARILY.
TRUE, OUR CONFIDENCE IS HIGHER WHEN WE CHOOSE 99%, BUT THE INTERVAL WILL BE MUCH LARGER, TOO.
99%
THE NUMBER OF ORDERS OF ICED TEA IS ALMOST CERTAINLY BETWEEN 0 AND 120!
OH SURE...
THAT'S NOT SURPRISING.
95%
THE NUMBER OF ORDERS OF ICED TEA IS PROBABLY BETWEEN 40 AND 80!
SWEET!
HOWEVER, IF THE CONFIDENCE COEFFICIENT IS TOO LOW, THE RESULT IS NOT CONVINCING.
HMM, I SEE.
NOW, SHALL WE CALCULATE THE CONFIDENCE INTERVAL FOR THE POPULATION OF DAYS WITH A HIGH TEMPERATURE OF 31°C?
YES, LET'S!

HERE'S HOW TO CALCULATE A 95% CONFIDENCE INTERVAL FOR ICED TEA ORDERS ON DAYS WITH A HIGH OF 31°C.

This is the confidence interval.

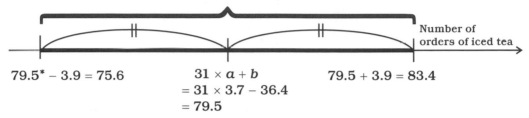

Number of orders of iced tea

$79.5^* - 3.9 = 75.6$

$31 \times a + b$
$= 31 \times 3.7 - 36.4$
$= 79.5$

$79.5 + 3.9 = 83.4$

Distance from the estimated mean is

$$\sqrt{F(1, n-2; .05) \times \left(\frac{1}{n} + \frac{(x_0 - \bar{x})^2}{S_{xx}} \right) \times \frac{S_e}{n-2}}$$

$$= \sqrt{F(1, 14-2; .05) \times \left(\frac{1}{14} + \frac{(31 - 29.1)^2}{129.7} \right) \times \frac{391.1}{14-2}}$$

$$= 3.9$$

where n is the number of data points in our sample and F is a ratio of two chi-squared distributions, as described on page 57.

TO CALCULATE A 99% CONFIDENCE INTERVAL, JUST CHANGE

$F(1, 14-2; .05) = 4.7$

TO

$F(1, 14-2; .01) = 9.3$

(REFER TO PAGE 58 FOR AN EXPLANATION OF $F(1, n-2; .05) = 4.7$, AND SO ON.)

* THE VALUE 79.5 WAS CALCULATED USING UNROUNDED NUMBERS.

SO WE ARE 95% SURE THAT, IF WE LOOK AT THE POPULATION OF DAYS WITH A HIGH OF 31°C, THE MEAN NUMBER OF ICED TEA ORDERS IS BETWEEN 76 AND 83.

EXACTLY!

STEP 6: MAKE A PREDICTION!

AT LAST, WE MAKE THE PREDICTION.

THE FINAL STEP!

TOMORROW'S WEATHER

15:22

☼ FAIR

HIGH 27°C
LOW 20°C
RAIN 0%

IF TOMORROW'S HIGH TEMPERATURE IS 27°C...

...HOW MANY ICED TEA ORDERS WILL WE GET AT THE SHOP TOMORROW?

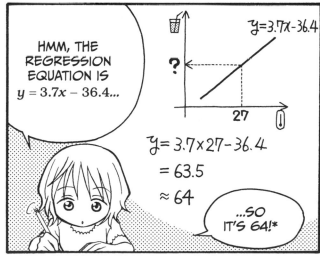

HMM, THE REGRESSION EQUATION IS
$y = 3.7x - 36.4$...

$y = 3.7x - 36.4$

?

27

$$y = 3.7 \times 27 - 36.4$$
$$= 63.5$$
$$\approx 64$$

...SO IT'S 64!*

BINGO!

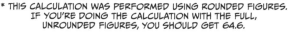

* THIS CALCULATION WAS PERFORMED USING ROUNDED FIGURES. IF YOU'RE DOING THE CALCULATION WITH THE FULL, UNROUNDED FIGURES, YOU SHOULD GET 64.6.

BUT WILL THERE BE EXACTLY 64 ORDERS?

HOW CAN WE POSSIBLY KNOW FOR SURE?

THAT'S A GREAT QUESTION.

WE SHOULD GET CLOSE TO 64 ORDERS BECAUSE THE VALUE OF R^2 IS 0.8225, BUT... HOW CLOSE?

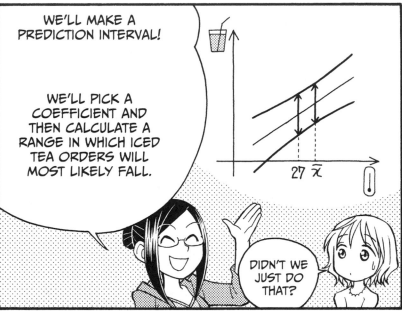

WE'LL MAKE A PREDICTION INTERVAL!

WE'LL PICK A COEFFICIENT AND THEN CALCULATE A RANGE IN WHICH ICED TEA ORDERS WILL MOST LIKELY FALL.

DIDN'T WE JUST DO THAT?

NOT QUITE. BEFORE, WE WERE PREDICTING THE MEAN NUMBER OF ICED TEA ORDERS FOR THE POPULATION OF DAYS WITH A CERTAIN HIGH TEMPERATURE, BUT NOW WE'RE PREDICTING THE LIKELY NUMBER OF ICED TEA ORDERS ON A GIVEN DAY WITH A CERTAIN TEMPERATURE.

I DON'T SEE THE DIFFERENCE.

WHAT'S THE POPULATION LIKE?

$y=ax+b$

3|

CONFIDENCE INTERVALS HELP US ASSESS THE POPULATION.

HOW MANY ORDERS WILL I GET?

$y=ax+b$

27

PREDICTION INTERVALS GIVE A PROBABLE RANGE OF FUTURE VALUES.

THE PREDICTION INTERVAL LOOKS LIKE A CONFIDENCE INTERVAL, BUT IT'S NOT THE SAME.

PREDICTION INTERVAL

27°C

MAXIMUM ORDERS

MINIMUM ORDERS

AS WITH A CONFIDENCE INTERVAL, WE NEED TO CHOOSE THE CONFIDENCE COEFFICIENT BEFORE WE CAN DO THE CALCULATION. AGAIN, 95% AND 99% ARE POPULAR.

I JUST CALCULATED AN INTERVAL, SO THIS SHOULD BE A SNAP.

THE CALCULATION IS VERY SIMILAR, WITH ONE IMPORTANT DIFFERENCE...

THE PREDICTION INTERVAL WILL BE WIDER BECAUSE IT COVERS THE RANGE OF ALL EXPECTED VALUES, NOT JUST WHERE THE MEAN SHOULD BE.

CONFIDENCE INTERVAL

PREDICTION INTERVAL

REGRESSION EQUATION

y

x

x

THE FUTURE IS ALWAYS SURPRISING.

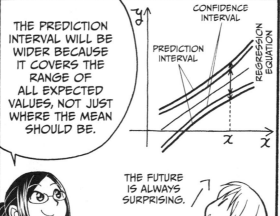

NOW, TRY CALCULATING THE PREDICTION INTERVAL FOR 27°C.

NO SWEAT!

HERE'S HOW WE CALCULATE A 95% PREDICTION INTERVAL FOR TOMORROW'S ICED TEA SALES.

This is the prediction interval.

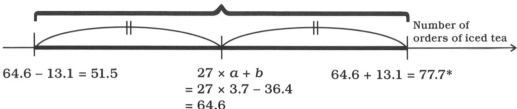

Number of orders of iced tea

$64.6 - 13.1 = 51.5$

$27 \times a + b$
$= 27 \times 3.7 - 36.4$
$= 64.6$

$64.6 + 13.1 = 77.7*$

Distance from the estimated value is

$$\sqrt{F\left(1, n-2; .05\right) \times \left(1 + \frac{1}{n} + \frac{\left(x_0 - \bar{x}\right)^2}{S_{xx}}\right) \times \frac{S_e}{n-2}}$$

$$= \sqrt{F\left(1, 14-2; .05\right) \times \left(1 + \frac{1}{14} + \frac{\left(27 - 29.1\right)^2}{129.7}\right) \times \frac{391.1}{14-2}}$$

$$= 13.1$$

THE ESTIMATED NUMBER OF TEA ORDERS WE CALCULATED EARLIER (ON PAGE 95) WAS ROUNDED, BUT WE'VE USED THE NUMBER OF TEA ORDERS ESTIMATED USING UNROUNDED NUMBERS, 64.6, HERE.

HERE WE USED THE F DISTRIBUTION TO FIND THE PREDICTION INTERVAL AND POPULATION REGRESSION. TYPICALLY, STATISTICIANS USE THE T DISTRIBUTION TO GET THE SAME RESULTS.

* THIS CALCULATION WAS PERFORMED USING THE ROUNDED NUMBERS SHOWN HERE. THE FULL, UNROUNDED CALCULATION RESULTS IN 77.6.

SO WE'RE 95% CONFIDENT THAT THE NUMBER OF ICED TEA ORDERS WILL BE BETWEEN 52 AND 78 WHEN THE HIGH TEMPERATURE FOR THAT DAY IS 27°C.

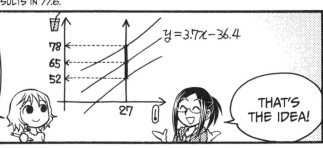

$y = 3.7x - 36.4$

THAT'S THE IDEA!

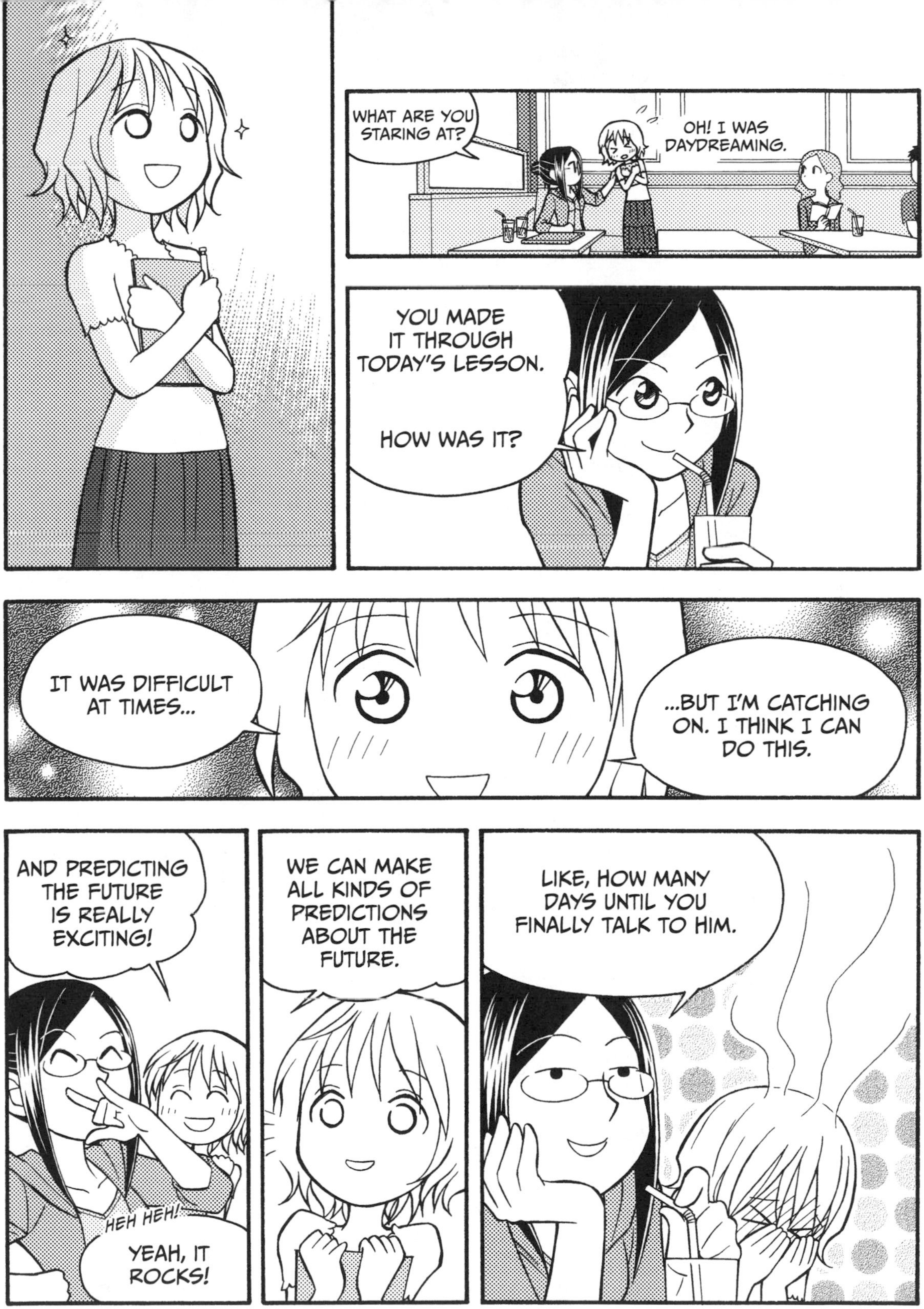
WHAT ARE YOU STARING AT?
OH! I WAS DAYDREAMING.
YOU MADE IT THROUGH TODAY'S LESSON.
HOW WAS IT?
IT WAS DIFFICULT AT TIMES...
...BUT I'M CATCHING ON. I THINK I CAN DO THIS.
AND PREDICTING THE FUTURE IS REALLY EXCITING!
HEH HEH!
YEAH, IT ROCKS!
WE CAN MAKE ALL KINDS OF PREDICTIONS ABOUT THE FUTURE.
LIKE, HOW MANY DAYS UNTIL YOU FINALLY TALK TO HIM.

WHICH STEPS ARE NECESSARY?

Remember the regression analysis procedure introduced on page 68?

1. Draw a scatter plot of the independent variable versus the dependent variable. If the dots line up, the variables may be correlated.

2. Calculate the regression equation.

3. Calculate the correlation coefficient (R) and assess our population and assumptions.

4. Conduct the analysis of variance.

5. Calculate the confidence intervals.

6. Make a prediction!

In this chapter, we walked through each of the six steps, but it isn't always necessary to do every step. Recall the example of Miu's age and height on page 25.

- Fact: There is only one Miu in this world.

- Fact: Miu's height when she was 10 years old was 137.5 cm.

Given these two facts, it makes no sense to say that "Miu's height when she was 10 years old follows a normal distribution with mean $Ax + B$ and standard deviation σ." In other words, it's nonsense to analyze the population of Miu's heights at 10 years old. She was just one height, and we know what her height was.

In regression analysis, we either analyze the entire population or, much more commonly, analyze a sample of the larger population. When you analyze a sample, you should perform all the steps. However, since Steps 4 and 5 assess how well the sample represents the population, you can skip them if you're using data from an entire population instead of just a sample.

NOTE   *We use the term* statistic *to describe a measurement of a characteristic from a sample, like a sample mean, and* parameter *to describe a measurement that comes from a population, like a population mean or coefficient.*

STANDARDIZED RESIDUAL

Remember that a *residual* is the difference between the *measured* value and the value *estimated* with the regression equation. The *standardized residual* is the residual divided by its estimated standard deviation. We use the standardized residual to assess whether a particular measurement deviates significantly from the trend. For example, say a group of thirsty joggers stopped by Norns on the 4th, meaning that though iced tea orders were expected to be about 76 based on that day's high temperature, customers actually placed 84 orders for iced tea. Such an event would result in a large standardized residual.

Standardized residuals are calculated by dividing each residual by an estimate of its standard deviation, which is calculated using the residual sum of squares. The calculation is a little complicated, and most statistics software does it automatically, so we won't go into the details of the calculation here.

Table 2-1 shows the standardized residual for the Norns data used in this chapter.

TABLE 2-1: CALCULATING THE STANDARDIZED RESIDUAL

	High temperature x	Measured number of orders of iced tea y	Estimated number of orders of iced tea $\hat{y} = 3.7x - 36.4$	Residual $y - \hat{y}$	Standardized residual
22nd (Mon.)	29	77	72.0	5.0	0.9
23rd (Tues.)	28	62	68.3	−6.3	−1.2
24th (Wed.)	34	93	90.7	2.3	0.5
25th (Thurs.)	31	84	79.5	4.5	0.8
26th (Fri.)	25	59	57.1	1.9	0.4
27th (Sat.)	29	64	72.0	−8.0	−1.5
28th (Sun.)	32	80	83.3	−3.3	−0.6
29th (Mon.)	31	75	79.5	−4.5	−0.8
30th (Tues.)	24	58	53.3	4.7	1.0
31st (Wed.)	33	91	87.0	4.0	0.8
1st (Thurs.)	25	51	57.1	−6.1	−1.2
2nd (Fri.)	31	73	79.5	−6.5	−1.2
3rd (Sat.)	26	65	60.8	4.2	0.8
4th (Sun.)	30	84	75.8	8.2	1.5

As you can see, the standardized residual on the 4th is 1.5. If iced tea orders had been 76, as expected, the standardized residual would have been 0.

Sometimes a measured value can deviate so much from the trend that it adversely affects the analysis. If the standardized residual is greater than 3 or less than −3, the measurement is considered an *outlier*. There are a number of ways to handle outliers, including removing them, changing them to a set value, or just keeping them in the analysis as is. To determine which approach is most appropriate, investigate the underlying cause of the outliers.

INTERPOLATION AND EXTRAPOLATION

If you look at the *x* values (high temperature) on page 64, you can see that the highest value is 34°C and the lowest value is 24°C. Using regression analysis, you can *interpolate* the number of iced tea orders on days with a high temperature between 24°C and 34°C and *extrapolate* the number of iced tea orders on days with a high below 24°C or above 34°C. In other words, extrapolation is the estimation of values that fall outside the range of your observed data.

Since we've only observed the trend between 24°C and 34°C, we don't know whether iced tea sales follow the same trend when the weather is extremely cold or extremely hot. Extrapolation is therefore less reliable than interpolation, and some statisticians avoid it entirely.

For everyday use, it's fine to extrapolate—as long as you're aware that your result isn't completely trustworthy. However, avoid using extrapolation in academic research or to estimate a value that's far beyond the scope of the measured data.

AUTOCORRELATION

The independent variable used in this chapter was high temperature; this is used to predict iced tea sales. In most places, it's unlikely that the high temperature will be 20°C one day and then shoot up to 30°C the next day. Normally, the temperature rises or drops gradually over a period of several days, so if the two variables are related, the number of iced tea orders should rise or drop gradually as well. Our assumption, however, has been that the deviation (error) values are random. Therefore, our predicted values do not change from day to day as smoothly as they might in real life.

When analyzing variables that may be affected by the passage of time, it's a good idea to check for autocorrelation. Autocorrelation occurs when the error is correlated over time, and it can indicate that you need to use a different type of regression model.

There's an index to describe autocorrelation—the *Durbin-Watson statistic*, which is calculated as follows:

$$d = \frac{\sum_{t=2}^{T} (e_t - e_{t-1})^2}{\sum_{t=1}^{T} e_t^2}$$

The equation can be read as "the sum of the square of each residual minus the previous residual, divided by the sum of each residual squared." You can calculate the value of the Durbin-Watson statistic for the example in this chapter:

$$\frac{(-6.3 - 5.0)^2 + (2.3 - (-6.3))^2 + \cdots + (8.2 - 4.2)^2}{5.0^2 + (-6.3)^2 + \cdots + 8.2^2} = 1.8$$

The exact critical value of the Durbin-Watson test differs for each analysis, and you can use a table to find it, but generally we use 1 as a cutoff: a result less than 1 may indicate the presence of autocorrelation. This result is close to 2, so we can conclude that there is no autocorrelation in our example.

NONLINEAR REGRESSION

On page 66, Risa said:

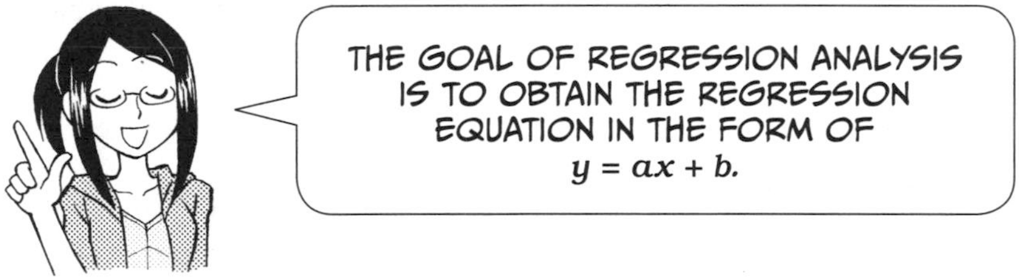

THE GOAL OF REGRESSION ANALYSIS IS TO OBTAIN THE REGRESSION EQUATION IN THE FORM OF
$y = ax + b.$

This equation is linear, but regression equations don't have to be linear. For example, these equations may also be used as regression equations:

- $y = \dfrac{a}{x} + b$
- $y = a\sqrt{x} + b$
- $y = ax^2 + bx + c$
- $y = a \times \log x + b$

The regression equation for Miu's age and height introduced on page 26 is actually in the form of $y = \dfrac{a}{x} + b$ rather than $y = ax + b$.

Of course, this raises the question of which type of equation you should choose when performing regression analysis on your own data. Below are some steps that can help you decide.

1. Draw a scatter plot of the data points, with the dependent variable values on the x-axis and the independent variable values on the y-axis. Examine the relationship between the variables suggested by the spread of the dots: Are they in roughly a straight line? Do they fall along a curve? If the latter, what is the shape of the curve?

2. Try the regression equation suggested by the shape in the variables plotted in Step 1. Plot the residuals (or standardized residuals) on the y-axis and the independent variable on the x-axis. The residuals should appear to be random, so if there is an obvious pattern in the residuals, like a curved shape, this suggests that the regression equation doesn't match the shape of the relationship.

3. If the residuals plot from Step 2 shows a pattern in the residuals, try a different regression equation and repeat Step 2. Try the shapes of several regression equations and pick one that appears to most closely match the data. It's usually best to pick the simplest equation that fits the data well.

TRANSFORMING NONLINEAR EQUATIONS INTO LINEAR EQUATIONS

There's another way to deal with nonlinear equations: simply turn them into linear equations. For an example, look at the equation for Miu's age and height (from page 26):

$$y = -\frac{326.6}{x} + 173.3$$

You can turn this into a linear equation. Remember:

$$\text{If } \frac{1}{x} = X, \text{ then } \frac{1}{X} = x.$$

So we'll define a new variable X, set it equal to $\frac{1}{x}$, and use X in the normal $y = aX + b$ regression equation. As shown on page 76, the value of a and b in the regression equation $y = aX + b$ can be calculated as follows:

$$\begin{cases} a = \dfrac{S_{xy}}{S_{xx}} \\ b = \bar{y} - \bar{X}a \end{cases}$$

We continue with the analysis as usual. See Table 2-2.

TALBE 2-2: CALCULATING THE REGRESSION EQUATION

Age x	$\dfrac{1}{age}$ $\dfrac{1}{x} = X$	Height y	$(X - \bar{X})$	$y - \bar{y}$	$(X - \bar{X})^2$	$(y - \bar{y})^2$	$(X - \bar{X})(y - \bar{y})$
4	0.2500	100.1	0.1428	−38.1625	0.0204	1456.3764	−5.4515
5	0.2000	107.2	0.0928	−31.0625	0.0086	964.8789	−2.8841
6	0.1667	114.1	0.0595	−24.1625	0.0035	583.8264	−1.4381
7	0.1429	121.7	0.0357	−16.5625	0.0013	274.3164	−0.5914
8	0.1250	126.8	0.0178	−11.4625	0.0003	131.3889	−0.2046
9	0.1111	130.9	0.0040	−7.3625	0.0000	54.2064	−0.0292
10	0.1000	137.5	−0.0072	−0.7625	0.0001	0.5814	−0.0055
11	0.0909	143.2	−0.0162	4.9375	0.0003	24.3789	−0.0802
12	0.0833	149.4	−0.0238	11.1375	0.0006	124.0439	−0.2653
13	0.0769	151.6	−0.0302	13.3375	0.0009	177.889	−0.4032
14	0.0714	154.0	−0.0357	15.7375	0.0013	247.6689	−0.5622
15	0.0667	154.6	−0.0405	16.3375	0.0016	266.9139	−0.6614
16	0.0625	155.0	−0.0447	16.7375	0.0020	280.1439	−0.7473
17	0.0588	155.1	−0.0483	16.8375	0.0023	283.5014	−0.8137
18	0.0556	155.3	−0.0516	17.0375	0.0027	290.2764	−0.8790
19	0.0526	155.7	−0.0545	17.4375	0.0030	304.0664	−0.9507
Sum 184	1.7144	2212.2	0.0000	0.0000	0.0489	5464.4575	−15.9563
Average 11.5	0.1072	138.3					

According to the table:

$$\begin{cases} a - \dfrac{S_{xy}}{S_{xx}} = \dfrac{-15.9563}{0.0489} = -326.6^* \\ b = \bar{y} - \bar{X}a = 138.2625 - 0.1072 \times (-326.6) = 173.3 \end{cases}$$

So the regression equation is this:

$$y = -326.6X + 173.3$$

$$\uparrow \qquad\qquad \uparrow$$

$$\text{height} \qquad \dfrac{1}{age}$$

---

* If your result is slightly different from 326.6, the difference might be due to rounding. If so, it should be very small.

which is the same as this:

$$\underset{\text{height}}{\underset{\uparrow}{y}} = -\frac{326.6}{\underset{\text{age}}{\underset{\uparrow}{x}}} + 173.3$$

We've transformed our original, nonlinear equation into a linear one!

3

MULTIPLE
REGRESSION
ANALYSIS

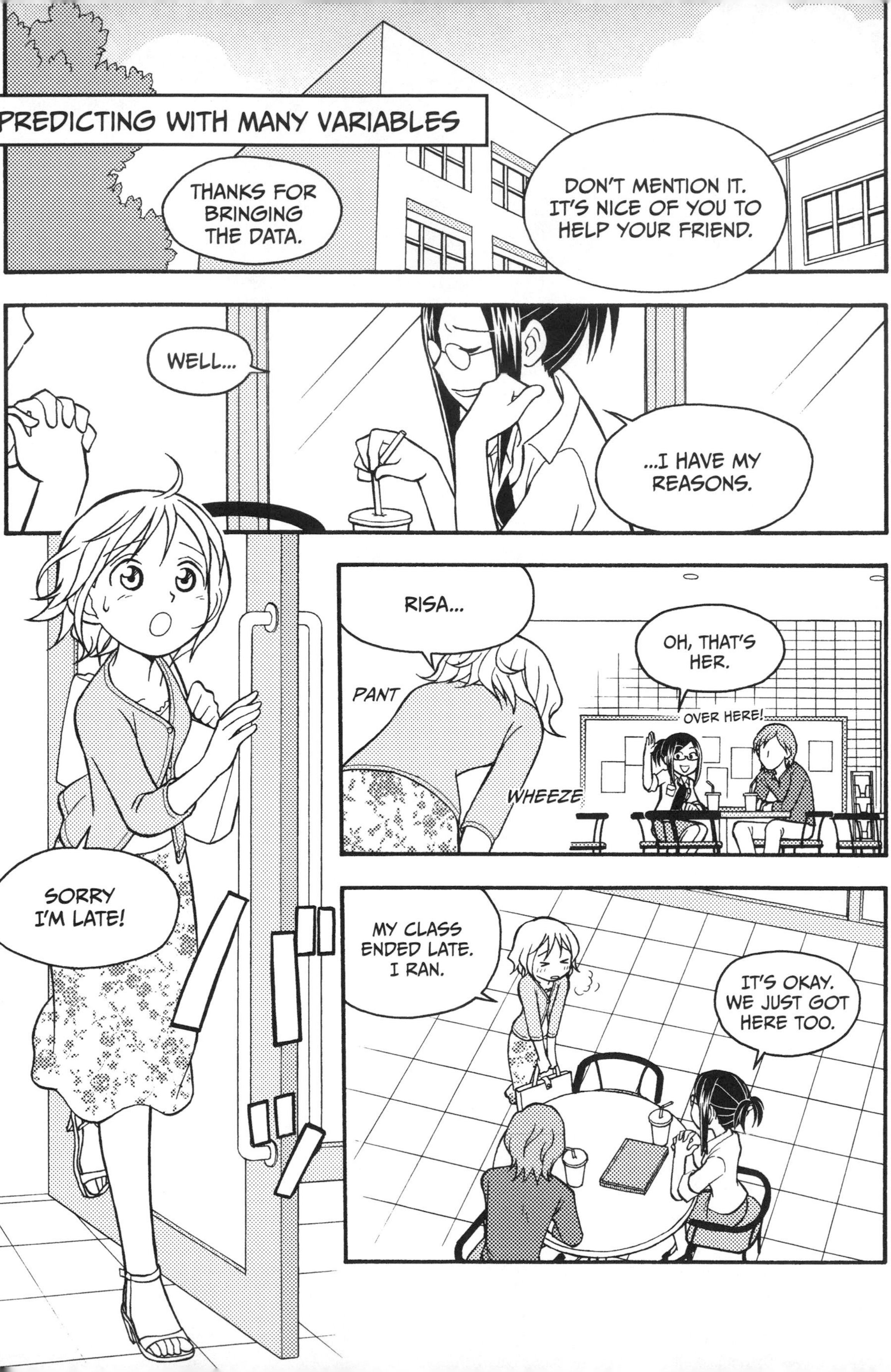
PREDICTING WITH MANY VARIABLES
THANKS FOR BRINGING THE DATA.
DON'T MENTION IT. IT'S NICE OF YOU TO HELP YOUR FRIEND.
WELL...
...I HAVE MY REASONS.
SORRY I'M LATE!
RISA...
PANT
WHEEZE
OH, THAT'S HER.
OVER HERE!
MY CLASS ENDED LATE. I RAN.
IT'S OKAY. WE JUST GOT HERE TOO.

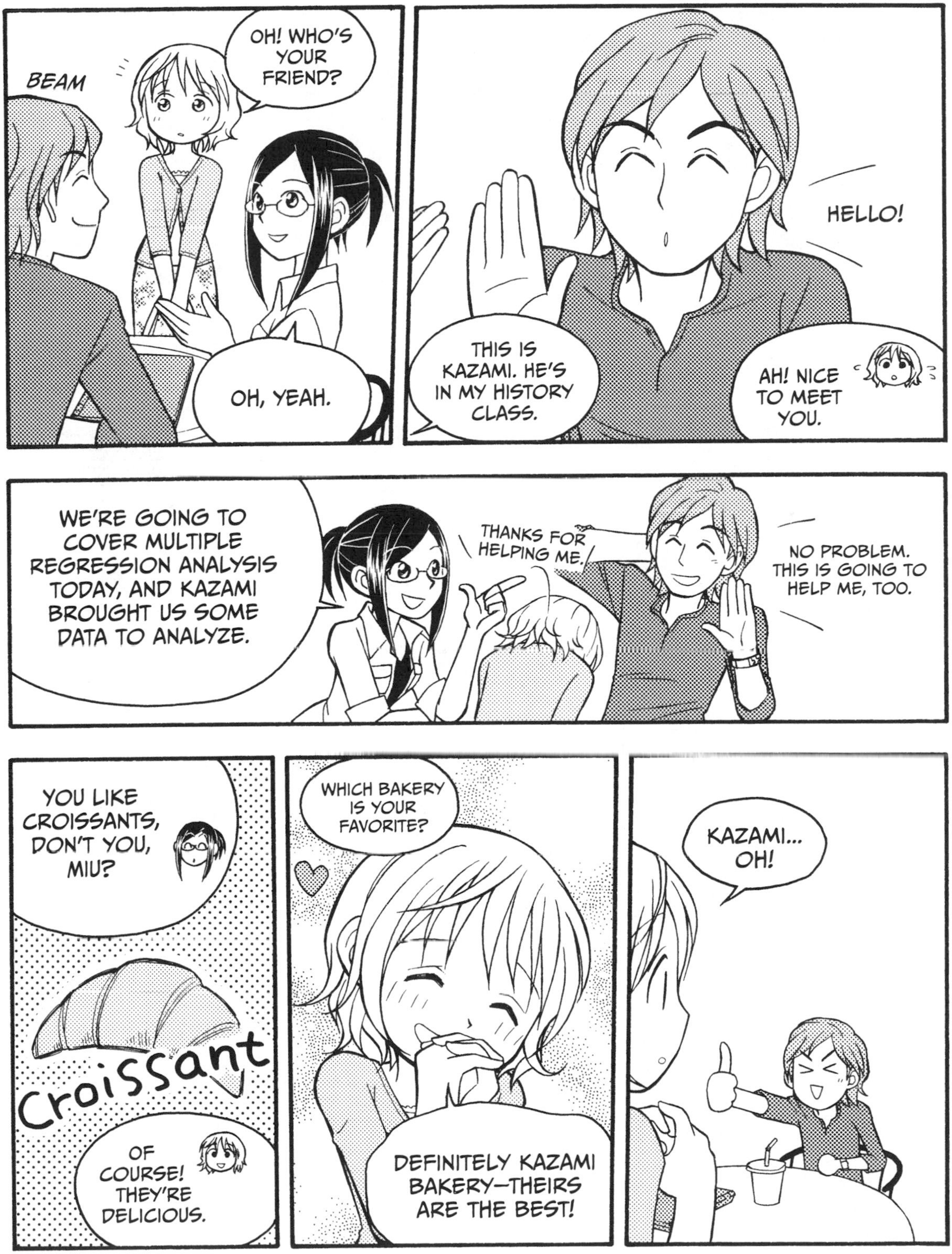
BEAM
OH! WHO'S YOUR FRIEND?
OH, YEAH.
HELLO!
THIS IS KAZAMI. HE'S IN MY HISTORY CLASS.
AH! NICE TO MEET YOU.
WE'RE GOING TO COVER MULTIPLE REGRESSION ANALYSIS TODAY, AND KAZAMI BROUGHT US SOME DATA TO ANALYZE.
THANKS FOR HELPING ME.
NO PROBLEM. THIS IS GOING TO HELP ME, TOO.
YOU LIKE CROISSANTS, DON'T YOU, MIU?
Croissant
OF COURSE! THEY'RE DELICIOUS.
WHICH BAKERY IS YOUR FAVORITE?
DEFINITELY KAZAMI BAKERY—THEIRS ARE THE BEST!
KAZAMI... OH!

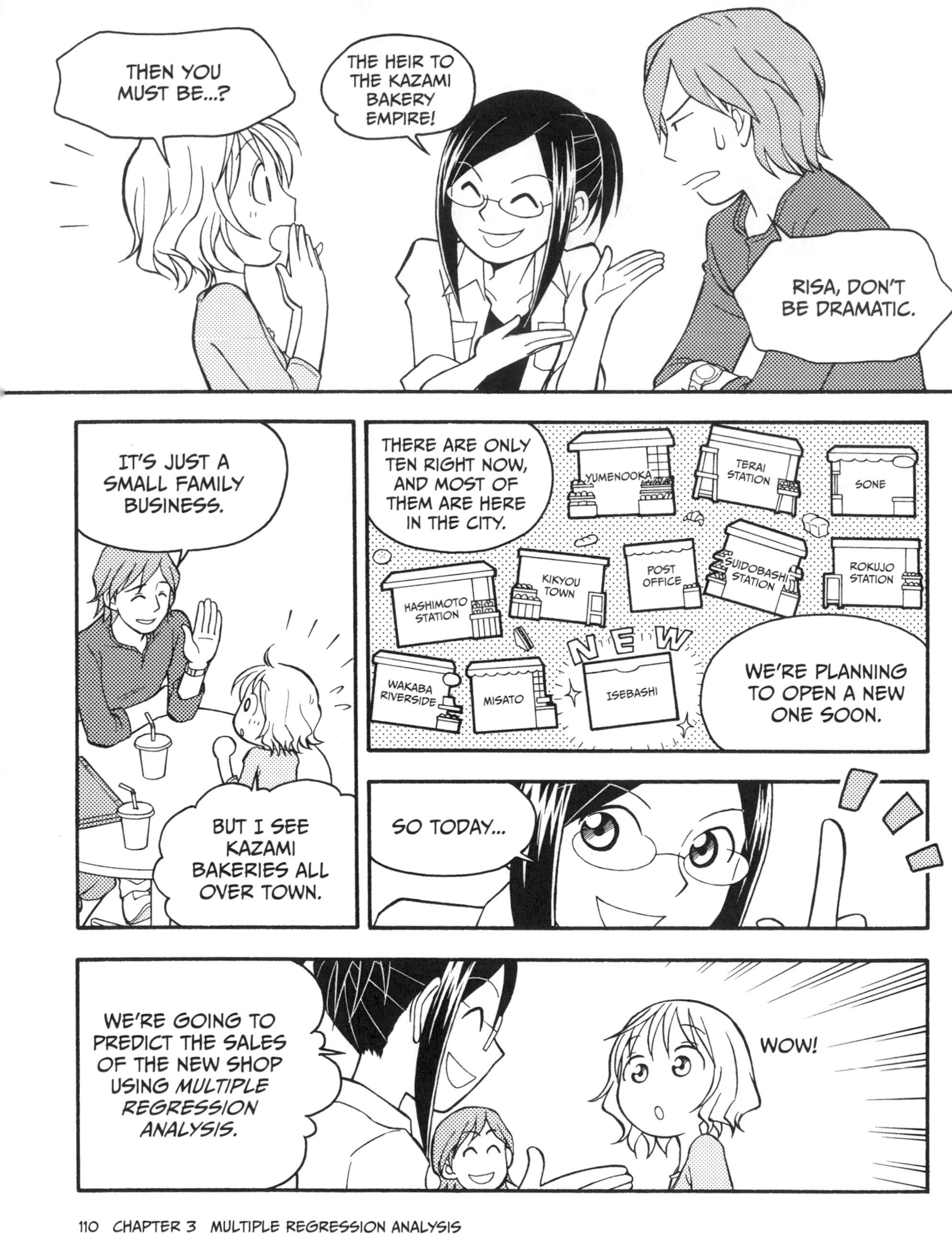
THEN YOU MUST BE...?
THE HEIR TO THE KAZAMI BAKERY EMPIRE!
RISA, DON'T BE DRAMATIC.
IT'S JUST A SMALL FAMILY BUSINESS.
BUT I SEE KAZAMI BAKERIES ALL OVER TOWN.
THERE ARE ONLY TEN RIGHT NOW, AND MOST OF THEM ARE HERE IN THE CITY.
YUMENOOKA
TERAI STATION
SONE
HASHIMOTO STATION
KIKYOU TOWN
POST OFFICE
SUIDOBASHI STATION
ROKUJO STATION
WAKABA RIVERSIDE
MISATO
NEW
ISEBASHI
WE'RE PLANNING TO OPEN A NEW ONE SOON.
SO TODAY...
WE'RE GOING TO PREDICT THE SALES OF THE NEW SHOP USING *MULTIPLE REGRESSION ANALYSIS*.
WOW!

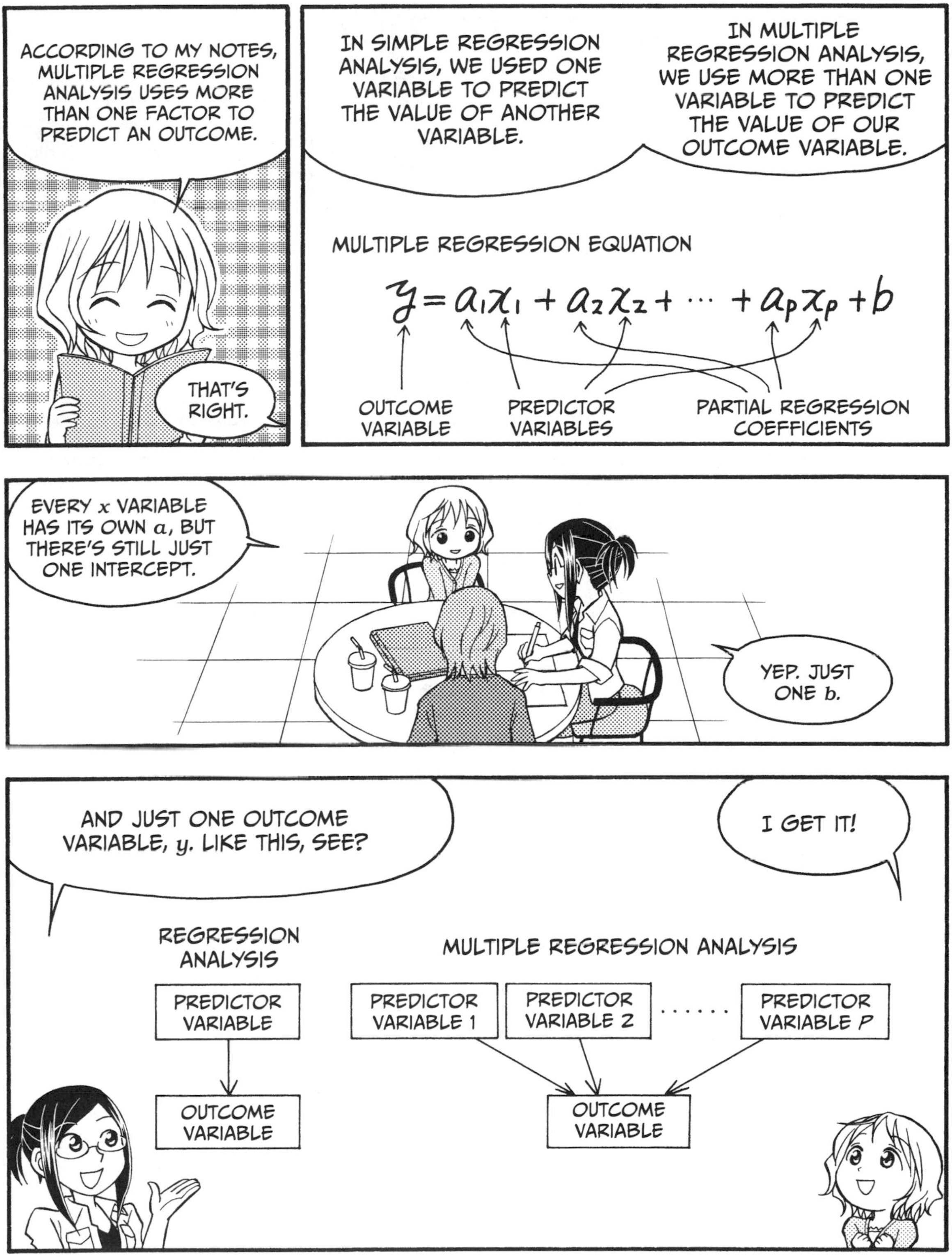

ACCORDING TO MY NOTES, MULTIPLE REGRESSION ANALYSIS USES MORE THAN ONE FACTOR TO PREDICT AN OUTCOME.

THAT'S RIGHT.

IN SIMPLE REGRESSION ANALYSIS, WE USED ONE VARIABLE TO PREDICT THE VALUE OF ANOTHER VARIABLE.

IN MULTIPLE REGRESSION ANALYSIS, WE USE MORE THAN ONE VARIABLE TO PREDICT THE VALUE OF OUR OUTCOME VARIABLE.

MULTIPLE REGRESSION EQUATION

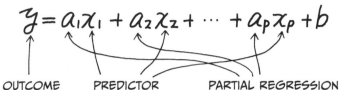

$$y = a_1 x_1 + a_2 x_2 + \cdots + a_p x_p + b$$

OUTCOME VARIABLE

PREDICTOR VARIABLES

PARTIAL REGRESSION COEFFICIENTS

EVERY x VARIABLE HAS ITS OWN a, BUT THERE'S STILL JUST ONE INTERCEPT.

YEP. JUST ONE b.

AND JUST ONE OUTCOME VARIABLE, y. LIKE THIS, SEE?

I GET IT!

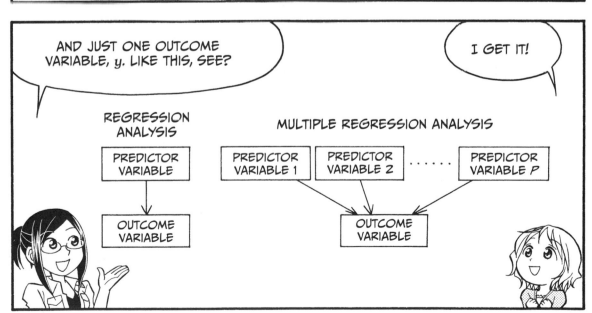

REGRESSION ANALYSIS

PREDICTOR VARIABLE

OUTCOME VARIABLE

MULTIPLE REGRESSION ANALYSIS

PREDICTOR VARIABLE 1

PREDICTOR VARIABLE 2

...... PREDICTOR VARIABLE P

OUTCOME VARIABLE

THE MULTIPLE REGRESSION EQUATION

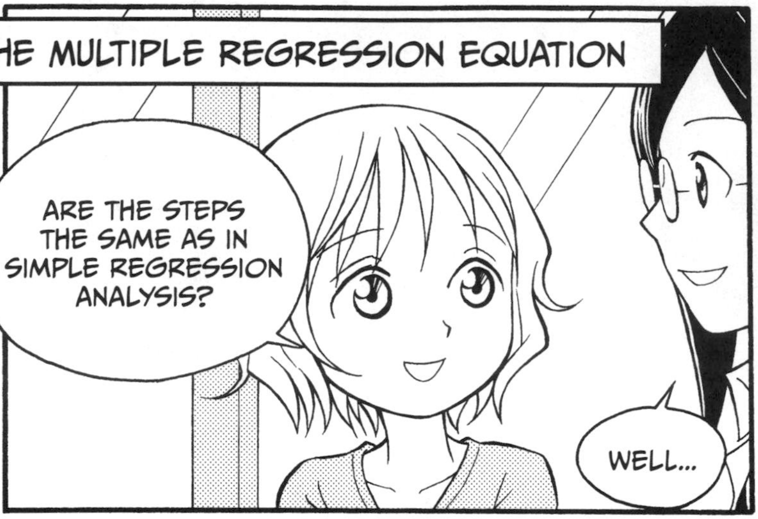
ARE THE STEPS THE SAME AS IN SIMPLE REGRESSION ANALYSIS?
WELL...

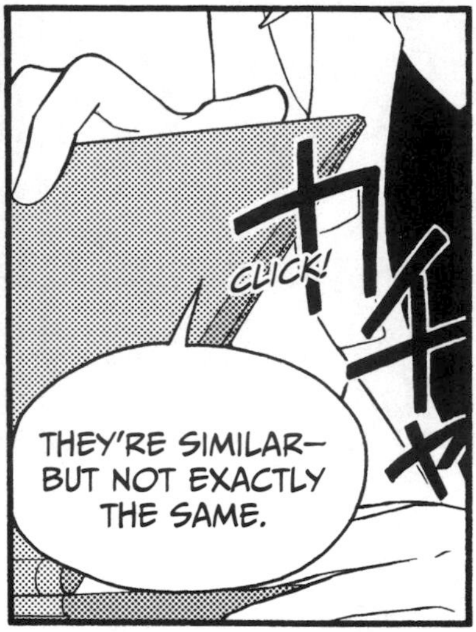
CLICK!
THEY'RE SIMILAR— BUT NOT EXACTLY THE SAME.

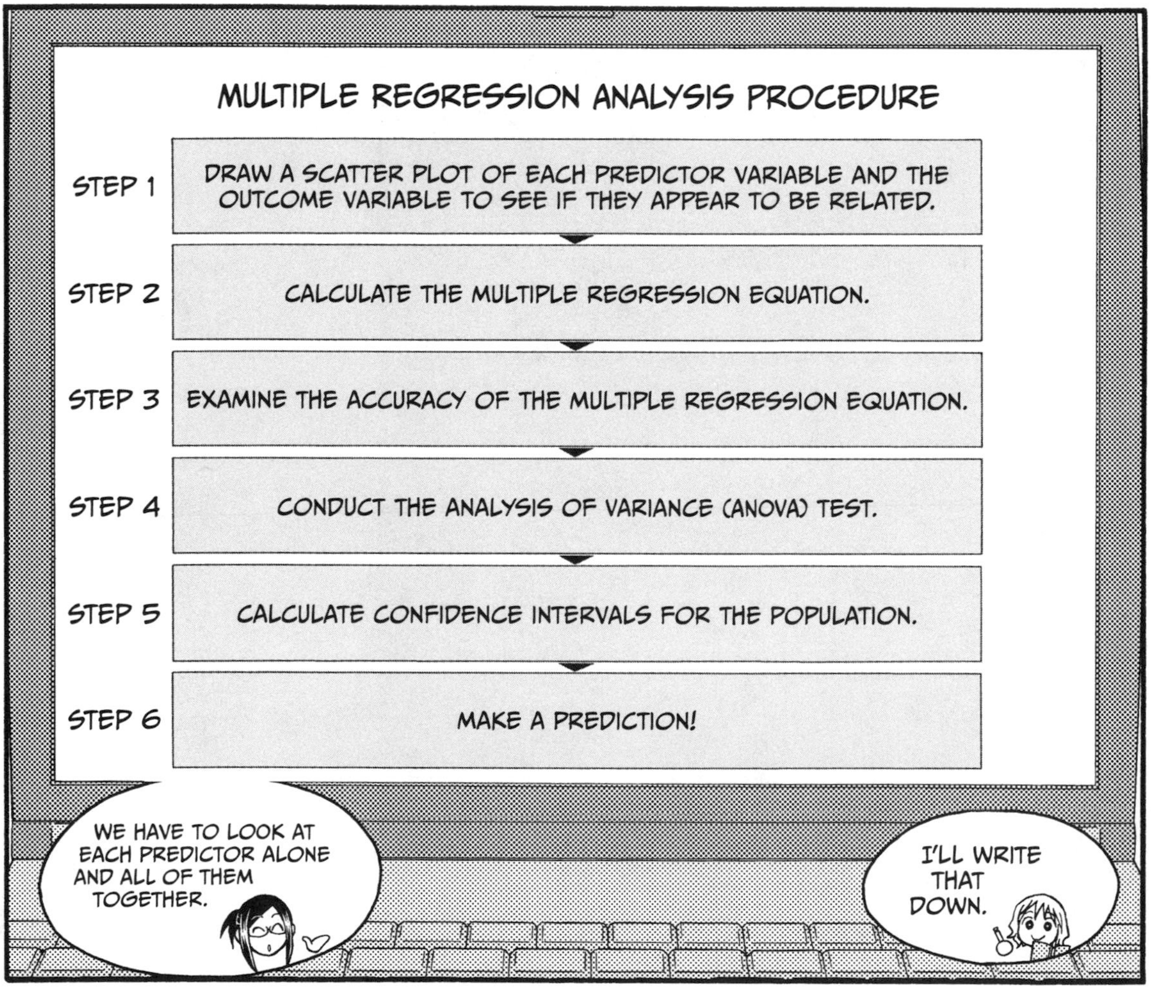
MULTIPLE REGRESSION ANALYSIS PROCEDURE
STEP 1 DRAW A SCATTER PLOT OF EACH PREDICTOR VARIABLE AND THE OUTCOME VARIABLE TO SEE IF THEY APPEAR TO BE RELATED.
STEP 2 CALCULATE THE MULTIPLE REGRESSION EQUATION.
STEP 3 EXAMINE THE ACCURACY OF THE MULTIPLE REGRESSION EQUATION.
STEP 4 CONDUCT THE ANALYSIS OF VARIANCE (ANOVA) TEST.
STEP 5 CALCULATE CONFIDENCE INTERVALS FOR THE POPULATION.
STEP 6 MAKE A PREDICTION!
WE HAVE TO LOOK AT EACH PREDICTOR ALONE AND ALL OF THEM TOGETHER.
I'LL WRITE THAT DOWN.

STEP 1: DRAW A SCATTER PLOT OF EACH PREDICTOR VARIABLE AND THE OUTCOME VARIABLE TO SEE IF THEY APPEAR TO BE RELATED.

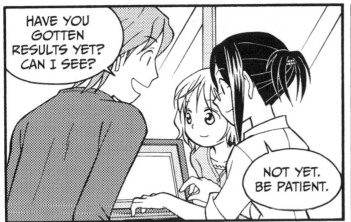

HAVE YOU GOTTEN RESULTS YET? CAN I SEE?
NOT YET. BE PATIENT.

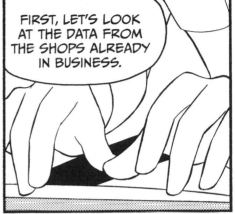

FIRST, LET'S LOOK AT THE DATA FROM THE SHOPS ALREADY IN BUSINESS.

Bakery	Floor space of the shop (tsubo*)	Distance to the nearest station (meters)	Monthly sales (¥10,000)
Yumenooka Shop	10	80	469
Terai Station Shop	8	0	366
Sone Shop	8	200	371
Hashimoto Station Shop	5	200	208
Kikyou Town Shop	7	300	246
Post Office Shop	8	230	297
Suidobashi Station Shop	7	40	363
Rokujo Station Shop	9	0	436
Wakaba Riverside Shop	6	330	198
Misato Shop	9	180	364

* 1 tsubo is about 36 square feet.

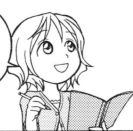

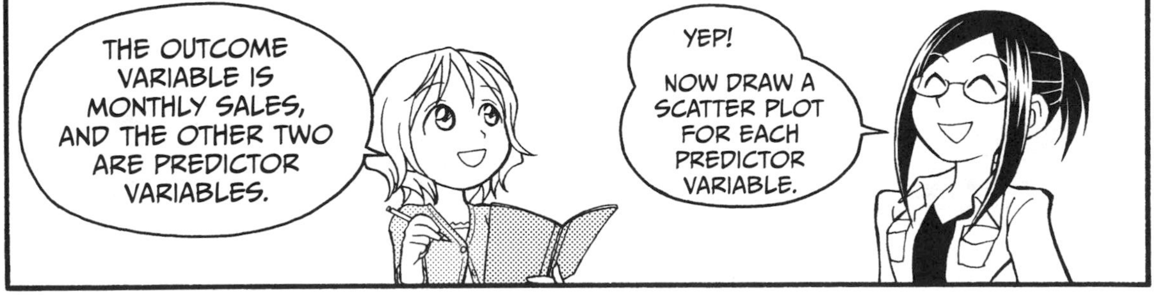

THE OUTCOME VARIABLE IS MONTHLY SALES, AND THE OTHER TWO ARE PREDICTOR VARIABLES.
YEP!
NOW DRAW A SCATTER PLOT FOR EACH PREDICTOR VARIABLE.

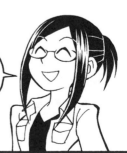

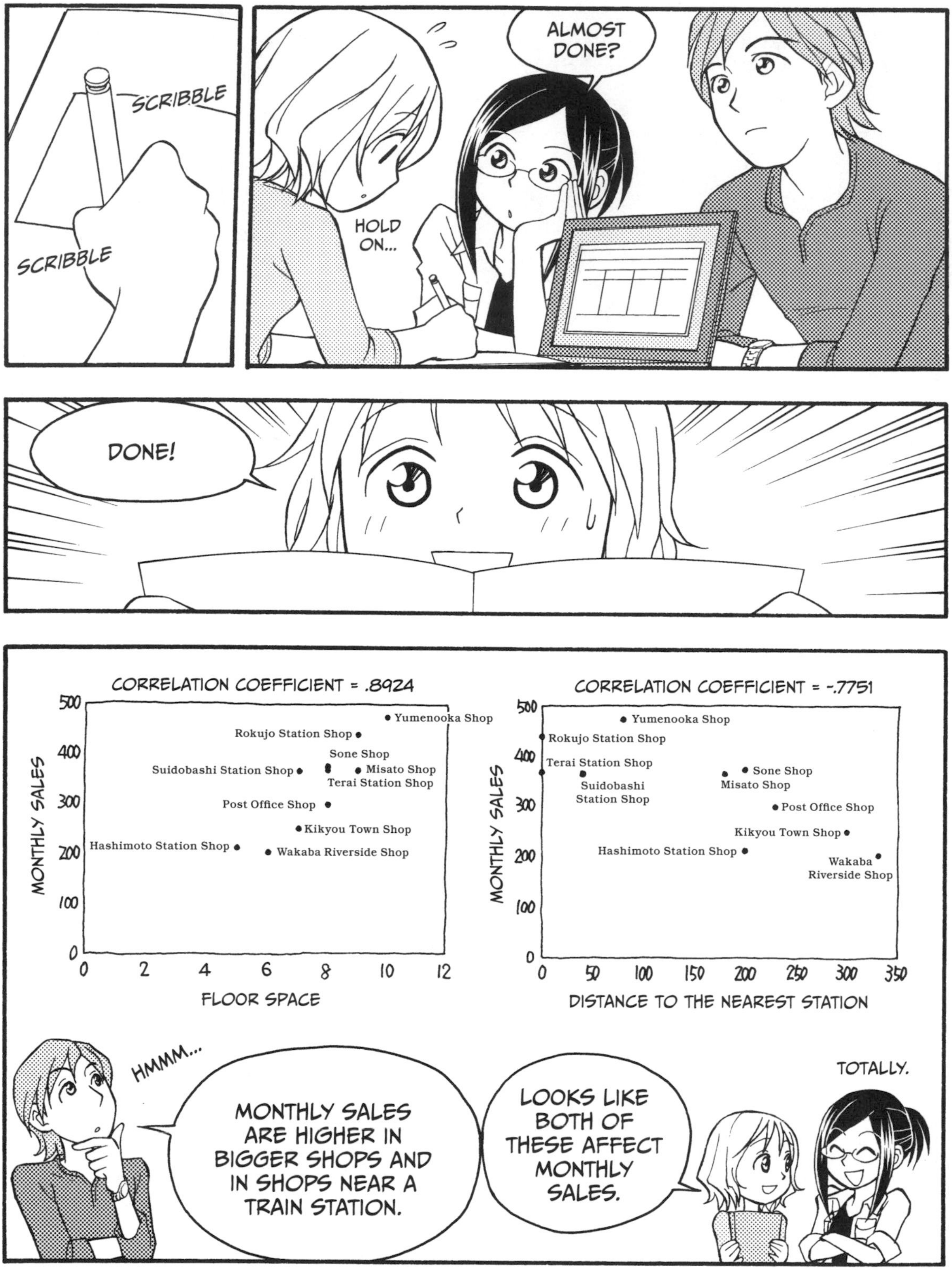
SCRIBBLE
SCRIBBLE
ALMOST DONE?
HOLD ON...
DONE!
CORRELATION COEFFICIENT = .8924
MONTHLY SALES
FLOOR SPACE
Yumenooka Shop
Rokujo Station Shop
Sone Shop
Suidobashi Station Shop
Misato Shop
Terai Station Shop
Post Office Shop
Kikyou Town Shop
Hashimoto Station Shop
Wakaba Riverside Shop
CORRELATION COEFFICIENT = -.7751
MONTHLY SALES
DISTANCE TO THE NEAREST STATION
Yumenooka Shop
Rokujo Station Shop
Terai Station Shop
Suidobashi Station Shop
Sone Shop
Misato Shop
Post Office Shop
Kikyou Town Shop
Hashimoto Station Shop
Wakaba Riverside Shop
HMMM...
MONTHLY SALES ARE HIGHER IN BIGGER SHOPS AND IN SHOPS NEAR A TRAIN STATION.
LOOKS LIKE BOTH OF THESE AFFECT MONTHLY SALES.
TOTALLY.

STEP 2: CALCULATE THE MULTIPLE REGRESSION EQUATION.

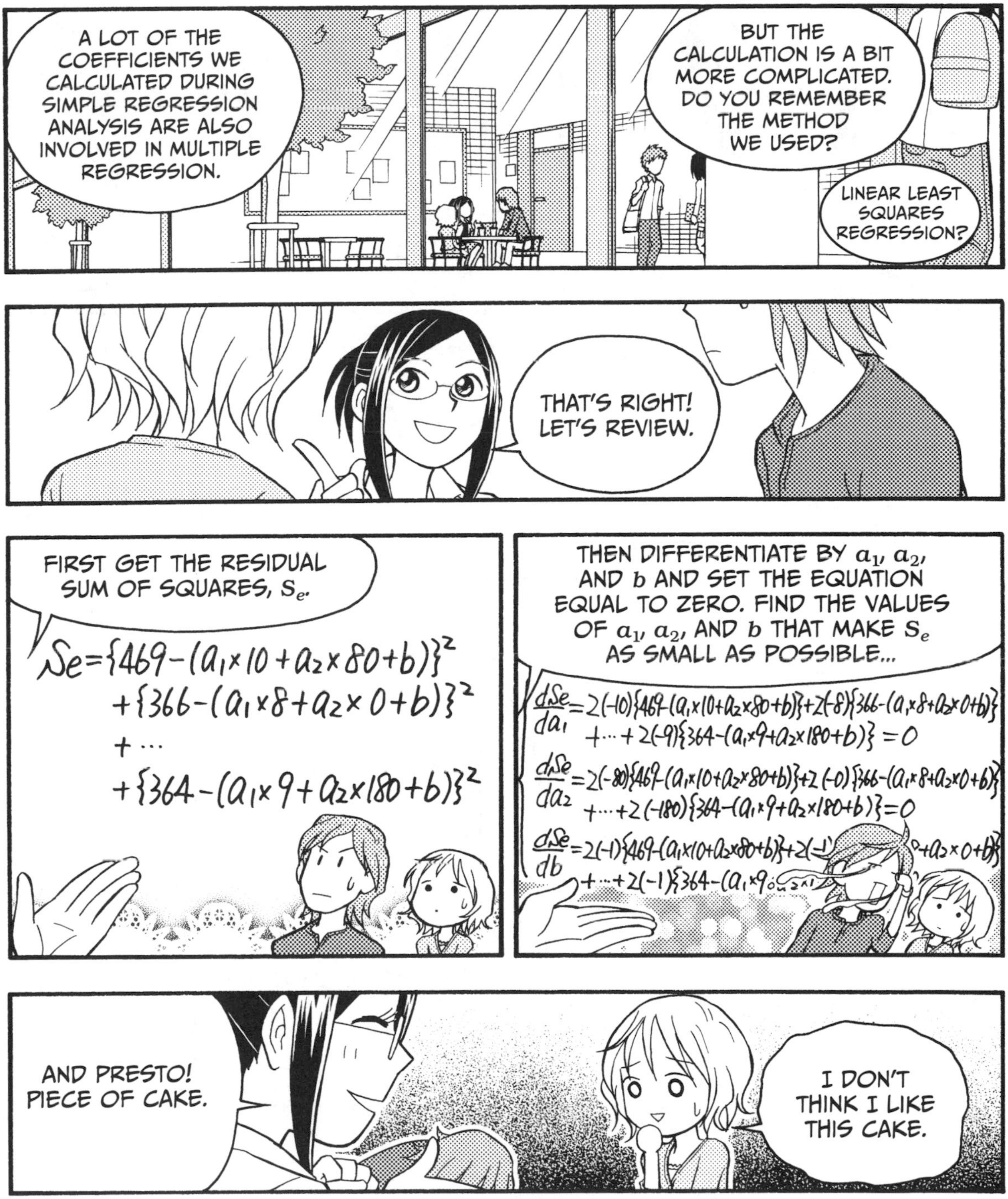

A LOT OF THE COEFFICIENTS WE CALCULATED DURING SIMPLE REGRESSION ANALYSIS ARE ALSO INVOLVED IN MULTIPLE REGRESSION.
BUT THE CALCULATION IS A BIT MORE COMPLICATED. DO YOU REMEMBER THE METHOD WE USED?
LINEAR LEAST SQUARES REGRESSION?
THAT'S RIGHT! LET'S REVIEW.
FIRST GET THE RESIDUAL SUM OF SQUARES, S_e.
$S_e = \{469 - (a_1 \times 10 + a_2 \times 80 + b)\}^2 + \{366 - (a_1 \times 8 + a_2 \times 0 + b)\}^2 + \cdots + \{364 - (a_1 \times 9 + a_2 \times 180 + b)\}^2$
THEN DIFFERENTIATE BY a_1, a_2, AND b AND SET THE EQUATION EQUAL TO ZERO. FIND THE VALUES OF a_1, a_2, AND b THAT MAKE S_e AS SMALL AS POSSIBLE...
$\frac{dS_e}{da_1} = 2(-10)\{469 - (a_1 \times 10 + a_2 \times 80 + b)\} + 2(-8)\{366 - (a_1 \times 8 + a_2 \times 0 + b)\} + \cdots + 2(-9)\{364 - (a_1 \times 9 + a_2 \times 180 + b)\} = 0$
$\frac{dS_e}{da_2} = 2(-80)\{469 - (a_1 \times 10 + a_2 \times 80 + b)\} + 2(-0)\{366 - (a_1 \times 8 + a_2 \times 0 + b)\} + \cdots + 2(-180)\{364 - (a_1 \times 9 + a_2 \times 180 + b)\} = 0$
$\frac{dS_e}{db} = 2(-1)\{469 - (a_1 \times 10 + a_2 \times 80 + b)\} + 2(-1)$
$+ a_2 \times 0 + b)\}$
$+ \cdots + 2(-1)\{364 - (a_1 \times 9$
AND PRESTO! PIECE OF CAKE.
I DON'T THINK I LIKE THIS CAKE.

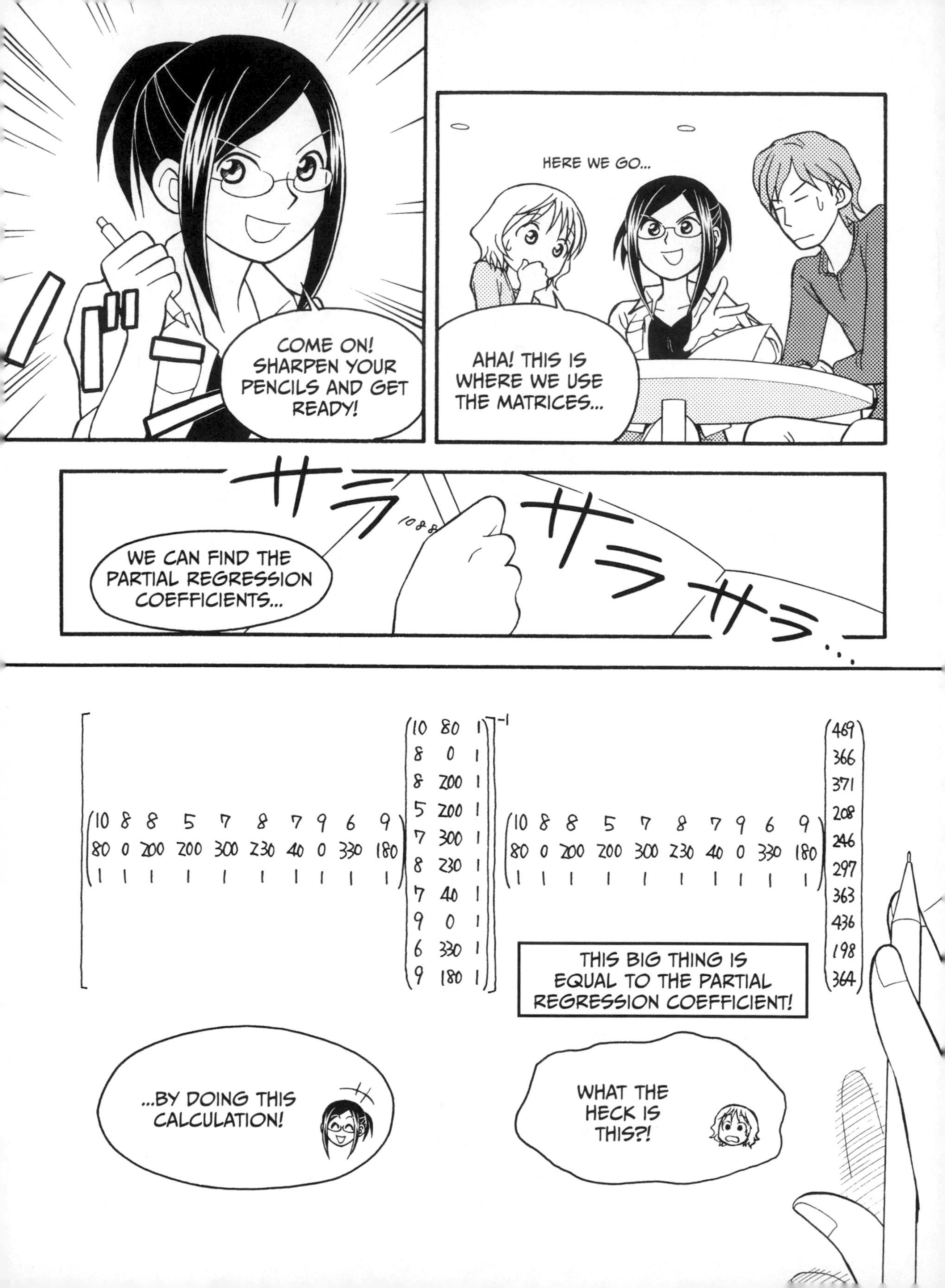
COME ON! SHARPEN YOUR PENCILS AND GET READY!
HERE WE GO...
AHA! THIS IS WHERE WE USE THE MATRICES...
サラ サラ サラ...
WE CAN FIND THE PARTIAL REGRESSION COEFFICIENTS...
$$\left[\begin{pmatrix}10&8&8&5&7&8&7&9&6&9\\80&0&200&200&300&230&40&0&330&180\\1&1&1&1&1&1&1&1&1&1\end{pmatrix}\begin{pmatrix}10&80&1\\8&0&1\\8&200&1\\5&200&1\\7&300&1\\8&230&1\\7&40&1\\9&0&1\\6&330&1\\9&180&1\end{pmatrix}\right]^{-1}\begin{pmatrix}10&8&8&5&7&8&7&9&6&9\\80&0&200&200&300&230&40&0&330&180\\1&1&1&1&1&1&1&1&1&1\end{pmatrix}\begin{pmatrix}469\\366\\371\\208\\246\\297\\363\\436\\198\\364\end{pmatrix}$$
THIS BIG THING IS EQUAL TO THE PARTIAL REGRESSION COEFFICIENT!
...BY DOING THIS CALCULATION!
WHAT THE HECK IS THIS?!

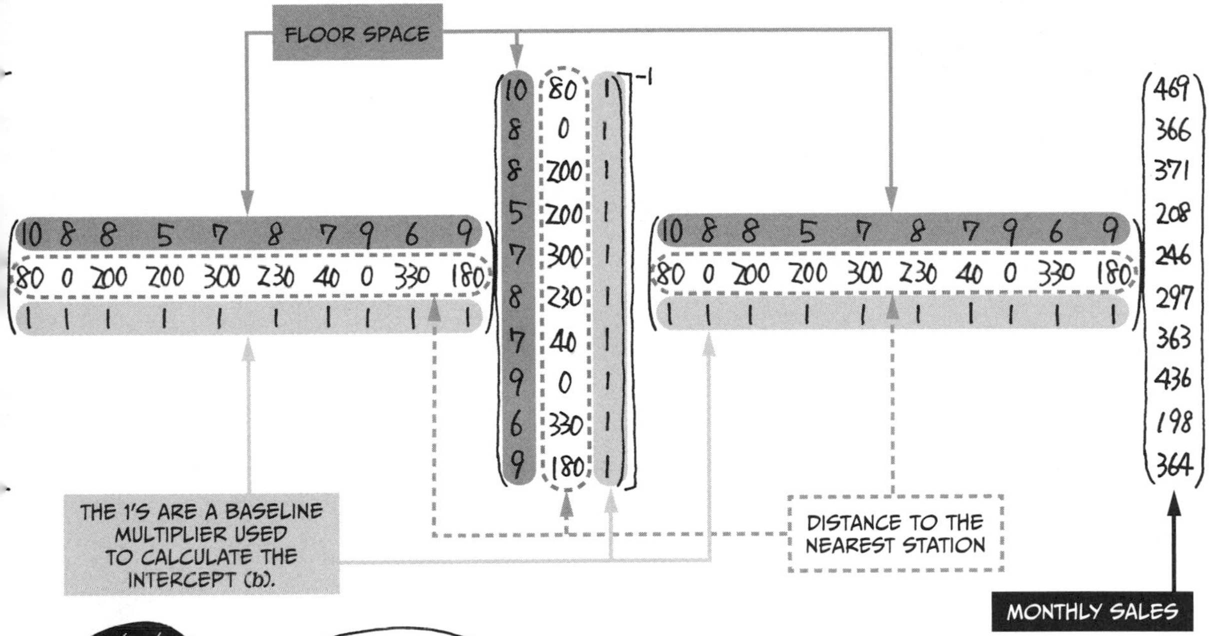
FLOOR SPACE
THE 1'S ARE A BASELINE MULTIPLIER USED TO CALCULATE THE INTERCEPT (b).
DISTANCE TO THE NEAREST STATION
MONTHLY SALES

$$\left[\begin{pmatrix} 10 & 8 & 8 & 5 & 7 & 8 & 7 & 9 & 6 & 9 \\ 80 & 0 & 200 & 200 & 300 & 230 & 40 & 0 & 330 & 180 \\ 1 & 1 & 1 & 1 & 1 & 1 & 1 & 1 & 1 & 1 \end{pmatrix}\begin{pmatrix} 10 & 80 & 1 \\ 8 & 0 & 1 \\ 8 & 200 & 1 \\ 5 & 200 & 1 \\ 7 & 300 & 1 \\ 8 & 230 & 1 \\ 7 & 40 & 1 \\ 9 & 0 & 1 \\ 6 & 330 & 1 \\ 9 & 180 & 1 \end{pmatrix}\right]^{-1}\begin{pmatrix} 10 & 8 & 8 & 5 & 7 & 8 & 7 & 9 & 6 & 9 \\ 80 & 0 & 200 & 200 & 300 & 230 & 40 & 0 & 330 & 180 \\ 1 & 1 & 1 & 1 & 1 & 1 & 1 & 1 & 1 & 1 \end{pmatrix}\begin{pmatrix} 469 \\ 366 \\ 371 \\ 208 \\ 246 \\ 297 \\ 363 \\ 436 \\ 198 \\ 364 \end{pmatrix}$$

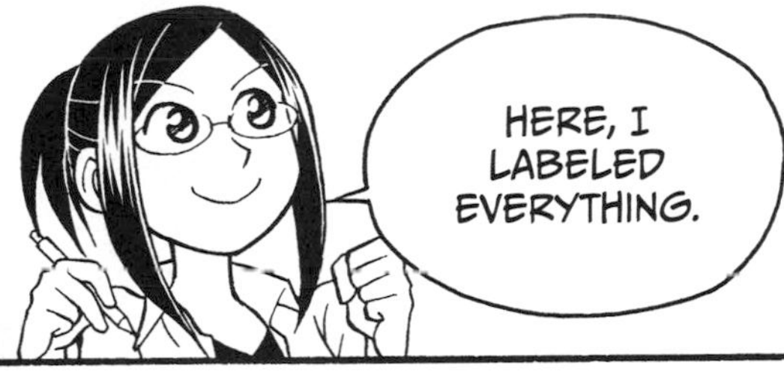
HERE, I LABELED EVERYTHING.

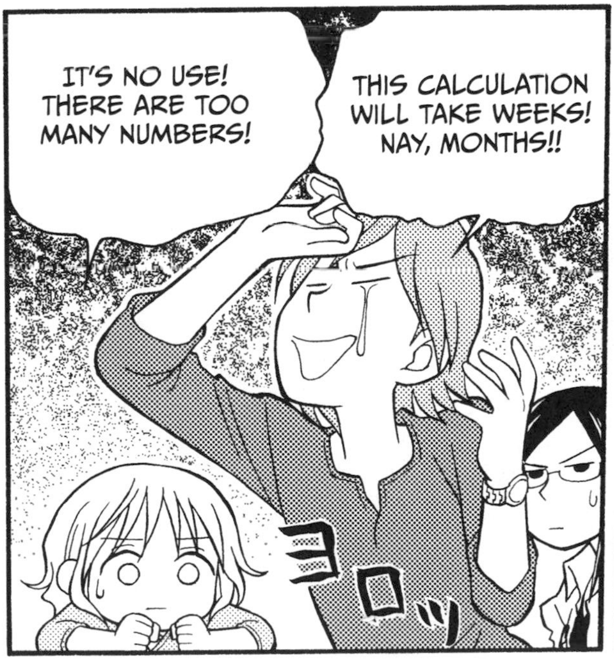
IT'S NO USE! THERE ARE TOO MANY NUMBERS!
THIS CALCULATION WILL TAKE WEEKS! NAY, MONTHS!!
ヨロッ

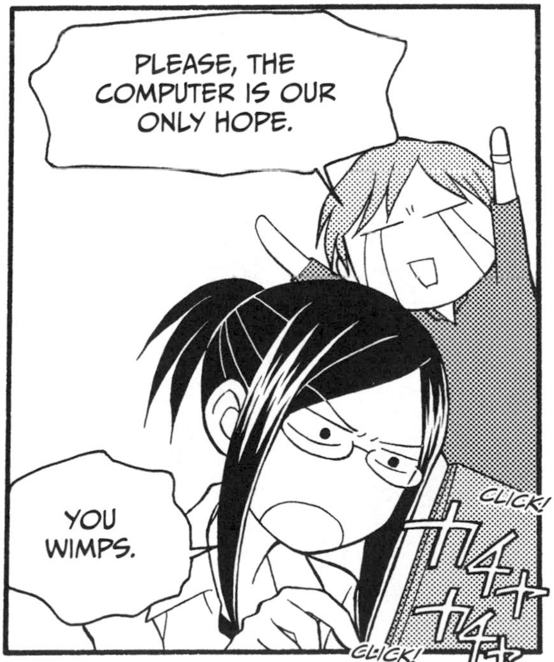
PLEASE, THE COMPUTER IS OUR ONLY HOPE.
YOU WIMPS.
CLICK!
CLICK!

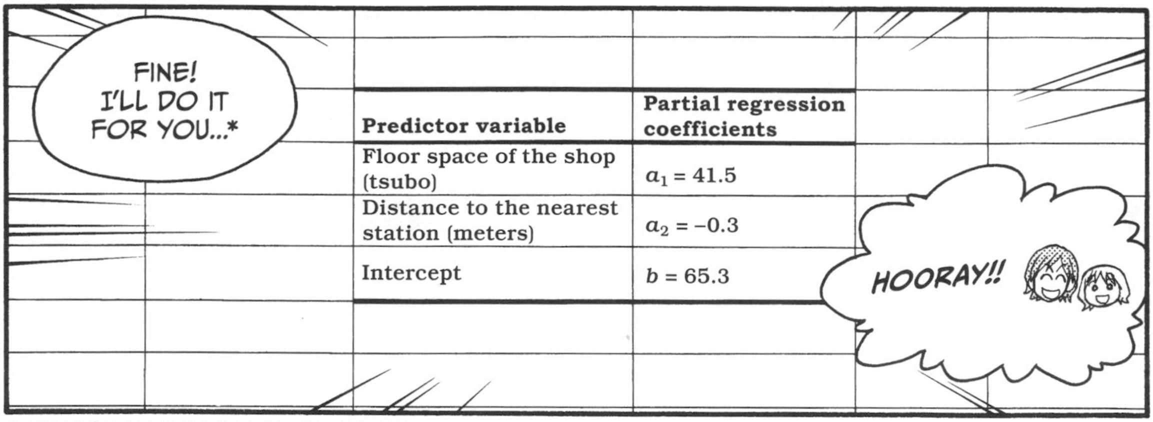
FINE! I'LL DO IT FOR YOU...*
HOORAY!!

Predictor variable	Partial regression coefficients
Floor space of the shop (tsubo)	$a_1 = 41.5$
Distance to the nearest station (meters)	$a_2 = -0.3$
Intercept	$b = 65.3$

* SEE PAGE 209 FOR THE FULL CALCULATION.

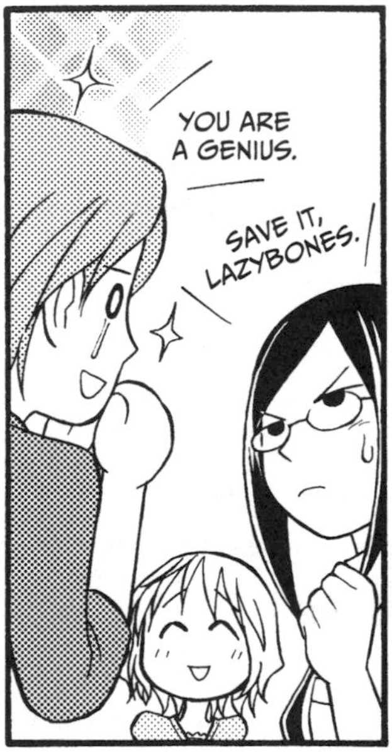
YOU ARE A GENIUS.
SAVE IT, LAZYBONES.

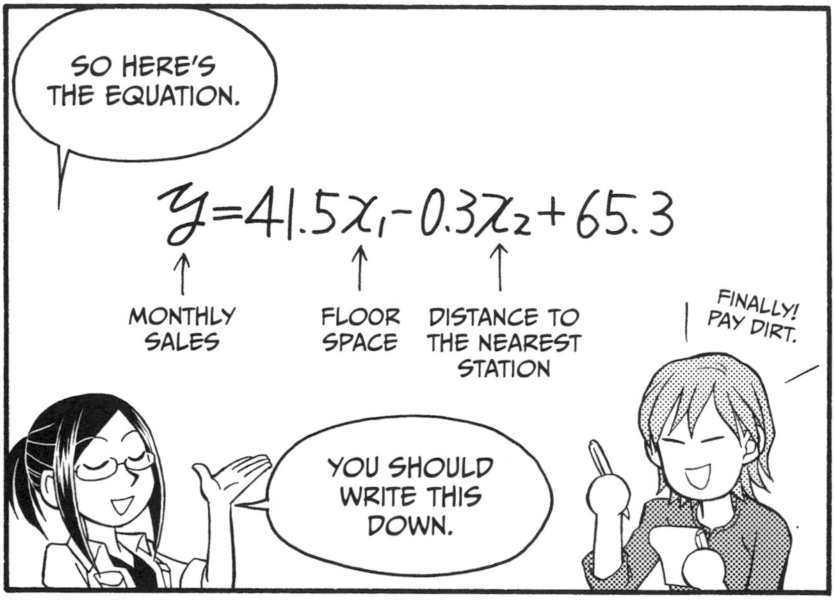
SO HERE'S THE EQUATION.
y=41.5x1-0.3x2+65.3
MONTHLY SALES
FLOOR SPACE
DISTANCE TO THE NEAREST STATION
FINALLY! PAY DIRT.
YOU SHOULD WRITE THIS DOWN.

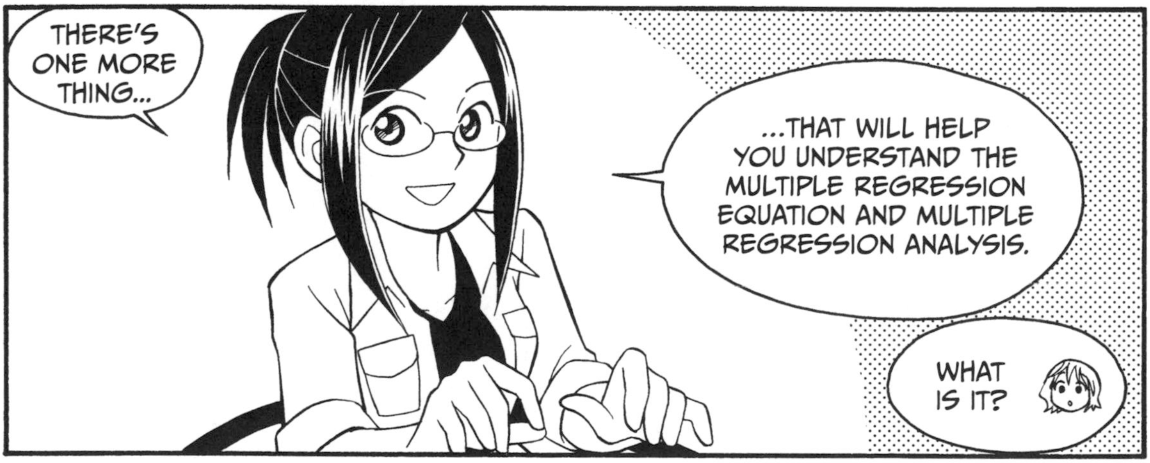
THERE'S ONE MORE THING...
...THAT WILL HELP YOU UNDERSTAND THE MULTIPLE REGRESSION EQUATION AND MULTIPLE REGRESSION ANALYSIS.
WHAT IS IT?

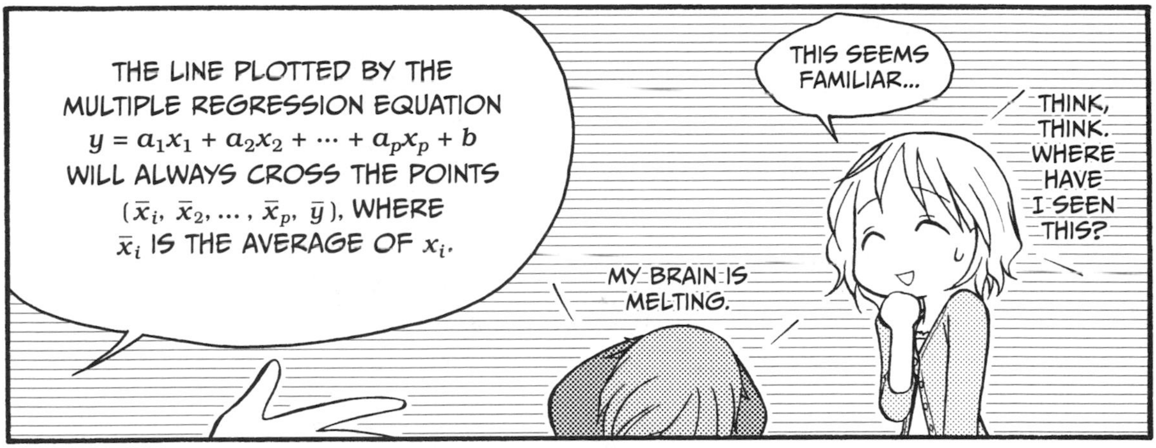
THE LINE PLOTTED BY THE MULTIPLE REGRESSION EQUATION $y = a_1x_1 + a_2x_2 + \cdots + a_px_p + b$ WILL ALWAYS CROSS THE POINTS $(\bar{x}_i, \bar{x}_2, \ldots, \bar{x}_p, \bar{y})$, WHERE $\bar{x}_i$ IS THE AVERAGE OF x_i.
THIS SEEMS FAMILIAR...
THINK, THINK. WHERE HAVE I SEEN THIS?
MY BRAIN IS MELTING.

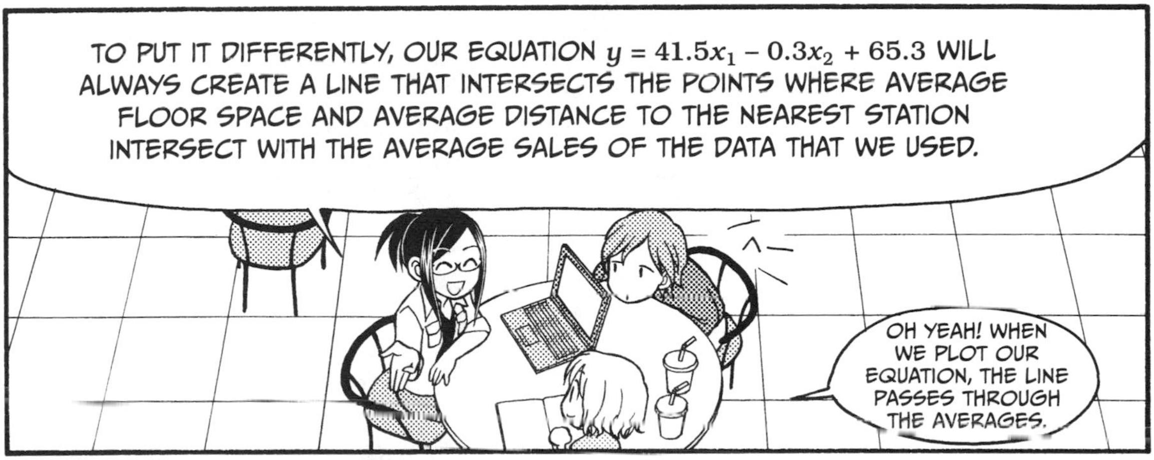
TO PUT IT DIFFERENTLY, OUR EQUATION $y = 41.5x_1 - 0.3x_2 + 65.3$ WILL ALWAYS CREATE A LINE THAT INTERSECTS THE POINTS WHERE AVERAGE FLOOR SPACE AND AVERAGE DISTANCE TO THE NEAREST STATION INTERSECT WITH THE AVERAGE SALES OF THE DATA THAT WE USED.
OH YEAH! WHEN WE PLOT OUR EQUATION, THE LINE PASSES THROUGH THE AVERAGES.

STEP 3: EXAMINE THE ACCURACY OF THE MULTIPLE REGRESSION EQUATION.

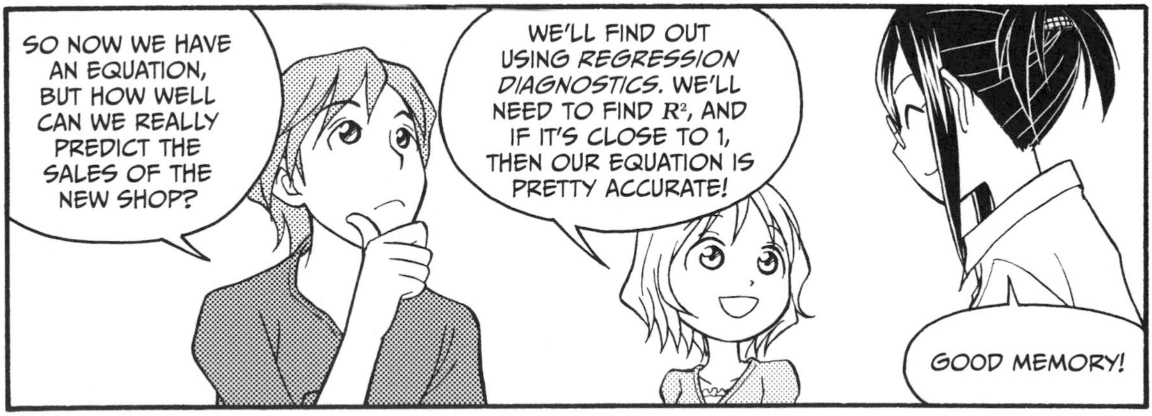
SO NOW WE HAVE AN EQUATION, BUT HOW WELL CAN WE REALLY PREDICT THE SALES OF THE NEW SHOP?
WE'LL FIND OUT USING *REGRESSION DIAGNOSTICS*. WE'LL NEED TO FIND R^2, AND IF IT'S CLOSE TO 1, THEN OUR EQUATION IS PRETTY ACCURATE!
GOOD MEMORY!

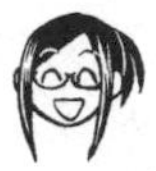
BEFORE WE FIND R^2, WE NEED TO FIND PLAIN OLD R, WHICH IN THIS CASE IS CALLED THE MULTIPLE CORRELATION COEFFICIENT. REMEMBER: R IS A WAY OF COMPARING THE ACTUAL MEASURED VALUES (y) WITH OUR ESTIMATED VALUES ($\hat{y}$).*

Bakery	Actual value y	Estimated value $\hat{y} = 41x_1 - 0.3x_2 + 65.3$	$y - \bar{y}$	$\hat{y} - \bar{\hat{y}}$	$(y - \bar{y})^2$	$(\hat{y} - \bar{\hat{y}})^2$	$(y - \bar{y})(\hat{y} - \bar{\hat{y}})$	$(y - \hat{y})^2$
Yumenooka	469	453.2	137.2	121.4	18823.8	14735.1	16654.4	250.0
Terai	366	397.4	34.2	65.6	1169.6	4307.5	2244.6	988.0
Sone	371	329.3	39.2	−2.5	1536.6	6.5	−99.8	1742.6
Hashimoto	208	204.7	−123.8	−127.1	15326.4	16150.7	15733.2	10.8
Kikyou	246	253.7	−85.8	−78.1	7361.6	6016.9	6705.0	58.6
Post Office	297	319.0	−34.8	−12.8	1211.0	163.1	444.4	485.3
Suidobashi	363	342.3	31.2	10.5	973.4	109.9	327.1	429.2
Rokujo	436	438.9	104.2	107.1	10857.6	11480.1	11164.5	8.7
Wakaba	198	201.9	−133.8	−129.9	17902.4	16870.5	17378.8	15.3
Misato	364	377.6	32.2	45.8	1036.8	2096.4	1474.3	184.6
Total	3318	3318	0	0	76199.6	72026.6	72026.6	4173.0
Average	331.8	331.8						
	↓ $\bar{y}$	↓ $\bar{\hat{y}}$			↓ S_{yy}	↓ $S_{\hat{y}\hat{y}}$	↓ $S_{y\hat{y}}$	↓ S_e

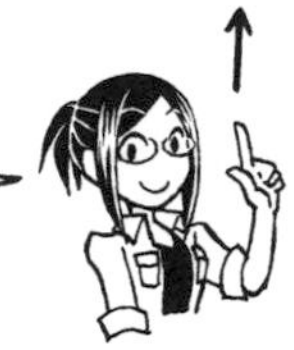
WE DON'T NEED S_e YET, BUT WE WILL USE IT LATER.

$$R = \frac{\text{sum of } (y - \bar{y})(\hat{y} - \bar{\hat{y}})}{\sqrt{\text{sum of } (y - \bar{y})^2 \times \text{sum of } (\hat{y} - \bar{\hat{y}})^2}} = \frac{S_{y\hat{y}}}{\sqrt{S_{yy} \times S_{\hat{y}\hat{y}}}}$$

$$= \frac{72026.6}{\sqrt{76199.6 \times 72026.6}} = .9722$$

$$R^2 = (.9722)^2 = .9452$$

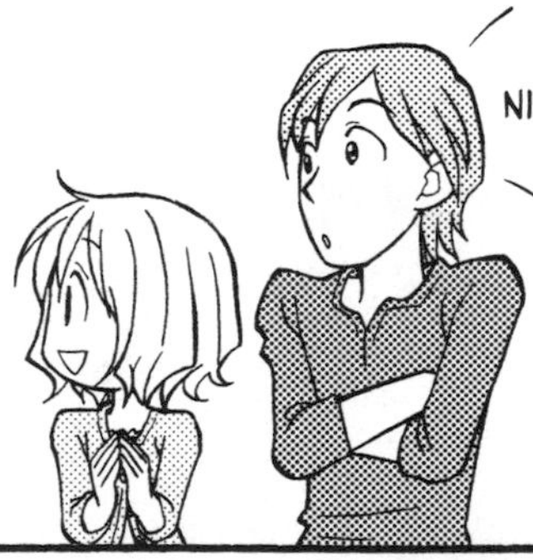
R^2 IS .9452.
NICE!

* AS IN CHAPTER 2, SOME OF THE FIGURES IN THIS CHAPTER ARE ROUNDED FOR READABILITY, BUT ALL CALCULATIONS ARE DONE USING THE FULL, UNROUNDED VALUES RESULTING FROM THE RAW DATA UNLESS OTHERWISE STATED.

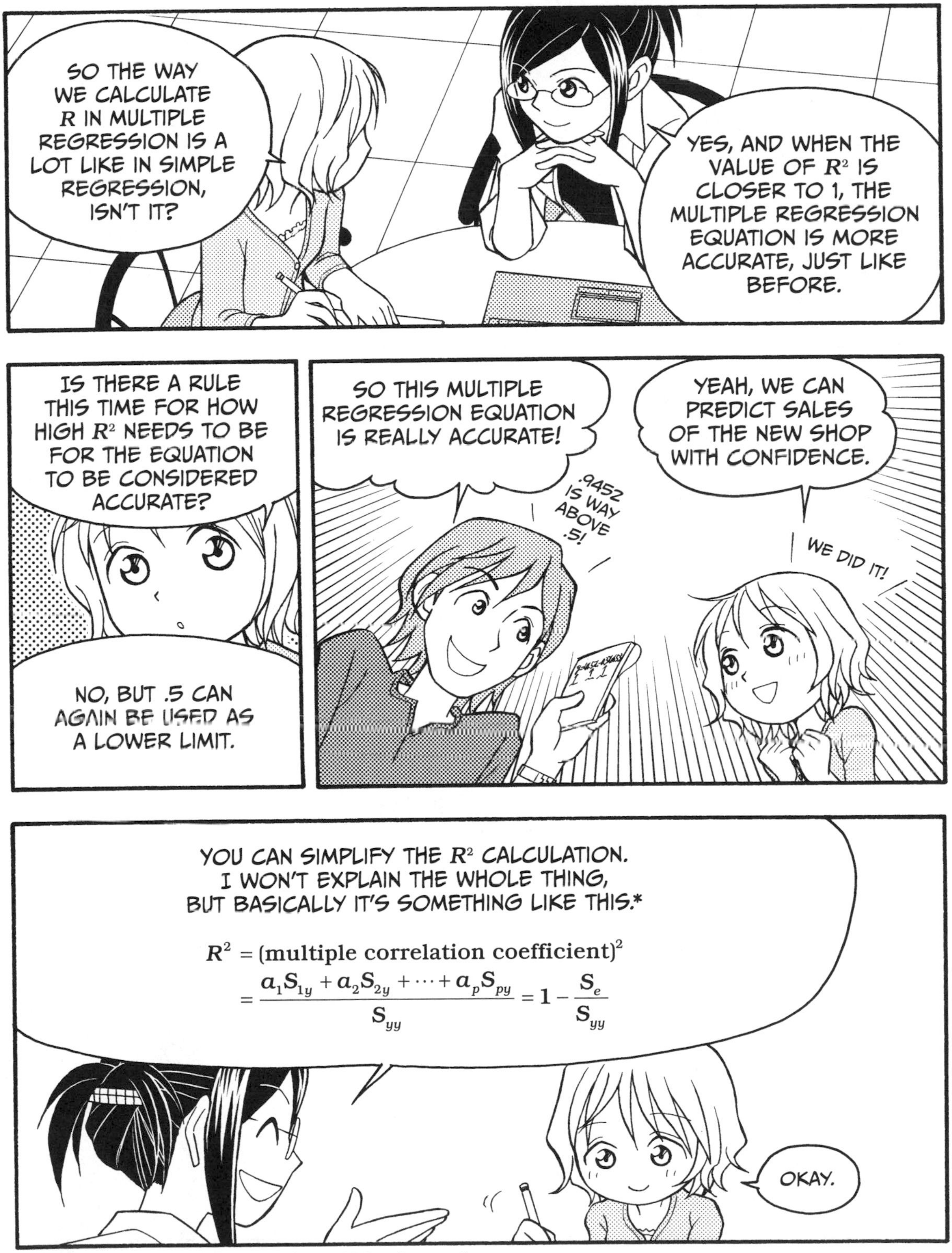
SO THE WAY WE CALCULATE R IN MULTIPLE REGRESSION IS A LOT LIKE IN SIMPLE REGRESSION, ISN'T IT?
YES, AND WHEN THE VALUE OF R^2 IS CLOSER TO 1, THE MULTIPLE REGRESSION EQUATION IS MORE ACCURATE, JUST LIKE BEFORE.
IS THERE A RULE THIS TIME FOR HOW HIGH R^2 NEEDS TO BE FOR THE EQUATION TO BE CONSIDERED ACCURATE?
NO, BUT .5 CAN AGAIN BE USED AS A LOWER LIMIT.
SO THIS MULTIPLE REGRESSION EQUATION IS REALLY ACCURATE!
.9452 IS WAY ABOVE .5!
YEAH, WE CAN PREDICT SALES OF THE NEW SHOP WITH CONFIDENCE.
WE DID IT!
YOU CAN SIMPLIFY THE R^2 CALCULATION. I WON'T EXPLAIN THE WHOLE THING, BUT BASICALLY IT'S SOMETHING LIKE THIS.*
$R^2 = (\text{multiple correlation coefficient})^2$
$= \frac{a_1 S_{1y} + a_2 S_{2y} + \cdots + a_p S_{py}}{S_{yy}} = 1 - \frac{S_e}{S_{yy}}$
OKAY.

* REFER TO PAGE 144 FOR AN EXPLANATION OF $S_{1y}, S_{2y}, \ldots, S_{py}$.

THE TROUBLE WITH R^2

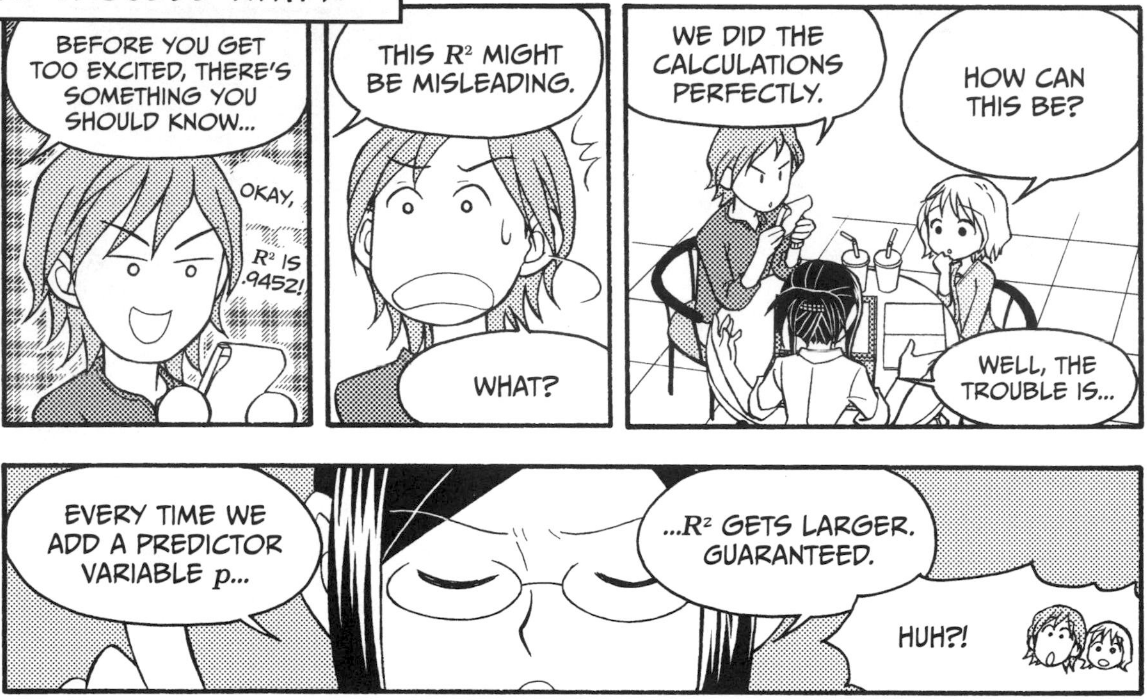
BEFORE YOU GET TOO EXCITED, THERE'S SOMETHING YOU SHOULD KNOW...
OKAY, R^2 IS .9452!
THIS R^2 MIGHT BE MISLEADING.
WHAT?
WE DID THE CALCULATIONS PERFECTLY.
HOW CAN THIS BE?
WELL, THE TROUBLE IS...
EVERY TIME WE ADD A PREDICTOR VARIABLE p...
...R^2 GETS LARGER. GUARANTEED.
HUH?!

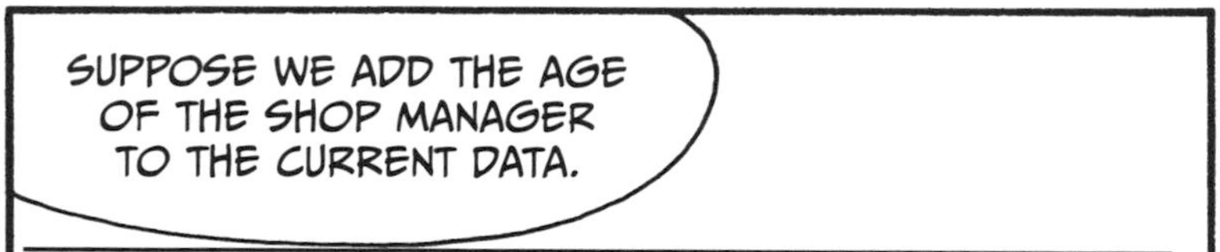
SUPPOSE WE ADD THE AGE OF THE SHOP MANAGER TO THE CURRENT DATA.

Bakery	Floor area of the shop (tsubo)	Distance to the nearest station (meters)	Shop manager's age (years)	Monthly sales (¥10,000)
Yumenooka Shop	10	80	42	469
Terai Station Shop	8	0	29	366
Sone Shop	8	200	33	371
Hashimoto Station Shop	5	200	41	208
Kikyou Town Shop	7	300	33	246
Post Office Shop	8	230	35	297
Suidobashi Shop	7	40	40	363
Rokujo Station Shop	9	0	46	436
Wakaba Riverside Shop	6	330	44	198
Misato Shop	9	180	34	364

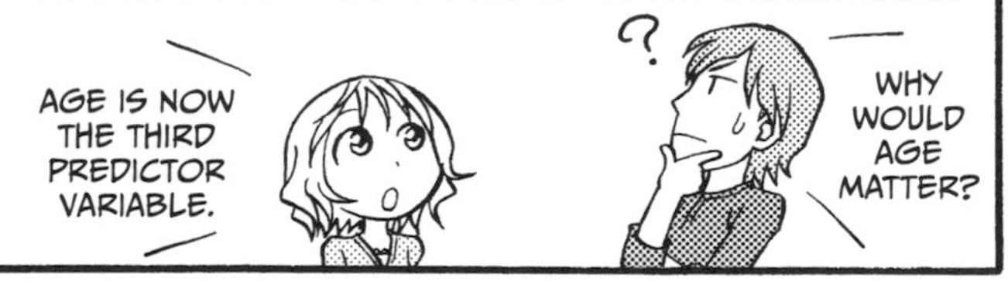
AGE IS NOW THE THIRD PREDICTOR VARIABLE.
?
WHY WOULD AGE MATTER?

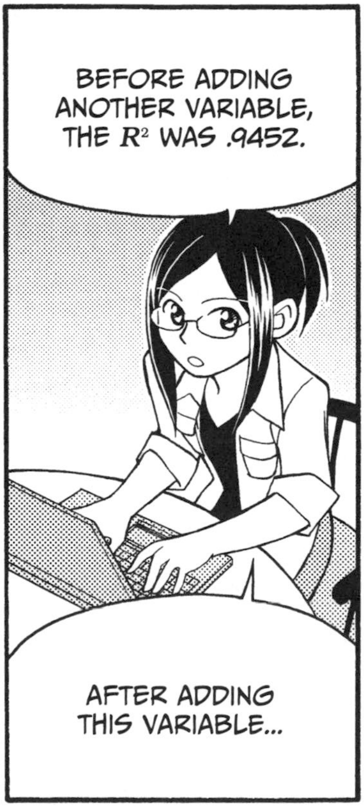
BEFORE ADDING ANOTHER VARIABLE, THE R^2 WAS .9452.
AFTER ADDING THIS VARIABLE...

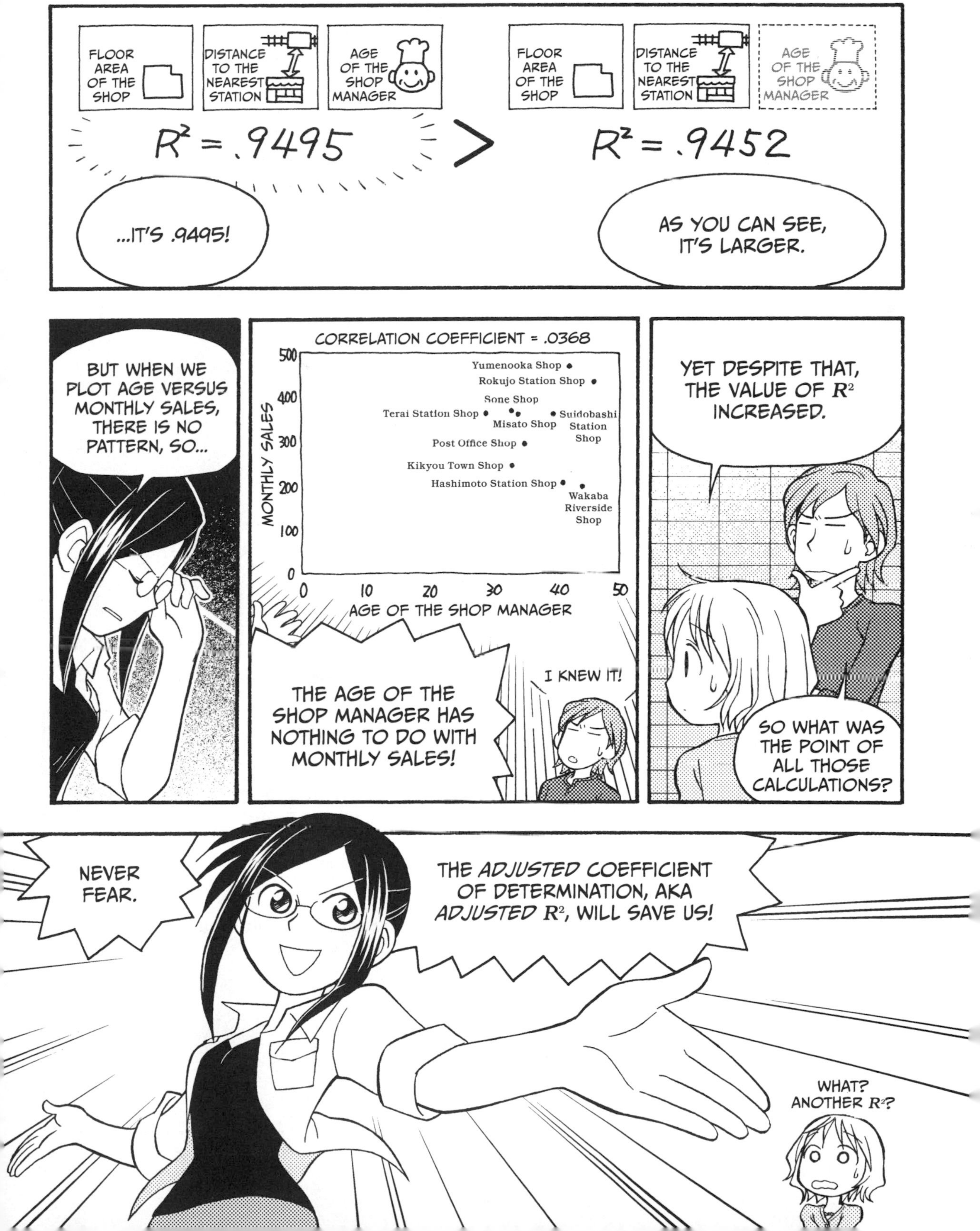

FLOOR AREA OF THE SHOP
DISTANCE TO THE NEAREST STATION
AGE OF THE SHOP MANAGER
FLOOR AREA OF THE SHOP
DISTANCE TO THE NEAREST STATION
AGE OF THE SHOP MANAGER
$R^2 = .9495 > R^2 = .9452$
...IT'S .9495!
AS YOU CAN SEE, IT'S LARGER.
BUT WHEN WE PLOT AGE VERSUS MONTHLY SALES, THERE IS NO PATTERN, SO...
CORRELATION COEFFICIENT = .0368
MONTHLY SALES
AGE OF THE SHOP MANAGER
Yumenooka Shop
Rokujo Station Shop
Sone Shop
Terai Station Shop
Misato Shop
Suidobashi Station Shop
Post Office Shop
Kikyou Town Shop
Hashimoto Station Shop
Wakaba Riverside Shop
THE AGE OF THE SHOP MANAGER HAS NOTHING TO DO WITH MONTHLY SALES!
I KNEW IT!
YET DESPITE THAT, THE VALUE OF R^2 INCREASED.
SO WHAT WAS THE POINT OF ALL THOSE CALCULATIONS?
NEVER FEAR.
THE *ADJUSTED* COEFFICIENT OF DETERMINATION, AKA *ADJUSTED* R^2, WILL SAVE US!
WHAT? ANOTHER R^2?

THE VALUE OF ADJUSTED R^2 ($\bar{R}^2$) CAN BE OBTAINED BY USING THIS FORMULA.

$$\bar{R}^2 = 1 - \frac{\left(\dfrac{S_e}{\text{sample size} - \text{number of predictor variables} - 1}\right)}{\left(\dfrac{S_{yy}}{\text{sample size} - 1}\right)}$$

TOO MANY LAYERS!

IT'S WORSE THAN R^2!

MIU, COULD YOU FIND THE VALUE OF ADJUSTED R^2 WITH AND WITHOUT THE AGE OF THE SHOP MANAGER?

UM, I THINK SO...

GO, MIU!

LET'S SEE...

WHEN THE PREDICTOR VARIABLES ARE ONLY FLOOR AREA AND DISTANCE...

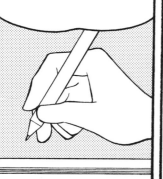

...R^2 IS .9452.

SO ADJUSTED R^2 IS:

$$1 - \frac{\left(\dfrac{S_e}{\text{sample size} - \text{number of predictor variables} - 1}\right)}{\left(\dfrac{S_{yy}}{\text{sample size} - 1}\right)}$$

$$= 1 - \frac{\left(\dfrac{4173.0}{10 - 2 - 1}\right)}{\left(\dfrac{76199.6}{10 - 1}\right)} = .9296$$

I GOT IT!

THE ANSWER IS .9296.

GREAT!

HOW ABOUT WHEN WE ALSO INCLUDE THE SHOP MANAGER'S AGE?

WE'VE ALREADY GOT R^2 FOR THAT, RIGHT?

| FLOOR AREA OF THE SHOP | DISTANCE TO THE NEAREST STATION | AGE OF THE SHOP MANAGER |

$R^2 = .9495$

YES, IT'S .9495.

SO ALL I HAVE TO GET IS THE VALUE OF ADJUSTED R^2...

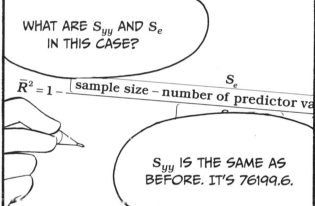

WHAT ARE S_{yy} AND S_e IN THIS CASE?

$$\bar{R}^2 = 1 - \frac{S_e}{\text{sample size} - \text{number of predictor va}}$$

S_{yy} IS THE SAME AS BEFORE. IT'S 76199.6.

WE'LL CHEAT AND CALCULATE S_e USING MY COMPUTER.

IT'S 3846.4.

PREDICTOR VARIABLES:

- FLOOR AREA
- DISTANCE
- MANAGER'S AGE

$$1 - \frac{\left(\dfrac{S_e}{\text{sample size} - \text{number of predictor variables} - 1}\right)}{\left(\dfrac{S_{yy}}{\text{sample size} - 1}\right)}$$

$$= 1 - \frac{\left(\dfrac{3846.4}{10 - 3 - 1}\right)}{\left(\dfrac{76199.6}{10 - 1}\right)} = \underline{.9243}$$

WAIT A MINUTE...

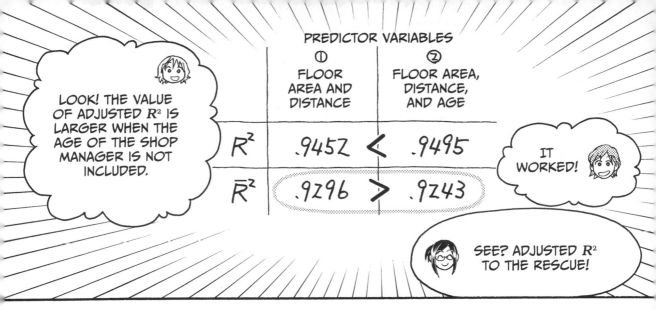

LOOK! THE VALUE OF ADJUSTED R^2 IS LARGER WHEN THE AGE OF THE SHOP MANAGER IS NOT INCLUDED.

IT WORKED!

SEE? ADJUSTED R^2 TO THE RESCUE!

	PREDICTOR VARIABLES	
	① FLOOR AREA AND DISTANCE	② FLOOR AREA, DISTANCE, AND AGE
R^2	.9452 <	.9495
$\bar{R}^2$	.9296 >	.9243

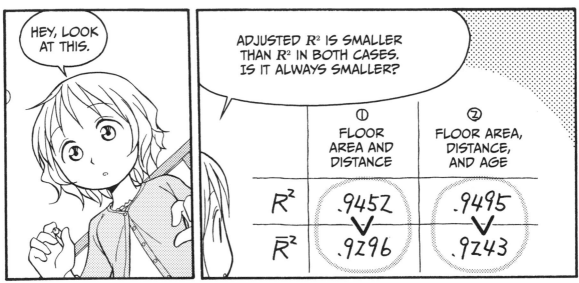

HEY, LOOK AT THIS.

ADJUSTED R^2 IS SMALLER THAN R^2 IN BOTH CASES. IS IT ALWAYS SMALLER?

	① FLOOR AREA AND DISTANCE	② FLOOR AREA, DISTANCE, AND AGE
R^2	.9452 ∨	.9495 ∨
$\bar{R}^2$	.9296	.9243

GOOD EYE! YES, IT IS ALWAYS SMALLER.

IS THAT GOOD?

IT MEANS THAT ADJUSTED R^2 IS A HARSHER JUDGE OF ACCURACY, SO WHEN WE USE IT, WE CAN BE MORE CONFIDENT IN OUR MULTIPLE REGRESSION EQUATION.

ADJUSTED R^2 IS AWESOME.

HYPOTHESIS TESTING WITH MULTIPLE REGRESSION

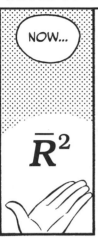

NOW...

$\bar{R}^2$

SINCE WE'RE HAPPY WITH ADJUSTED R^2, WE'LL TEST OUR ASSUMPTIONS ABOUT THE POPULATION.

WE'LL DO HYPOTHESIS AND REGRESSION COEFFICIENT TESTS, RIGHT?

YES, BUT IN MULTIPLE REGRESSION ANALYSIS, WE HAVE *PARTIAL REGRESSION COEFFICIENTS,* INSTEAD.

DO YOU REMEMBER HOW WE DID THE HYPOTHESIS TESTING BEFORE?

I THINK SO. WE TESTED WHETHER THE POPULATION MATCHED THE EQUATION AND THEN CHECKED THAT *A* DIDN'T EQUAL ZERO.

RIGHT! IT'S BASICALLY THE SAME WITH MULTIPLE REGRESSION.

~ ALTERNATIVE HYPOTHESIS ~

IF THE FLOOR AREA OF THE SHOP IS x_1 TSUBO AND THE DISTANCE TO THE NEAREST STATION IS x_2 METERS, THE MONTHLY SALES FOLLOW A NORMAL DISTRIBUTION WITH MEAN $A_1 x_1 + A_2 x_2 + B$ AND STANDARD DEVIATION σ.

NOW, WE HAVE MORE THAN ONE x AND MORE THAN ONE *A*. AT LEAST ONE OF THESE *A*'S MUST NOT EQUAL ZERO.

I SEE!

STEP 4: CONDUCT THE ANALYSIS OF VARIANCE (ANOVA) TEST.

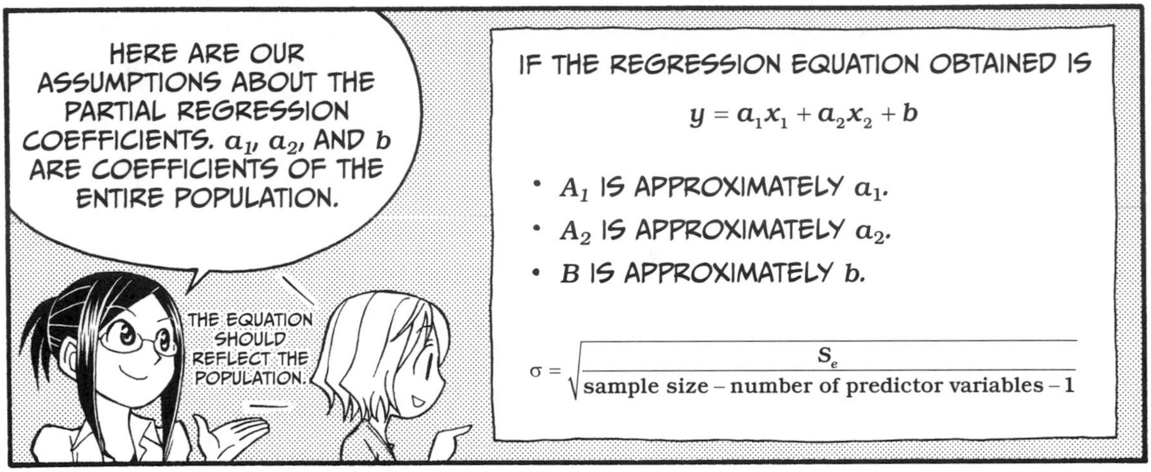
HERE ARE OUR ASSUMPTIONS ABOUT THE PARTIAL REGRESSION COEFFICIENTS. a_1, a_2, AND b ARE COEFFICIENTS OF THE ENTIRE POPULATION.
THE EQUATION SHOULD REFLECT THE POPULATION.
IF THE REGRESSION EQUATION OBTAINED IS
$y = a_1x_1 + a_2x_2 + b$
• A_1 IS APPROXIMATELY a_1.
• A_2 IS APPROXIMATELY a_2.
• B IS APPROXIMATELY b.
$\sigma = \sqrt{\frac{S_e}{\text{sample size} - \text{number of predictor variables} - 1}}$

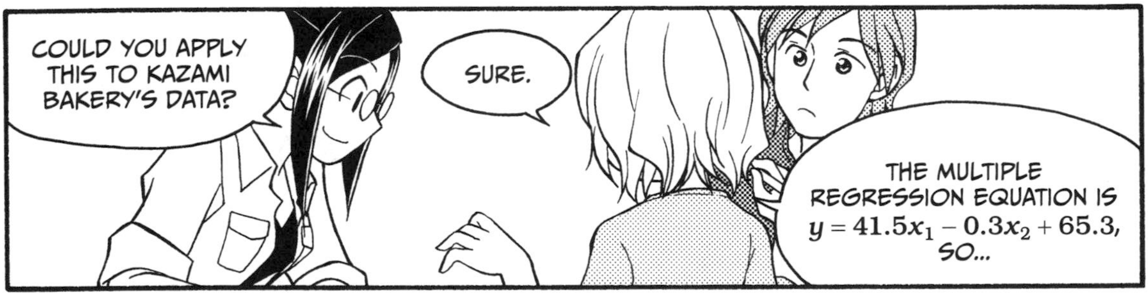
COULD YOU APPLY THIS TO KAZAMI BAKERY'S DATA?
SURE.
THE MULTIPLE REGRESSION EQUATION IS $y = 41.5x_1 - 0.3x_2 + 65.3$, SO...

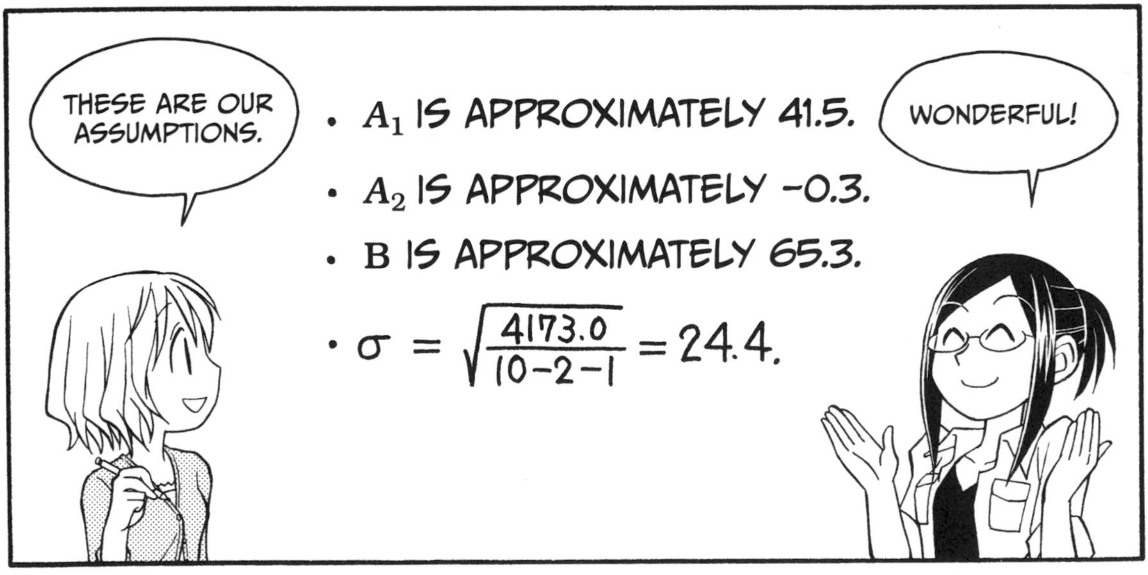
THESE ARE OUR ASSUMPTIONS.
• A_1 IS APPROXIMATELY 41.5.
• A_2 IS APPROXIMATELY -0.3.
• B IS APPROXIMATELY 65.3.
• $\sigma = \sqrt{\frac{4173.0}{10-2-1}} = 24.4$.
WONDERFUL!

NOW WE NEED TO TEST OUR MODEL USING AN *F*-TEST.

THERE ARE TWO TYPES.

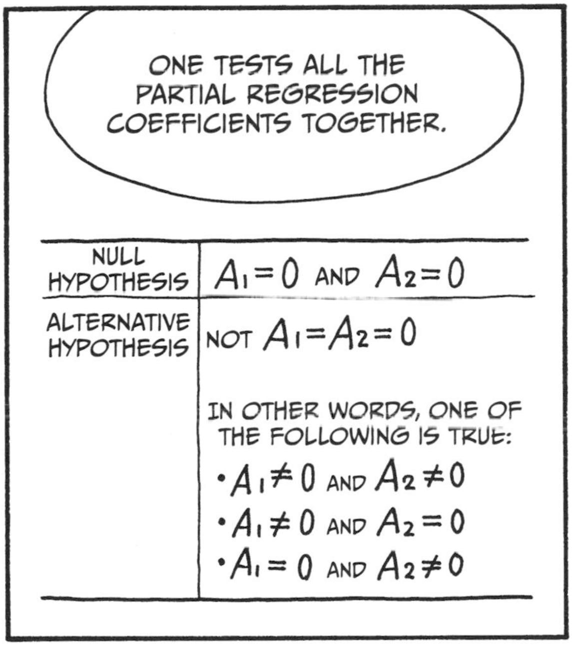

ONE TESTS ALL THE PARTIAL REGRESSION COEFFICIENTS TOGETHER.

NULL HYPOTHESIS	$A_1 = 0$ AND $A_2 = 0$
ALTERNATIVE HYPOTHESIS	NOT $A_1 = A_2 = 0$

IN OTHER WORDS, ONE OF THE FOLLOWING IS TRUE:

- $A_1 \neq 0$ AND $A_2 \neq 0$
- $A_1 \neq 0$ AND $A_2 = 0$
- $A_1 = 0$ AND $A_2 \neq 0$

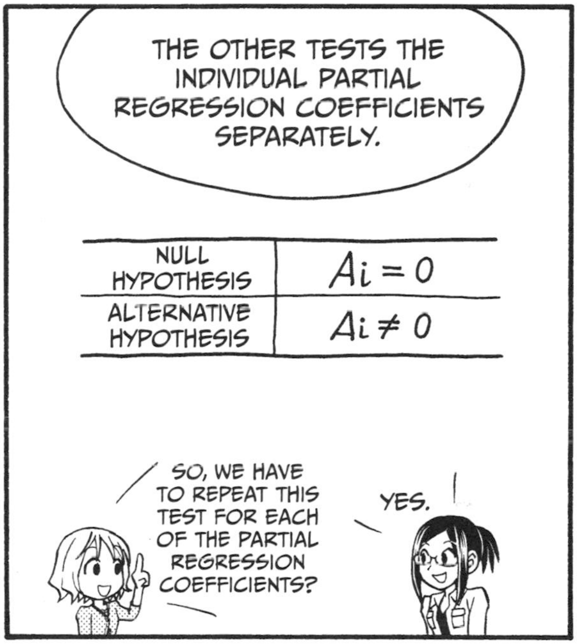

THE OTHER TESTS THE INDIVIDUAL PARTIAL REGRESSION COEFFICIENTS SEPARATELY.

NULL HYPOTHESIS	$A_i = 0$
ALTERNATIVE HYPOTHESIS	$A_i \neq 0$

SO, WE HAVE TO REPEAT THIS TEST FOR EACH OF THE PARTIAL REGRESSION COEFFICIENTS?

YES.

LET'S SET THE SIGNIFICANCE LEVEL TO .05. ARE YOU READY TO TRY DOING THESE TESTS?

YES, LET'S!

FIRST, WE'LL TEST ALL THE PARTIAL
REGRESSION COEFFICIENTS TOGETHER.

THE STEPS OF ANOVA

Step 1	Define the population.	The population is all Kazami Bakery shops.
Step 2	Set up a null hypothesis and an alternative hypothesis.	Null hypothesis is $A_1 = 0$ and $A_2 = 0$. Alternative hypothesis is that A_1 or A_2 or both $\neq 0$.
Step 3	Select which hypothesis test to conduct.	We'll use an F-test.
Step 4	Choose the significance level.	We'll use a significance level of .05.
Step 5	Calculate the test statistic from the sample data.	The test statistic is: $$\dfrac{\dfrac{S_{yy} - S_e}{\text{number of predictor variables}}}{\dfrac{S_e}{\text{sample size} - \text{number of predictor variables} - 1}} =$$ $$\dfrac{76199.6 - 4173.0}{2} \div \dfrac{4173.0}{10 - 2 - 1} = 60.4$$ The test statistic, 60.4, will follow an F distribution with first degree of freedom 2 (the number of predictor variables) and second degree of freedom 7 (sample size minus the number of predictor variables minus 1), if the null hypothesis is true.
Step 6	Determine whether the p-value for the test statistic obtained in Step 5 is smaller than the significance level.	At significance level .05, with d_1 being 2 and d_2 being 7 $(10 - 2 - 1)$, the critical value is 4.7374. Our test statistic is 60.4.
Step 7	Decide whether you can reject the null hypothesis.	Since our test statistic is greater than the critical value, we reject the null hypothesis.

NEXT, WE'LL TEST THE INDIVIDUAL PARTIAL REGRESSION COEFFICIENTS. I WILL DO THIS FOR A_1 AS AN EXAMPLE.

THE STEPS OF ANOVA

Step 1	Define the population.	The population is all Kazami Bakery shops.
Step 2	Set up a null hypothesis and an alternative hypothesis.	Null hypothesis is $A_1 = 0$. Alternative hypothesis is $A_1 \neq 0$.
Step 3	Select which hypothesis test to conduct.	We'll use an F-test.
Step 4	Choose the significance level.	We'll use a significance level of .05.
Step 5	Calculate the test statistic from the sample data.	The test statistic is:

$$\frac{a_1^2}{S_{11}} \div \frac{S_e}{\text{sample size} - \text{number of predictor variables} - 1} =$$

$$\frac{41.5^2}{0.0657} \div \frac{4173.0}{10 - 2 - 1} = 44$$

The test statistic will follow an F distribution with first degree of freedom 1 and second degree of freedom 7 (sample size minus the number of predictor variables minus 1), if the null hypothesis is true. (The value of S_{11} will be explained on the next page.)

Step 6	Determine whether the p-value for the test statistic obtained in Step 5 is smaller than the significance level.	At significance level .05, with d_1 being 1 and d_2 being 7, the critical value is 5.5914. Our test statistic is 44.
Step 7	Decide whether you can reject the null hypothesis.	Since our test statistic is greater than the critical value, we reject the null hypothesis.

REGARDLESS OF THE RESULT OF STEP 7, IF THE VALUE OF THE TEST STATISTIC

$$\frac{a_1^2}{S_{11}} \div \frac{S_e}{\text{sample size} - \text{number of predictor variables} - 1}$$

IS 2 OR MORE, WE STILL CONSIDER THE PREDICTOR VARIABLE CORRESPONDING TO THAT PARTIAL REGRESSION COEFFICIENT TO BE USEFUL FOR PREDICTING THE OUTCOME VARIABLE.

FINDING S_{11} AND S_{22}

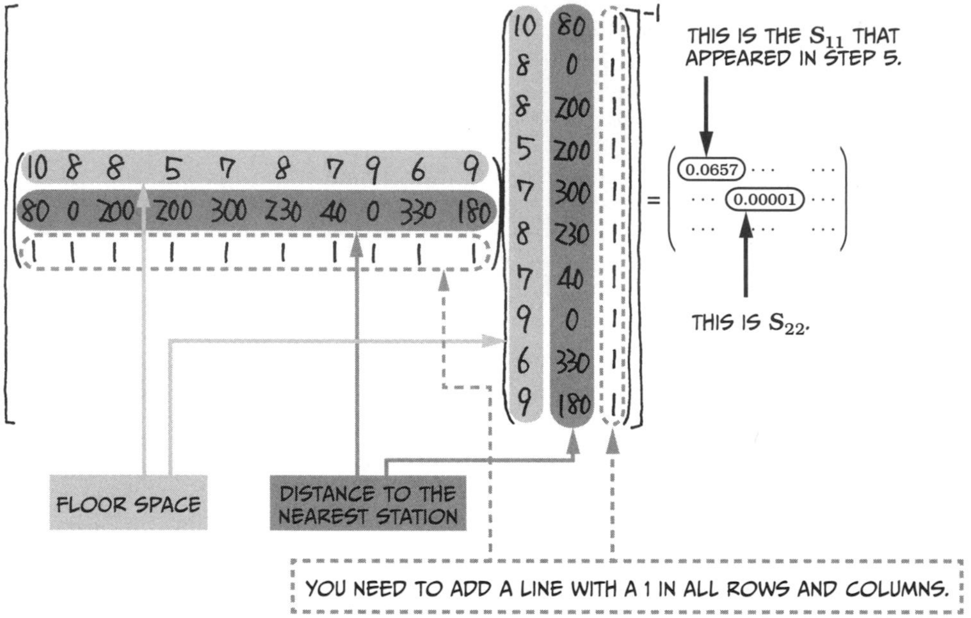

THIS IS THE S_{11} THAT APPEARED IN STEP 5.
THIS IS S_{22}.
0.0657
0.00001
FLOOR SPACE
DISTANCE TO THE NEAREST STATION
YOU NEED TO ADD A LINE WITH A 1 IN ALL ROWS AND COLUMNS.

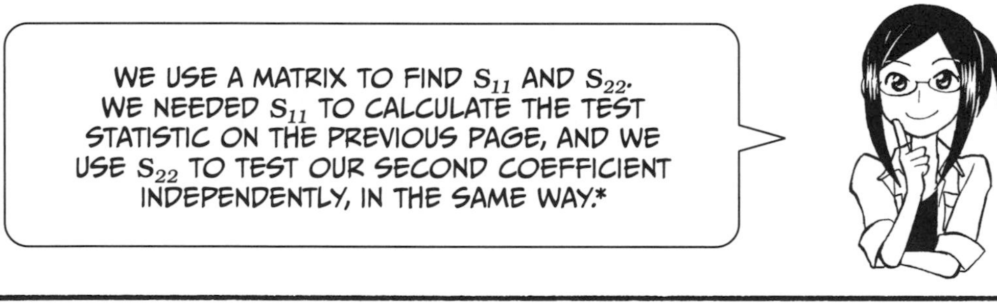

WE USE A MATRIX TO FIND S_{11} AND S_{22}. WE NEEDED S_{11} TO CALCULATE THE TEST STATISTIC ON THE PREVIOUS PAGE, AND WE USE S_{22} TO TEST OUR SECOND COEFFICIENT INDEPENDENTLY, IN THE SAME WAY.*

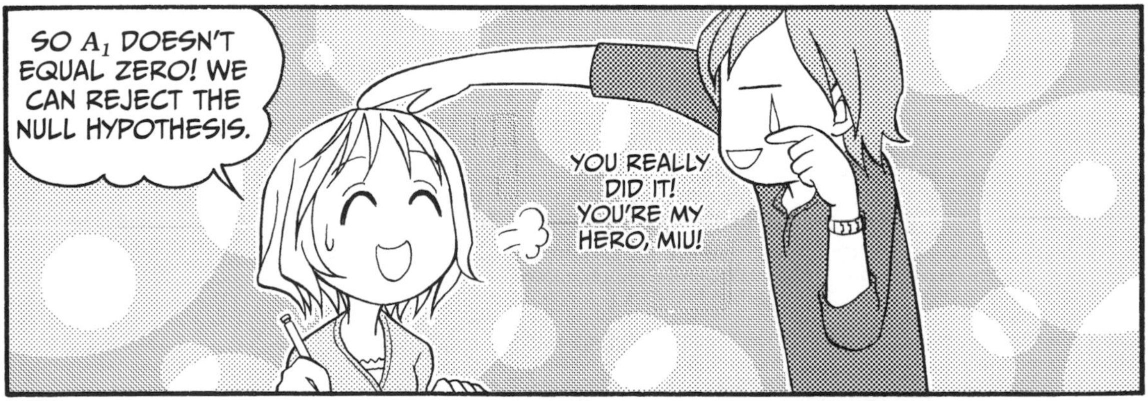

SO A_1 DOESN'T EQUAL ZERO! WE CAN REJECT THE NULL HYPOTHESIS.
YOU REALLY DID IT! YOU'RE MY HERO, MIU!

* SOME PEOPLE USE THE t DISTRIBUTION INSTEAD OF THE F DISTRIBUTION WHEN EXPLAINING THE "TEST OF PARTIAL REGRESSION COEFFICIENTS." YOUR FINAL RESULT WILL BE THE SAME NO MATTER WHICH METHOD YOU CHOOSE.

STEP 5: CALCULATE CONFIDENCE INTERVALS FOR THE POPULATION.

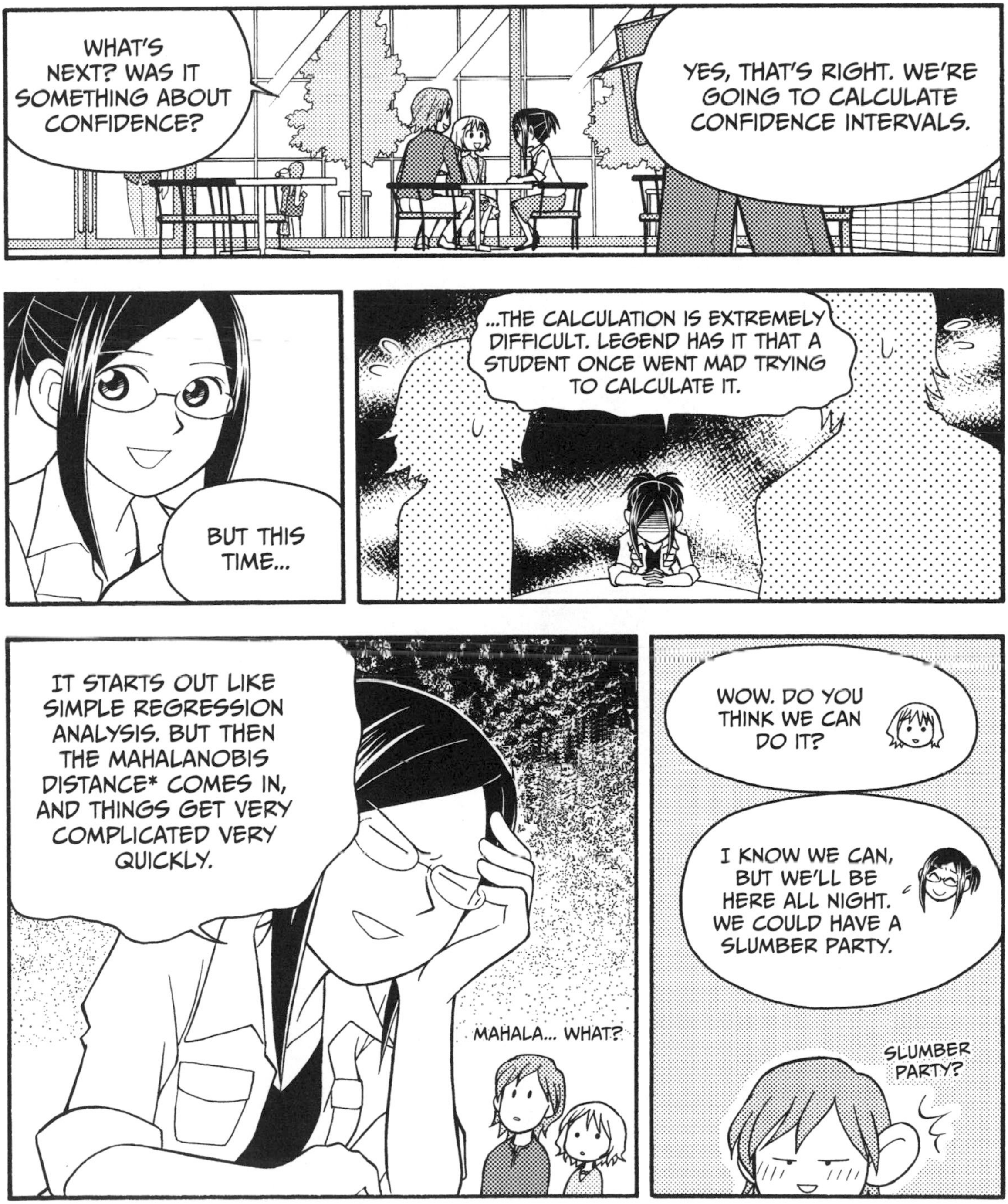
WHAT'S NEXT? WAS IT SOMETHING ABOUT CONFIDENCE?
YES, THAT'S RIGHT. WE'RE GOING TO CALCULATE CONFIDENCE INTERVALS.
BUT THIS TIME...
...THE CALCULATION IS EXTREMELY DIFFICULT. LEGEND HAS IT THAT A STUDENT ONCE WENT MAD TRYING TO CALCULATE IT.
IT STARTS OUT LIKE SIMPLE REGRESSION ANALYSIS. BUT THEN THE MAHALANOBIS DISTANCE* COMES IN, AND THINGS GET VERY COMPLICATED VERY QUICKLY.
MAHALA... WHAT?
WOW. DO YOU THINK WE CAN DO IT?
I KNOW WE CAN, BUT WE'LL BE HERE ALL NIGHT. WE COULD HAVE A SLUMBER PARTY.
SLUMBER PARTY?

* THE MATHEMATICIAN P.C. MAHALANOBIS INVENTED A WAY TO USE MULTIVARIATE DISTANCES TO COMPARE POPULATIONS.

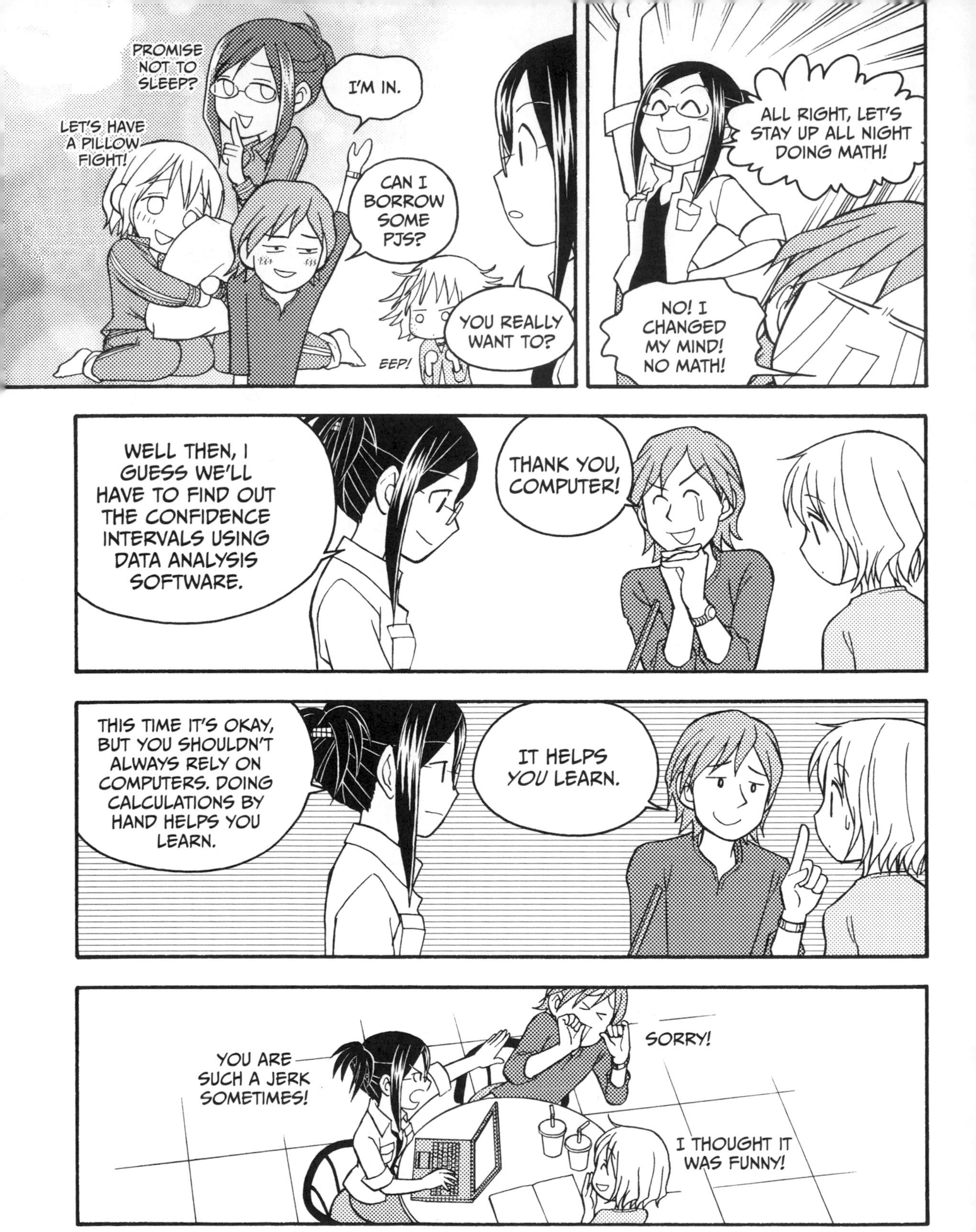
PROMISE NOT TO SLEEP?
LET'S HAVE A PILLOW FIGHT!
I'M IN.
CAN I BORROW SOME PJS?
YOU REALLY WANT TO?
EEP!
ALL RIGHT, LET'S STAY UP ALL NIGHT DOING MATH!
NO! I CHANGED MY MIND! NO MATH!
WELL THEN, I GUESS WE'LL HAVE TO FIND OUT THE CONFIDENCE INTERVALS USING DATA ANALYSIS SOFTWARE.
THANK YOU, COMPUTER!
THIS TIME IT'S OKAY, BUT YOU SHOULDN'T ALWAYS RELY ON COMPUTERS. DOING CALCULATIONS BY HAND HELPS YOU LEARN.
IT HELPS *YOU* LEARN.
YOU ARE SUCH A JERK SOMETIMES!
SORRY!
I THOUGHT IT WAS FUNNY!

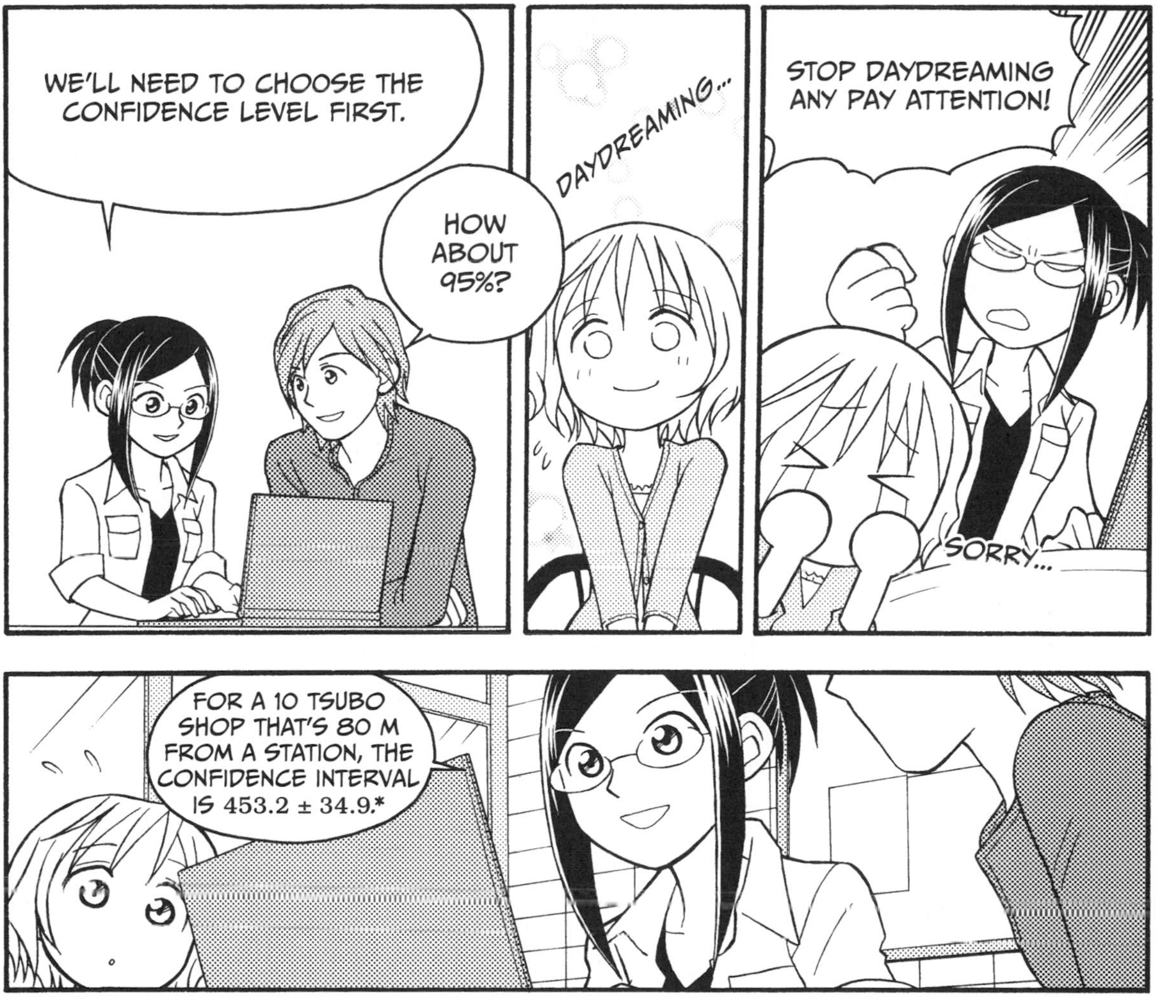
WE'LL NEED TO CHOOSE THE CONFIDENCE LEVEL FIRST.
HOW ABOUT 95%?
DAYDREAMING...
STOP DAYDREAMING ANY PAY ATTENTION!
SORRY...
FOR A 10 TSUBO SHOP THAT'S 80 M FROM A STATION, THE CONFIDENCE INTERVAL IS 453.2 ± 34.9.*

* THIS CALCULATION IS EXPLAINED IN MORE DETAIL ON PAGE 146.

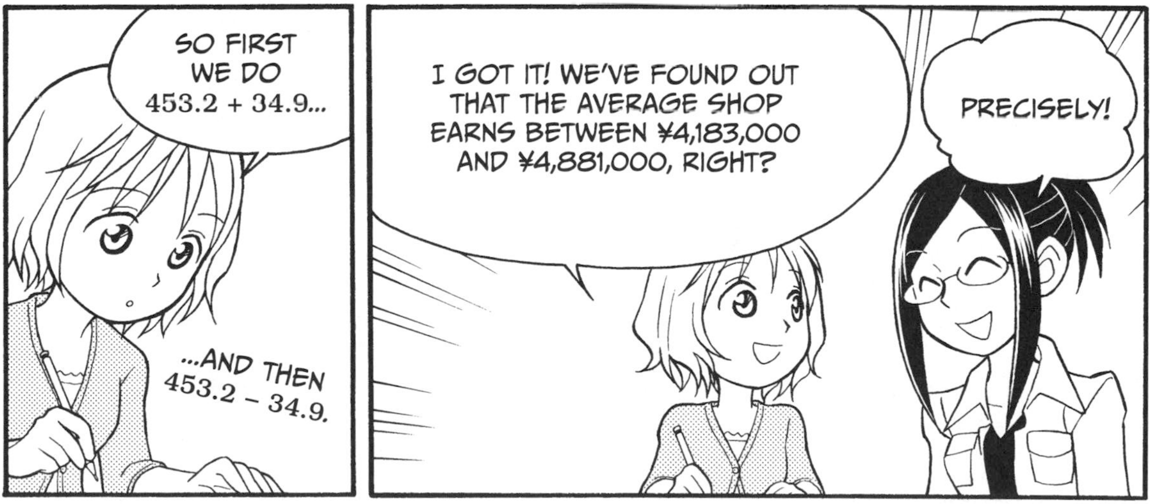
SO FIRST WE DO 453.2 + 34.9...
...AND THEN 453.2 − 34.9.
I GOT IT! WE'VE FOUND OUT THAT THE AVERAGE SHOP EARNS BETWEEN ¥4,183,000 AND ¥4,881,000, RIGHT?
PRECISELY!

HERE IS THE DATA FOR THE NEW SHOP WE'RE PLANNING TO OPEN.

	Floor space of the shop (tsubo)	Distance to the nearest station (meters)
Isebashi Shop	10	110

A SHOP IN ISEBASHI? THAT'S CLOSE TO MY HOUSE!

CAN YOU PREDICT THE SALES, MIU?

YEP!

$$y = 41.5x_1 - 0.3x_2 + 65.3$$
$$= 41.5 \times 10 - 0.3 \times 110 + 65.3$$
$$= \underline{447.3^*}$$

¥4,473,000 PER MONTH!

* THIS CALCULATION WAS MADE USING ROUNDED NUMBERS. IF YOU USE THE FULL, UNROUNDED NUMBERS, THE RESULT WILL BE 442.96.

YOU'RE A GENIUS, MIU! I SHOULD NAME THE SHOP AFTER YOU.

YOU SHOULD PROBABLY NAME IT AFTER RISA...

BUT HOW COULD WE KNOW THE EXACT SALES OF A SHOP THAT HASN'T BEEN BUILT? SHOULD WE CALCULATE A PREDICTION INTERVAL?

ABSOLUTELY.

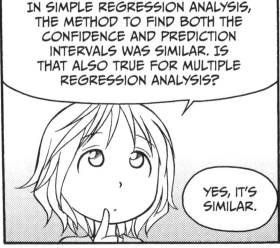

IN SIMPLE REGRESSION ANALYSIS, THE METHOD TO FIND BOTH THE CONFIDENCE AND PREDICTION INTERVALS WAS SIMILAR. IS THAT ALSO TRUE FOR MULTIPLE REGRESSION ANALYSIS?

YES, IT'S SIMILAR.

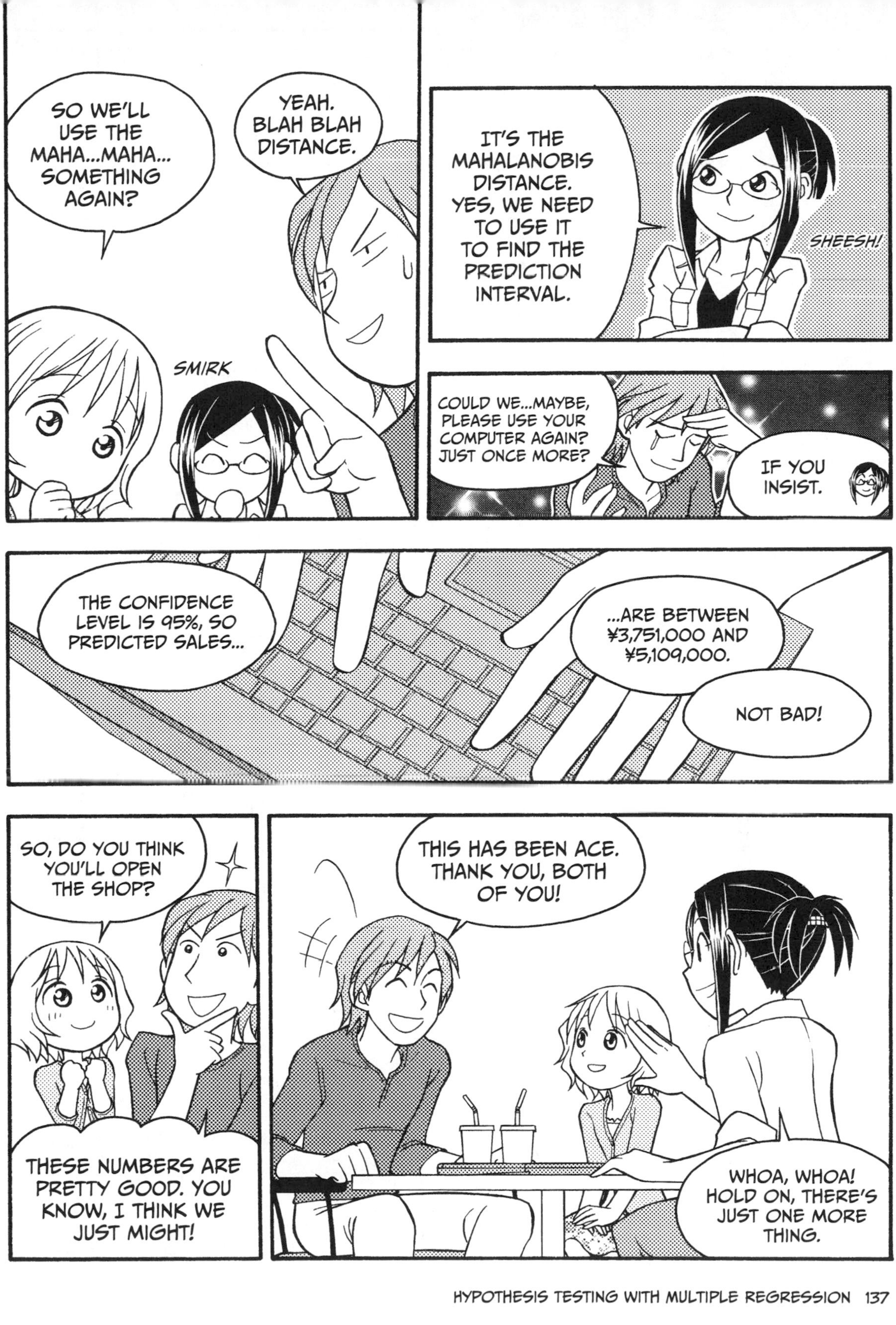
SO WE'LL USE THE MAHA...MAHA... SOMETHING AGAIN?
YEAH. BLAH BLAH DISTANCE.
SMIRK
IT'S THE MAHALANOBIS DISTANCE. YES, WE NEED TO USE IT TO FIND THE PREDICTION INTERVAL.
SHEESH!
COULD WE...MAYBE, PLEASE USE YOUR COMPUTER AGAIN? JUST ONCE MORE?
IF YOU INSIST.
THE CONFIDENCE LEVEL IS 95%, SO PREDICTED SALES...
...ARE BETWEEN ¥3,751,000 AND ¥5,109,000.
NOT BAD!
SO, DO YOU THINK YOU'LL OPEN THE SHOP?
THESE NUMBERS ARE PRETTY GOOD. YOU KNOW, I THINK WE JUST MIGHT!
THIS HAS BEEN ACE. THANK YOU, BOTH OF YOU!
WHOA, WHOA! HOLD ON, THERE'S JUST ONE MORE THING.

CHOOSING THE BEST COMBINATION OF PREDICTOR VARIABLES

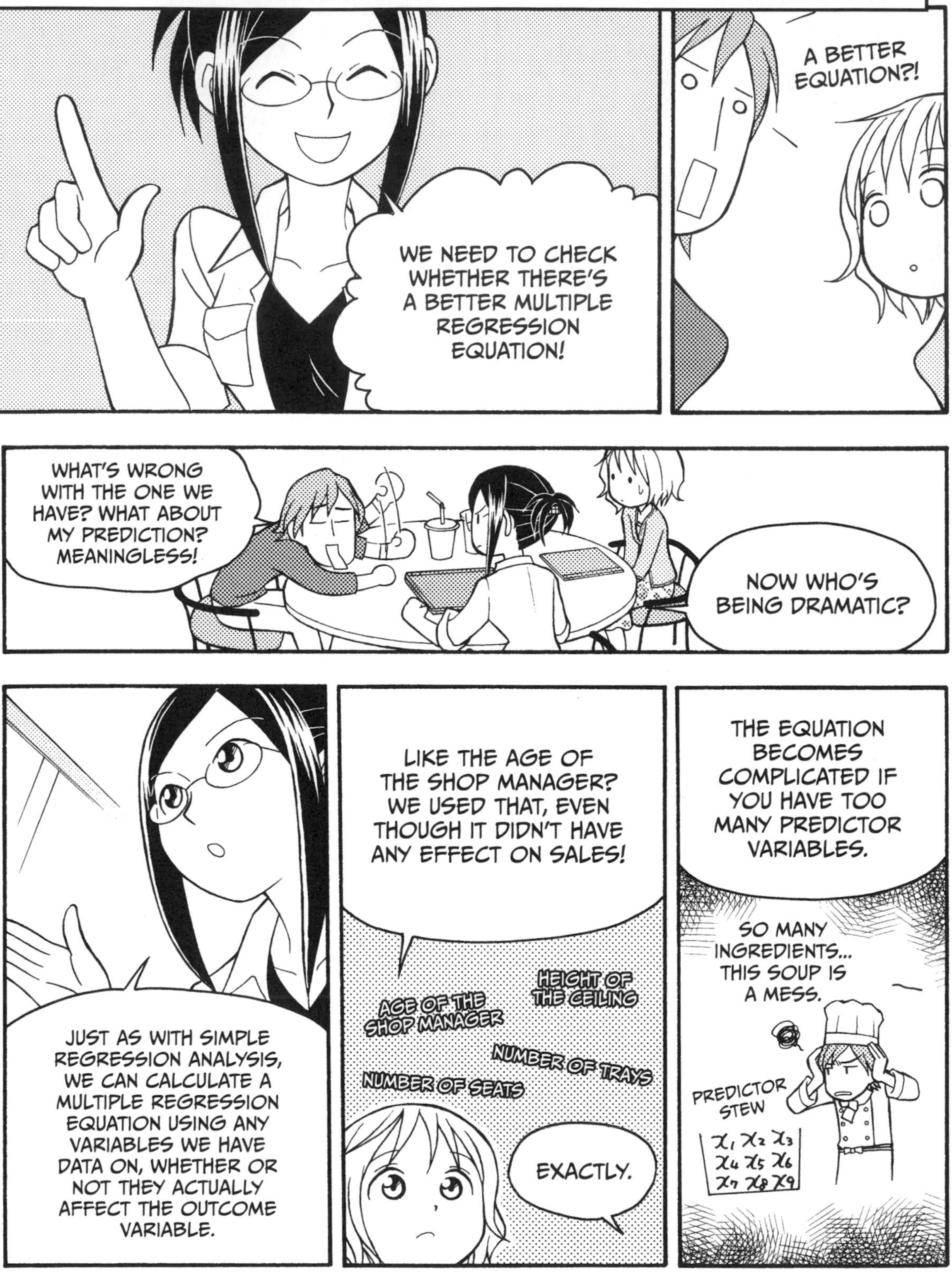
WE NEED TO CHECK WHETHER THERE'S A BETTER MULTIPLE REGRESSION EQUATION!
A BETTER EQUATION?!
WHAT'S WRONG WITH THE ONE WE HAVE? WHAT ABOUT MY PREDICTION? MEANINGLESS!
NOW WHO'S BEING DRAMATIC?
JUST AS WITH SIMPLE REGRESSION ANALYSIS, WE CAN CALCULATE A MULTIPLE REGRESSION EQUATION USING ANY VARIABLES WE HAVE DATA ON, WHETHER OR NOT THEY ACTUALLY AFFECT THE OUTCOME VARIABLE.
LIKE THE AGE OF THE SHOP MANAGER? WE USED THAT, EVEN THOUGH IT DIDN'T HAVE ANY EFFECT ON SALES!
AGE OF THE SHOP MANAGER
HEIGHT OF THE CEILING
NUMBER OF TRAYS
NUMBER OF SEATS
EXACTLY.
THE EQUATION BECOMES COMPLICATED IF YOU HAVE TOO MANY PREDICTOR VARIABLES.
SO MANY INGREDIENTS... THIS SOUP IS A MESS.
PREDICTOR STEW
x_1 x_2 x_3
x_4 x_5 x_6
x_7 x_8 x_9

THE BEST MULTIPLE REGRESSION EQUATION BALANCES ACCURACY AND COMPLEXITY BY INCLUDING ONLY THE PREDICTOR VARIABLES NEEDED TO MAKE THE BEST PREDICTION.

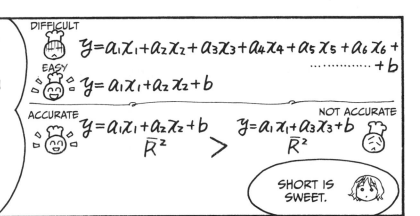

DIFFICULT

$$y = a_1 x_1 + a_2 x_2 + a_3 x_3 + a_4 x_4 + a_5 x_5 + a_6 x_6 + \cdots + b$$

EASY

$$y = a_1 x_1 + a_2 x_2 + b$$

ACCURATE

$$y = a_1 x_1 + a_2 x_2 + b \quad \bar{R}^2$$

$$>$$

$$y = a_1 x_1 + a_3 x_3 + b \quad \bar{R}^2$$

NOT ACCURATE

SHORT IS SWEET.

THERE ARE SEVERAL WAYS TO FIND THE EQUATION THAT GIVES YOU THE MOST BANG FOR YOUR BUCK.

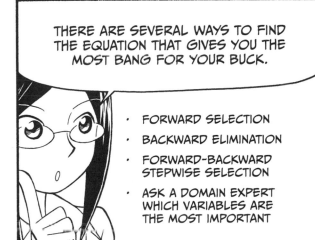

- FORWARD SELECTION
- BACKWARD ELIMINATION
- FORWARD-BACKWARD STEPWISE SELECTION
- ASK A DOMAIN EXPERT WHICH VARIABLES ARE THE MOST IMPORTANT

THESE ARE SOME COMMON WAYS.

THE METHOD WE'LL USE TODAY IS SIMPLER THAN ANY OF THOSE. IT'S CALLED *BEST SUBSETS REGRESSION*, OR SOMETIMES, THE *ROUND-ROBIN METHOD*.

ROUND ROBIN?

A FAT BIRD?

WHAT THE HECK IS THAT?

$$x_1 \quad x_2 \quad x_3$$

I'LL SHOW YOU. SUPPOSE x_1, x_2, AND x_3 ARE POTENTIAL PREDICTOR VARIABLES.

FIRST, WE'D CALCULATE THE MULTIPLE REGRESSION EQUATION FOR EVERY COMBINATION OF PREDICTOR VARIABLES!

- X_1
- X_2
- X_3

- X_1 AND X_2
- X_2 AND X_3
- X_1 AND X_3

- X_1 AND X_2 AND X_3

HAHA. THIS SURE IS ROUND-ABOUT.

LET'S REPLACE x_1, x_2, AND x_3 WITH SHOP SIZE, DISTANCE TO A STATION, AND MANAGER'S AGE.

IS OUR EQUATION THE WINNER?

WE'LL MAKE A TABLE THAT SHOWS THE PARTIAL REGRESSION COEFFICIENTS AND ADJUSTED R^2...

...LIKE THIS.

PRESTO!

Predictor variables	a_1	a_2	a_3	b	$\bar{R}^2$
1	54.9			−91.3	.07709
2		−0.6		424.8	.5508
3			0.6	309.1	.0000
1 and 2	41.5	−0.3		65.3	.9296
1 and 3	55.6		2.0	−170.1	.7563
2 and 3		−0.6	−0.4	438.9	.4873
1 and 2 and 3	42.2	−0.3	1.1	17.7	.9243

1 IS FLOOR AREA, 2 IS DISTANCE TO A STATION, AND 3 IS MANAGER'S AGE. WHEN 1 AND 2 ARE USED, ADJUSTED R^2 IS HIGHEST.

SO OUR EQUATION IS THE BEST! WE ROCK.

NOW WE REALLY KNOW THAT $y = 41.5x_1 - 0.3x_2 + 65.3$ DOES A GOOD JOB AT PREDICTING THE SALES AT THE NEW SHOP.

THAT'S RIGHT! GOOD WORK, FOLKS!

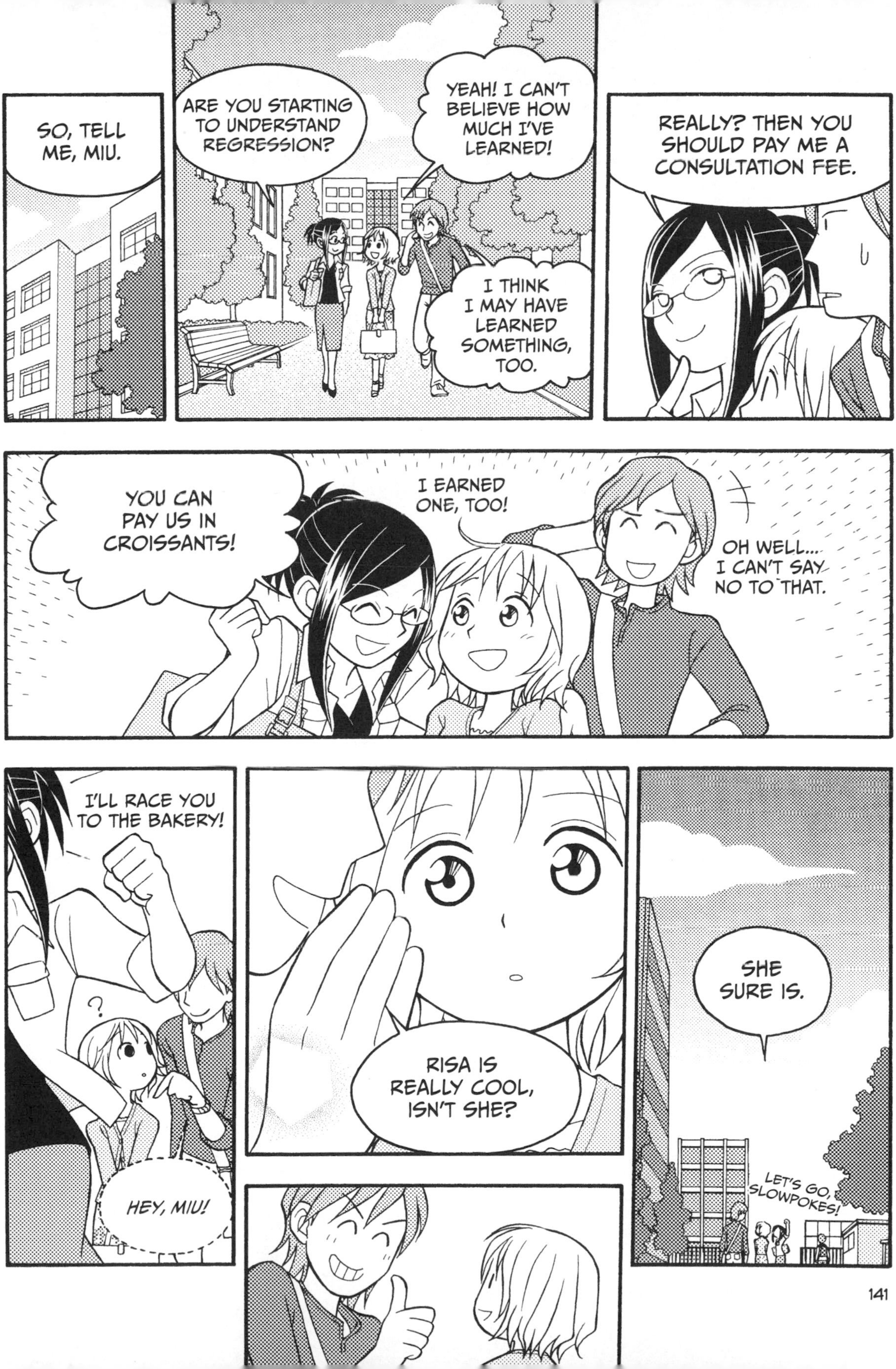
SO, TELL ME, MIU.
ARE YOU STARTING TO UNDERSTAND REGRESSION?
YEAH! I CAN'T BELIEVE HOW MUCH I'VE LEARNED!
I THINK I MAY HAVE LEARNED SOMETHING, TOO.
REALLY? THEN YOU SHOULD PAY ME A CONSULTATION FEE.
YOU CAN PAY US IN CROISSANTS!
I EARNED ONE, TOO!
OH WELL... I CAN'T SAY NO TO THAT.
I'LL RACE YOU TO THE BAKERY!
?
HEY, MIU!
RISA IS REALLY COOL, ISN'T SHE?
SHE SURE IS.
LET'S GO, SLOWPOKES!

ASSESSING POPULATIONS WITH
MULTIPLE REGRESSION ANALYSIS

Let's review the procedure of multiple regression analysis, shown on page 112.

1. Draw a scatter plot of each predictor variable and the outcome variable to see if they appear to be related.

2. Calculate the multiple regression equation.

3. Examine the accuracy of the multiple regression equation.

4. Conduct the analysis of variance (ANOVA) test.

5. Calculate confidence intervals for the population.

6. Make a prediction!

As in Chapter 2, we've talked about Steps 1 through 6 as if they were all mandatory. In reality, Steps 4 and 5 can be skipped for the analysis of some data sets.

Kazami Bakery currently has only 10 stores, and of those 10 stores, only one (Yumenooka Shop) has a floor area of 10 tsubo[1] and is 80 m to the nearest station. However, Risa calculated a confidence interval for the population of stores that were 10 tsubo and 80 m from a station. Why would she do that?

Well, it's possible that Kazami Bakery could open another 10-tsubo store that's also 80 m from a train station. If the chain keeps growing, there could be dozens of Kazami shops that fit that description. When Risa did that analysis, she was assuming that more 10-tsubo stores 80 m from a station might open someday.

The usefulness of this assumption is disputable. Yumenooka Shop has more sales than any other shop, so maybe the Kazami family will decide to open more stores just like that one. However, the bakery's next store, Isebashi Shop, will be 10 tsubo but 110 m from a station. In fact, it probably wasn't necessary to analyze such a specific population of stores. Risa could have skipped from calculating adjusted R^2 to making the prediction, but being a good friend, she wanted to show Miu all the steps.

---

1. Remember that 1 tsubo is about 36 square feet.

STANDARDIZED RESIDUALS

As in simple regression analysis, we calculate standardized residuals in multiple regression analysis when assessing how well the equation fits the actual sample data that's been collected.

Table 3-1 shows the residuals and standardized residuals for the Kazami Bakery data used in this chapter. An example calculation is shown for the Misato Shop.

TABLE 3-1: STANDARDIZED RESIDUALS OF THE KAZAMI BAKERY EXAMPLE

Bakery	Floor area of the shop x_1	Distance to the nearest station x_2	Monthly sales y	Monthly sales $\hat{y} = 41.5x_1 - 0.3x_2 + 65.3$	Residual $y - \hat{y}$	Standardized residual
Yumenooka Shop	10	80	469	453.2	15.8	0.8
Terai Station Shop	8	0	366	397.4	−31.4	−1.6
Sone Shop	8	200	371	329.3	41.7	1.8
Hashimoto Station Shop	5	200	208	204.7	3.3	0.2
Kikyou Town Shop	7	300	246	253.7	−7.7	−0.4
Post Office Shop	8	230	297	319.0	−22.0	1.0
Suidobashi Station Shop	7	40	363	342.3	20.7	1.0
Rokujo Station Shop	9	0	436	438.9	−2.9	−0.1
Wakaba Riverside Shop	6	330	198	201.9	−3.9	−0.2
Misato Shop	9	180	364	377.6	−13.6	−0.6

If a residual is positive, the measurement is higher than predicted by our equation, and if the residual is negative, the measurement is lower than predicted; if it's 0, the measurement and our prediction are the same. The absolute value of the residual tells us how well the equation predicted what actually happened. The larger the absolute value, the greater the difference between the measurement and the prediction.

If the absolute value of the standardized residual is greater than 3, the data point can be considered an *outlier*. Outliers are measurements that don't follow the general trend. In this case, an outlier could be caused by a store closure, by road construction around a store, or by a big event held at one of the bakeries—anything that would significantly affect sales. When you detect an outlier, you should investigate the data point to see if it needs to be removed and the regression equation calculated again.

MAHALANOBIS DISTANCE

The *Mahalanobis distance* was introduced in 1936 by mathematician and scientist P.C. Mahalanobis, who also founded the Indian Statistical Institute. Mahalanobis distance is very useful in statistics because it considers an entire set of data, rather than looking at each measurement in isolation. It's a way of calculating distance that, unlike the more common Euclidean concept of distance, takes into account the correlation between measurements to determine the similarity of a sample to an established data set. Because these calculations reflect a more complex relationship, a linear equation will not suffice. Instead, we use matrices, which condense a complex array of information into a more manageable form that can then be used to calculate all of these distances at once.

On page 137, Risa used her computer to find the prediction interval using the Mahalanobis distance. Let's work through that calculation now and see how she arrived at a prediction interval of ¥3,751,000 and ¥5,109,000 at a confidence level of 95%.

STEP 1
Obtain the inverse matrix of

$$\begin{pmatrix} S_{11} & S_{12} & \cdots & S_{1p} \\ S_{21} & S_{22} & \cdots & S_{2p} \\ \vdots & \vdots & \ddots & \vdots \\ S_{p1} & S_{p2} & \cdots & S_{pp} \end{pmatrix}, \text{ which is } \begin{pmatrix} S_{11} & S_{12} & \cdots & S_{1p} \\ S_{21} & S_{22} & \cdots & S_{2p} \\ \vdots & \vdots & \ddots & \vdots \\ S_{p1} & S_{p2} & \cdots & S_{pp} \end{pmatrix}^{-1} = \begin{pmatrix} S^{11} & S^{12} & \cdots & S^{1p} \\ S^{21} & S^{22} & \cdots & S^{2p} \\ \vdots & \vdots & \ddots & \vdots \\ S^{p1} & S^{p2} & \cdots & S^{pp} \end{pmatrix}.$$

The first matrix is the covariance matrix as calculated on page 132. The diagonal of this matrix (S_{11}, S_{22}, and so on) is the variance within a certain variable.

The inverse of this matrix, the second and third matrices shown here, is also known as the *concentration matrix* for the different predictor variables: floor area and distance to the nearest station.

For example, S_{22} is the variance of the values of the distance to the nearest station. S_{25} would be the covariance of the distance to the nearest station and some fifth predictor variable.

The values of S_{11} and S_{22} on page 132 were obtained through this series of calculations.

The values of S_{ii} and S_{ij} in

$$
\begin{pmatrix}
S_{11} & S_{12} & \cdots & S_{1p} \\
S_{21} & S_{22} & \cdots & S_{2p} \\
\vdots & \vdots & \ddots & \vdots \\
S_{p1} & S_{p2} & \cdots & S_{pp}
\end{pmatrix}^{-1}
$$

and the values of S_{ii} and S_{ij} obtained from conducting individual tests of the partial regression coefficients are always the same. That is, the values of S_{ii} and S_{ij} found through partial regression will be equivalent to the values of S_{ii} and S_{ij} found by calculating the inverse matrix.

STEP 2

Next we need to calculate the square of Mahalanobis distance for a given point using the following equation:

$$
D_M^2(x) = (x - \bar{x})^T \left(S^{-1} \right)(x - \bar{x})
$$

The x values are taken from the predictors, $\bar{x}$ is the mean of a given set of predictors, and S^{-1} is the concentration matrix from Step 1. The Mahalanobis distance for the shop at Yumenooka is shown here:

$$
D^2 = \left\{
\begin{aligned}
&(x_1 - \bar{x}_1)(x_1 - \bar{x}_1)S^{11} + (x_1 - \bar{x}_1)(x_2 - \bar{x}_2)S^{12} + \cdots + (x_1 - \bar{x}_1)(x_p - \bar{x}_p)S^{1p} \\
&+(x_2 - \bar{x}_2)(x_1 - \bar{x}_1)S^{21} + (x_2 - \bar{x}_2)(x_2 - \bar{x}_2)S^{22} + \cdots + (x_2 - \bar{x}_2)(x_p - \bar{x}_p)S^{2p} \\
&\cdots\cdots\cdots\cdots\cdots\cdots\cdots\cdots\cdots\cdots\cdots\cdots\cdots\cdots\cdots\cdots\cdots\cdots\cdots\cdots\cdots\cdots\cdots \\
&+(x_p - \bar{x}_p)(x_1 - \bar{x}_1)S^{p1} + (x_p - \bar{x}_p)(x_2 - \bar{x}_2)S^{p2} + \cdots + (x_p - \bar{x}_p)(x_p - \bar{x}_p)S^{pp}
\end{aligned}
\right\}(\text{number of individuals} - 1)
$$

$$
D^2 = \left\{
\begin{aligned}
&(x_1 - \bar{x}_1)(x_1 - \bar{x}_1)S^{11} + (x_1 - \bar{x}_1)(x_2 - \bar{x}_2)S^{12} \\
&+(x_2 - \bar{x}_2)(x_1 - \bar{x}_1)S^{21} + (x_2 - \bar{x}_2)(x_2 - \bar{x}_2)S^{22}
\end{aligned}
\right\}(\text{number of individuals} - 1)
$$

$$
= \left\{
\begin{aligned}
&(10 - 7.7)(10 - 7.7) \times 0.0657 + (10 - 7.7)(80 - 156) \times 0.0004 \\
&+(80 - 156)(10 - 7.7) \times 0.0004 + (80 - 156)(80 - 156) \times 0.00001
\end{aligned}
\right\}(10 - 1)
$$

$$
= 2.4
$$

STEP 3

Now we can calculate the confidence interval, as illustrated here:

This is the confidence interval.
Monthly sales

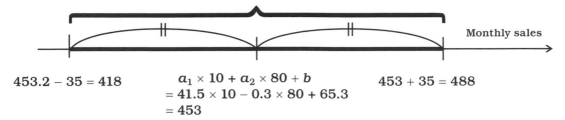

$453.2 - 35 = 418$

$$a_1 \times 10 + a_2 \times 80 + b$$
$$= 41.5 \times 10 - 0.3 \times 80 + 65.3$$
$$= 453$$

$453 + 35 = 488$

The minimum value of the confidence interval is the same distance from the mean as the maximum value of the interval. In other words, the confidence interval "straddles" the mean equally on each side. We calculate the distance from the mean as shown below (D^2 stands for Mahalanobis distance, and x represents the total number of predictors, not a value of some predictor):

$$\sqrt{F(1, \text{sample size} - x - 1; .05) \times \left(\frac{1}{\text{sample size}} + \frac{D^2}{\text{sample size} - 1}\right) \times \frac{S_e}{\text{sample size} - x - 1}}$$
$$= \sqrt{F(1, 10 - 2 - 1; .05) \times \left(\frac{1}{10} + \frac{2.4}{10 - 1}\right) \times \frac{4173.0}{10 - 2 - 1}}$$
$$= 35$$

As with simple regression analysis, when obtaining the prediction interval, we add 1 to the second term:

$$\sqrt{F(1, \text{sample size} - x - 1; .05) \times \left(1 + \frac{1}{\text{sample size}} + \frac{D^2}{\text{sample size} - 1}\right) \times \frac{S_e}{\text{sample size} - x - 1}}$$

If the confidence rate is 99%, just change the .05 to .01:

$$F(1, \text{sample size} - x - 1; .05) = F(1, 10 - 2 - 1; .05) = 5.6$$
$$F(1, \text{sample size} - x - 1; .01) = F(1, 10 - 2 - 1; .01) = 12.2$$

You can see that if you want to be more confident that the prediction interval will include the actual outcome, the interval needs to be larger.

USING CATEGORICAL DATA IN MULTIPLE REGRESSION ANALYSIS

Recall from Chapter 1 that categorical data is data that can't be measured with numbers. For example, the color of a store manager's eyes is categorical (and probably a terrible predictor variable for monthly sales). Although categorical variables can be *represented* by numbers (1 = blue, 2 = green), they are discrete—there's no such thing as "green and a half." Also, one cannot say that 2 (green eyes) is greater than 1 (blue eyes). So far we've been using the numerical data (which can be meaningfully represented by continuous numerical values—110 m from the train station is further than 109.9 m) shown in Table 3-2, which also appears on page 113.

TABLE 3-2: KAZAMI BAKERY EXAMPLE DATA

Bakery	Floor space of the shop (tsubo)	Distance to the nearest station (meters)	Monthly sales (¥10,000)
Yumenooka Shop	10	80	469
Terai Station Shop	8	0	366
Sone Shop	8	200	371
Hashimoto Station Shop	5	200	208
Kikyou Town Shop	7	300	246
Post Office Shop	8	230	297
Suidobashi Station Shop	7	40	363
Rokujo Station Shop	9	0	436
Wakaba Riverside Shop	6	330	198
Misato Shop	9	180	364

The predictor variable *floor area* is measured in tsubo, *distance to the nearest station* in meters, and *monthly sales* in yen. Clearly, these are all numerically measurable. In multiple regression analysis, the outcome variable must be a measurable, numerical variable, but the predictor variables can be

- all numerical variables,
- some numerical and some categorical variables, or
- all categorical variables.

Tables 3-3 and 3-4 both show valid data sets. In the first, categorical and numerical variables are both present, and in the second, all of the predictor variables are categorical.

TABLE 3-3: A COMBINATION OF CATEGORICAL AND NUMERICAL DATA

Bakery	Floor space of the shop (tsubo)	Distance to the nearest station (meters)	Free samples	Monthly sales (¥10,000)
Yumenooka Shop	10	80	1	469
Terai Station Shop	8	0	0	366
Sone Shop	8	200	1	371
Hashimoto Station Shop	5	200	0	208
Kikyou Town Shop	7	300	0	246
Post Office Shop	8	230	0	297
Suidobashi Station Shop	7	40	0	363
Rokujo Station Shop	9	0	1	436
Wakaba Riverside Shop	6	330	0	198
Misato Shop	9	180	1	364

In Table 3-3 we've included the categorical predictor variable *free samples*. Some Kazami Bakery locations put out a tray of free samples (1), and others don't (0). When we include this data in the analysis, we get the multiple regression equation

$$y = 30.6x_1 - 0.4x_2 + 39.5x_3 + 135.9$$

where y represents monthly sales, x_1 represents floor area, x_2 represents distance to the nearest station, and x_3 represents free samples.

TABLE 3-4: CATEGORICAL PREDICTOR DATA ONLY

Bakery	Floor space of the shop (tsubo)	Distance to the nearest station (meters)	Samples every day	Samples on the weekend only	Monthly sales (¥10,000)
Yumenooka Shop	1	0	1	0	469
Terai Station Shop	1	0	0	0	366
Sone Shop	1	1	1	0	371
Hashimoto Station Shop	0	1	0	0	208
Kikyou Town Shop	0	1	0	0	246
Post Office Shop	1	1	0	0	297
Suidobashi Station Shop	0	0	0	0	363
Rokujo Station Shop	1	0	1	1	436
Wakaba Riverside Shop	0	1	0	0	198
Misato Shop	1	0	1	1	364

Floor space: Less than 8 tsubo = 0; 8 tsubo or more = 1

Distance: Less than 200 m = 0; 200 m or more = 1

Samples: Does not offer samples = 0; Offers samples = 1

In Table 3-4, we've converted numerical data (floor space and distance to a station) to categorical data by creating some general categories. Using this data, we calculate the multiple regression equation

$$y = 50.2x_1 - 110.1x_2 + 13.4x_3 + 75.1x_4 + 336.4$$

where y represents monthly sales, x_1 represents floor area, x_2 represents distance to the nearest station, x_3 represents samples every day, and x_4 represents samples on the weekend only.

MULTICOLLINEARITY

Multicollinearity occurs when two of the predictor variables are strongly correlated with each other. When this happens, it's hard to distinguish between the effects of these variables on the outcome variable, and this can have the following effects on your analysis:

· Less accurate estimate of the impact of a given variable on the outcome variable

· Unusually large standard errors of the regression coefficients

· Failure to reject the null hypothesis

· *Overfitting*, which means that the regression equation describes a relationship between the outcome variable and random error, rather than the predictor variable

The presence of multicollinearity can be assessed by using an index such as *tolerance* or the inverse of tolerance, known as the *variance inflation factor (VIF)*. Generally, a tolerance of less than 0.1 or a VIF greater than 10 is thought to indicate significant multicollinearity, but sometimes more conservative thresholds are used.

When you're just starting out with multiple regression analysis, you don't need to worry too much about this. Just keep in mind that multicollinearity can cause problems when it's severe. Therefore, when predictor variables are correlated to each other strongly, it may be better to remove one of the highly correlated variables and then reanalyze the data.

DETERMINING THE RELATIVE INFLUENCE OF PREDICTOR VARIABLES ON THE OUTCOME VARIABLE

Some people use multiple regression analysis to examine the relative influence of each predictor variable on the outcome variable. This is a fairly common and accepted use of multiple regression analysis, but it's not always a wise use.

The story below illustrates how one researcher used multiple regression analysis to assess the relative impact of various factors on the overall satisfaction of people who bought a certain type of candy.

Mr. Torikoshi is a product development researcher in a confectionery company. He recently developed a new soda-flavored candy, Magic Fizz, that fizzes when wet. The candy is selling astonishingly well. To find out what makes it so popular, the company gave free samples of the candy to students at the local university and asked them to rate the product using the following questionnaire.

MAGIC FIZZ QUESTIONNAIRE

Please let us know what you thought of Magic Fizz by answering the following questions. Circle the answer that best represents your opinion.

Flavor	1. Unsatisfactory 2. Satisfactory 3. Exceptional
Texture	1. Unsatisfactory 2. Satisfactory 3. Exceptional
Fizz sensation	1. Unsatisfactory 2. Satisfactory 3. Exceptional
Package design	1. Unsatisfactory 2. Satisfactory 3. Exceptional
Overall satisfaction	1. Unsatisfactory 2. Satisfactory 3. Exceptional

Twenty students returned the questionnaires, and the results are compiled in Table 3-5. Note that unlike in the Kazami Bakery example, the values of the outcome variable—overall satisfaction—are already known. In the bakery problem, the goal was to predict the outcome variable (profit) of a not-yet-existing store based on the trends shown by existing stores. In this case, the purpose of this analysis is to examine the relative effects of the different predictor variables in order to learn how each of the predictors (flavor, texture, sensation, design) affects the outcome (satisfaction).

TABLE 3-5: RESULTS OF MAGIC FIZZ QUESTIONNAIRE

Respondent	Flavor	Texture	Fizz sensation	Package design	Overall satisfaction
1	2	2	3	2	2
2	1	1	3	1	3
3	2	2	1	1	3
3	2	2	1	1	1
4	3	3	3	2	2
5	1	1	2	2	1
6	1	1	1	1	1
7	3	3	1	3	3
8	3	3	1	2	2
9	3	3	1	2	3
10	1	1	3	1	1
11	2	3	2	1	3
12	2	1	1	1	1
13	3	3	3	1	3
14	3	3	1	3	3
15	3	2	1	1	2
16	1	1	3	3	1
17	2	2	2	1	1
18	1	1	1	3	1
19	3	1	3	3	3
20	3	3	3	3	3

Each of the variables was normalized before the multiple regression equation was calculated. Normalization reduces the effect of error or scale, allowing a researcher to compare two variables more accurately. The resulting equation is

$$y = 0.41x_1 + 0.32x_2 + 0.26x_3 + 0.11x_4$$

where y represents overall satisfaction, x_1 represents flavor, x_2 represents texture, x_3 represents fizz sensation, and x_4 represents package design.

If you compare the partial regression coefficients for the four predictor variables, you can see that the coefficient for flavor is the largest. Based on that fact, Mr. Torikoshi concluded that the flavor has the strongest influence on overall satisfaction.

Mr. Torikoshi's reasoning does make sense. The outcome variable is equal to the sum of the predictor variables multiplied by their partial regression coefficients. If you multiply a predictor variable by a higher number, it should have a greater impact on the final tally, right? Well, sometimes—but it's not always so simple.

Let's take a closer look at Mr. Torikoshi's reasoning as depicted here:

TEXTURE
PACKAGE DESIGN
FLAVOR
FIZZ SENSATION
OVERALL SATISFACTION

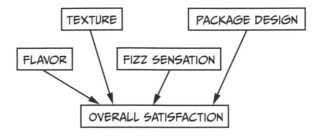

In other words, he is assuming that all the variables relate independently and directly to overall satisfaction. However, this is not necessarily true. Maybe in reality, the texture influences how satisfied people are with the flavor, like this:

TEXTURE
FIZZ SENSATION
FLAVOR
PACKAGE DESIGN
OVERALL SATISFACTION

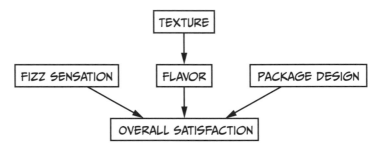

Structural equation modeling (*SEM*) is a better method for comparing the relative impact of various predictor variables on an outcome. This approach makes more flexible assumptions than linear regression does, and it can even be used to analyze data sets with multicollinearity. However, SEM is not a cure-all. It relies on the assumption that the data is relevant to answering the question asked.

SEM also assumes that the data is correctly modeled. It's worth noting that the questions in this survey ask each reviewer for a subjective interpretation. If Miu gave the candy two "satisfactory" and two "exceptional" marks, she might rate her overall satisfaction as either "satisfactory" or "exceptional." Which rating she picks might come down to what mood she is in that day!

Risa could rate the four primary categories the same as Miu, give a different overall satisfaction rating from Miu, and still be confident that she is giving an unbiased review. Because Miu and Risa had different thoughts on the final category, our data may not be correctly modeled. However, structural equation modeling can still yield useful results by telling us which variables have an impact on other variables rather than the final outcome.

4
LOGISTIC REGRESSION ANALYSIS

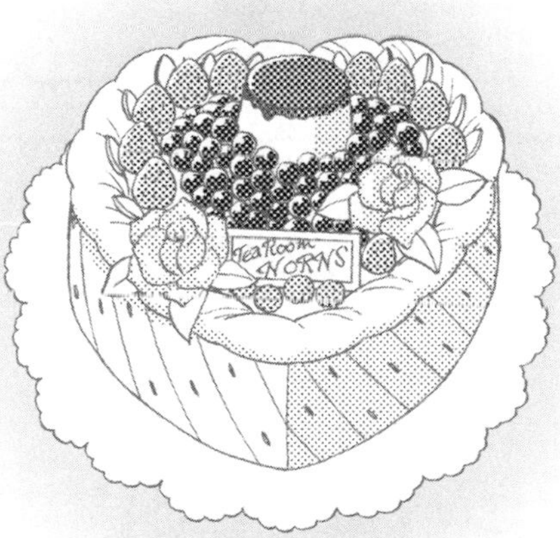
Tea Room NORNS

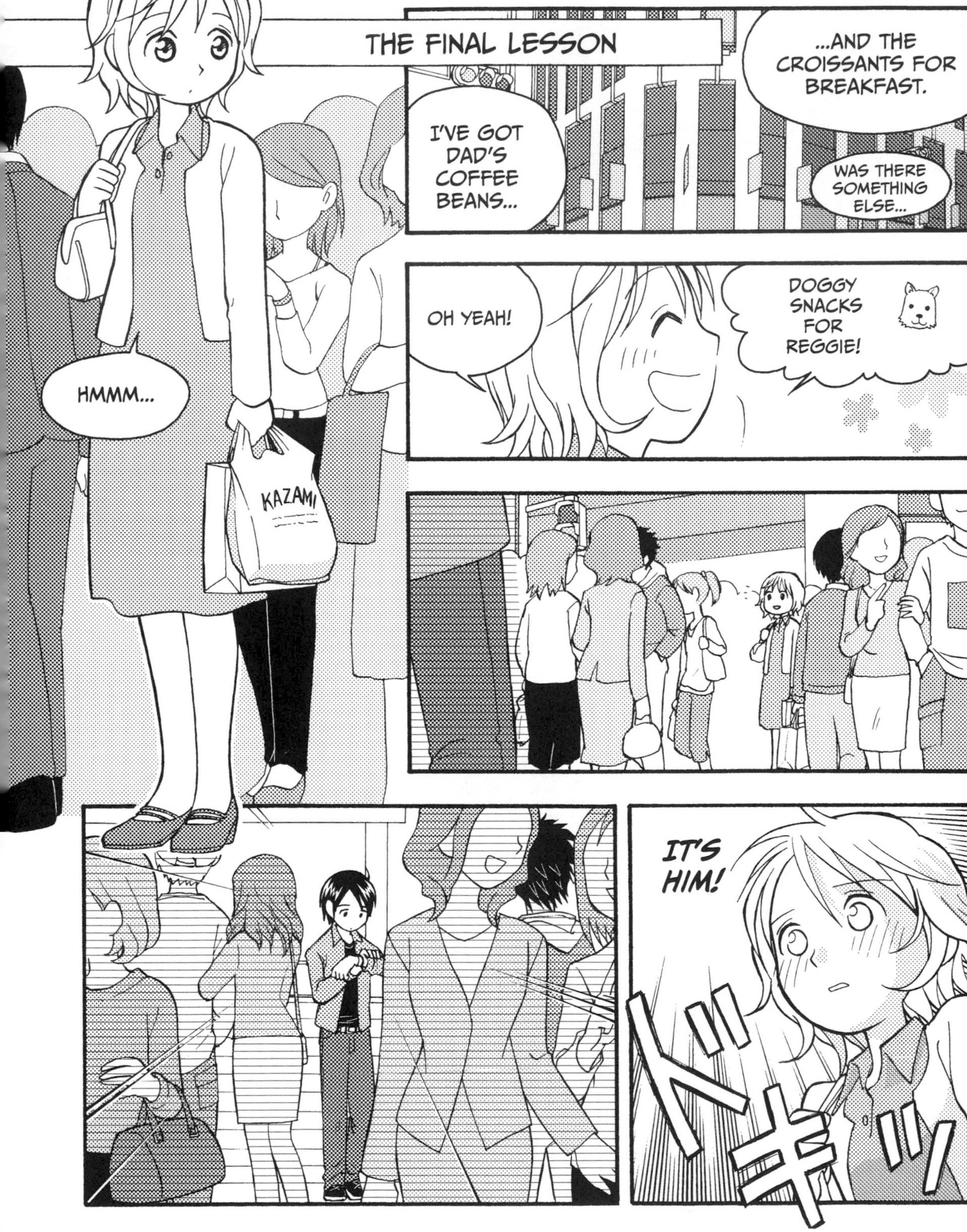
THE FINAL LESSON
HMMM...
I'VE GOT DAD'S COFFEE BEANS...
...AND THE CROISSANTS FOR BREAKFAST.
WAS THERE SOMETHING ELSE...
KAZAMI
OH YEAH!
DOGGY SNACKS FOR REGGIE!
IT'S HIM!
ドキッ

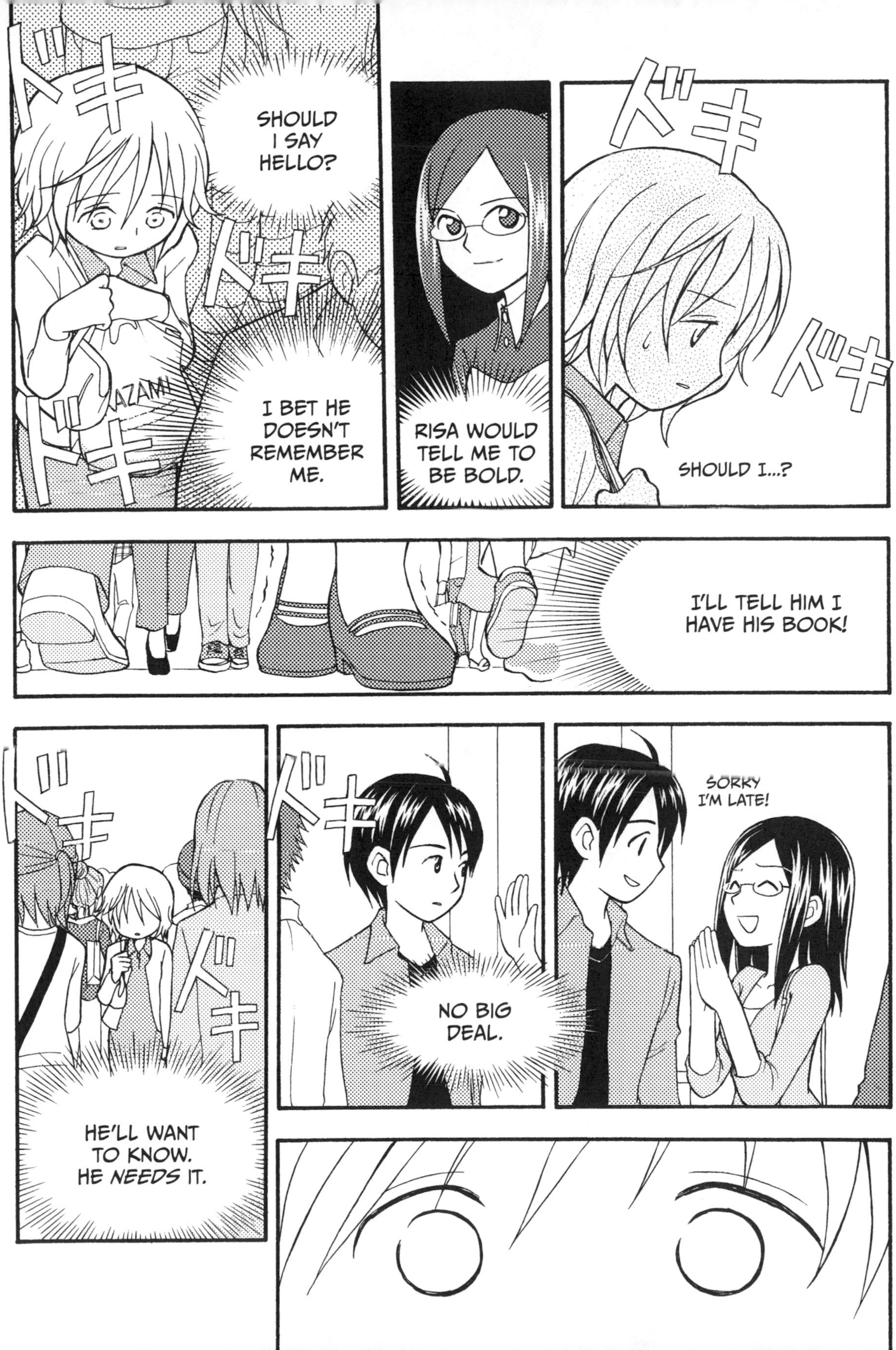
ドキ
SHOULD I SAY HELLO?
ドキ
ドキ
I BET HE DOESN'T REMEMBER ME.
RISA WOULD TELL ME TO BE BOLD.
ドキ
ドキ
SHOULD I...?
I'LL TELL HIM I HAVE HIS BOOK!
ドキ
ドキ
HE'LL WANT TO KNOW. HE *NEEDS* IT.
ドキ
NO BIG DEAL.
SORRY I'M LATE!

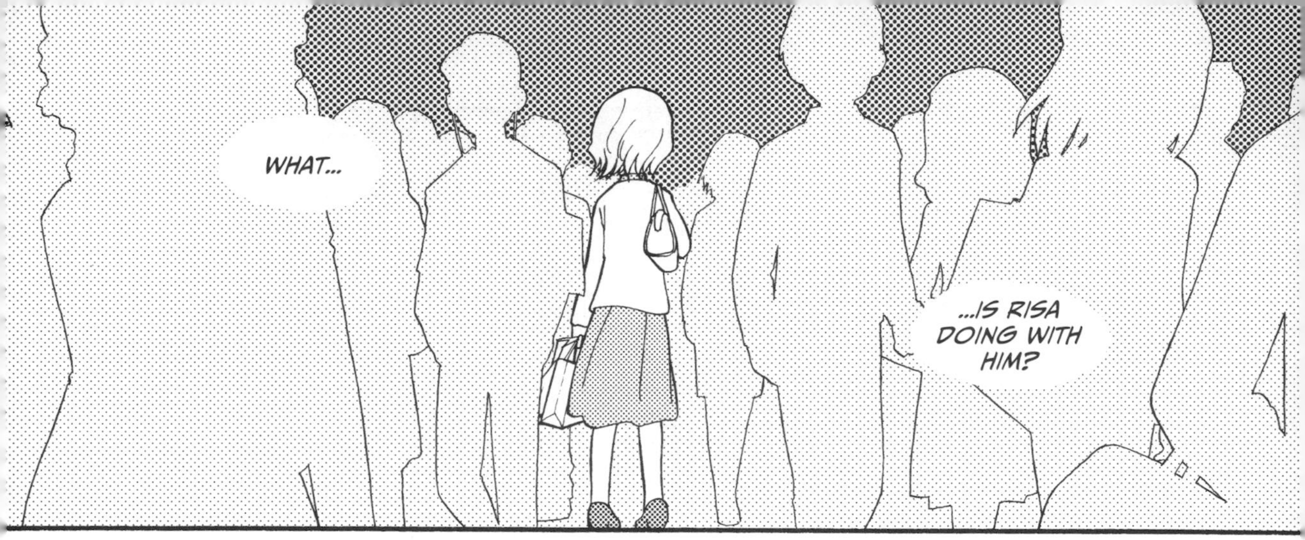
WHAT...
...IS RISA DOING WITH HIM?

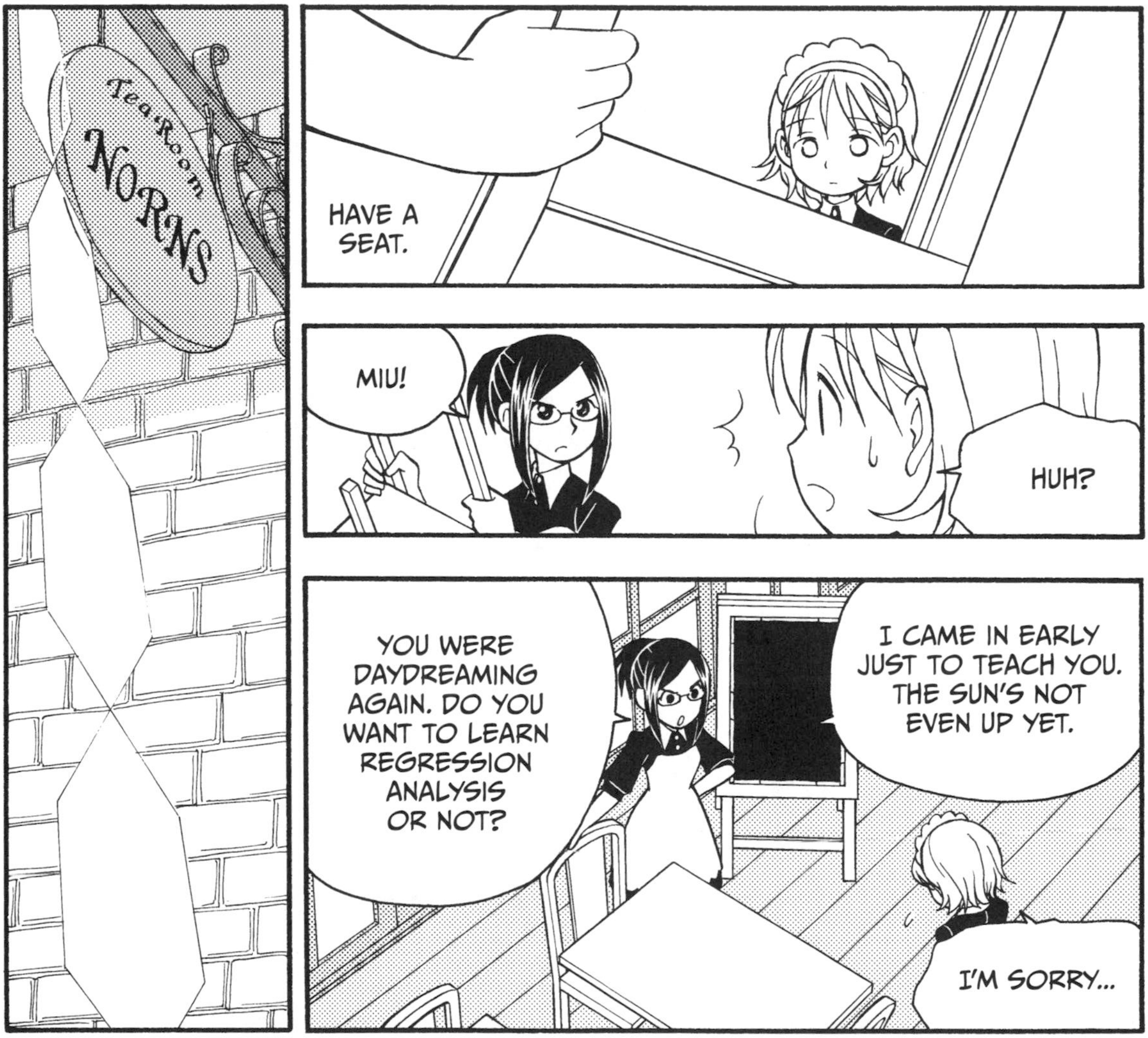
Tea-Room
NORNS
HAVE A SEAT.
MIU!
HUH?
YOU WERE DAYDREAMING AGAIN. DO YOU WANT TO LEARN REGRESSION ANALYSIS OR NOT?
I CAME IN EARLY JUST TO TEACH YOU. THE SUN'S NOT EVEN UP YET.
I'M SORRY...

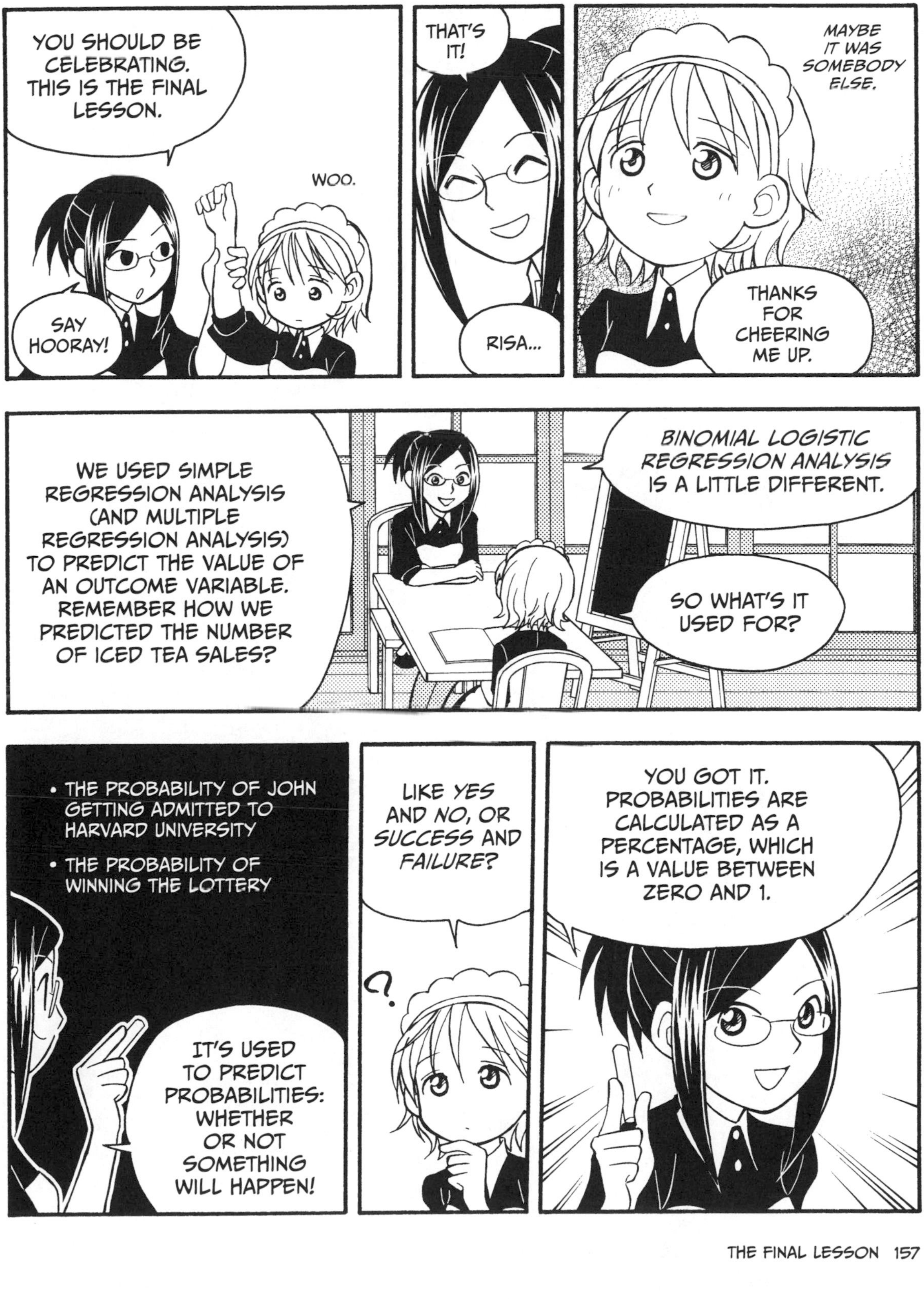
YOU SHOULD BE CELEBRATING. THIS IS THE FINAL LESSON.
WOO.
SAY HOORAY!
THAT'S IT!
RISA...
MAYBE IT WAS SOMEBODY ELSE.
THANKS FOR CHEERING ME UP.
WE USED SIMPLE REGRESSION ANALYSIS (AND MULTIPLE REGRESSION ANALYSIS) TO PREDICT THE VALUE OF AN OUTCOME VARIABLE. REMEMBER HOW WE PREDICTED THE NUMBER OF ICED TEA SALES?
BINOMIAL LOGISTIC REGRESSION ANALYSIS IS A LITTLE DIFFERENT.
SO WHAT'S IT USED FOR?
• THE PROBABILITY OF JOHN GETTING ADMITTED TO HARVARD UNIVERSITY
• THE PROBABILITY OF WINNING THE LOTTERY
IT'S USED TO PREDICT PROBABILITIES: WHETHER OR NOT SOMETHING WILL HAPPEN!
LIKE *YES* AND *NO*, OR *SUCCESS* AND *FAILURE*?
?
YOU GOT IT. PROBABILITIES ARE CALCULATED AS A PERCENTAGE, WHICH IS A VALUE BETWEEN ZERO AND 1.

BUT 70% IS GREATER THAN 1, ISN'T IT?

ACTUALLY, 70% IS EQUAL TO .7. WHEN WE DO LOGISTIC REGRESSION ANALYSIS, THE ANSWER WILL BE LESS THAN 1.

TO EXPRESS IT AS A PERCENTAGE, MULTIPLY BY 100 AND USE THE PERCENT SYMBOL.

THE LOGISTIC REGRESSION EQUATION...

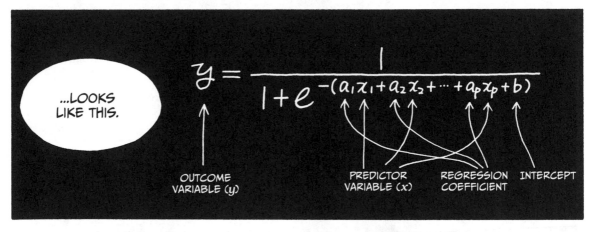

...LOOKS LIKE THIS.

$$y = \frac{1}{1 + e^{-(a_1 x_1 + a_2 x_2 + \cdots + a_p x_p + b)}}$$

OUTCOME VARIABLE (y)

PREDICTOR VARIABLE (x)

REGRESSION COEFFICIENT

INTERCEPT

THAT'S ONE BIG EXPONENT! THIS LOOKS COMPLICATED...

DON'T WORRY, I'LL SHOW YOU A SIMPLER WAY TO WRITE IT. WE'LL TAKE IT STEP-BY-STEP, AND IT WON'T SEEM SO TOUGH.

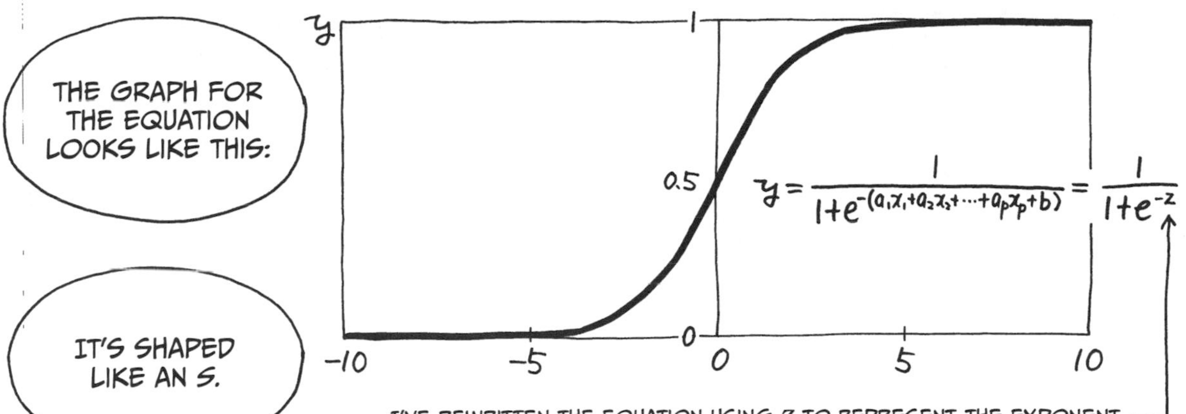
THE GRAPH FOR THE EQUATION LOOKS LIKE THIS:
IT'S SHAPED LIKE AN S.
y
1
0.5
0
−10
−5
0
5
10
$y = \frac{1}{1+e^{-(a_1x_1+a_2x_2+\cdots+a_px_p+b)}} = \frac{1}{1+e^{-z}}$

I'VE REWRITTEN THE EQUATION USING Z TO REPRESENT THE EXPONENT.
f(Z) IS THE PROBABILITY OF OUR OUTCOME!

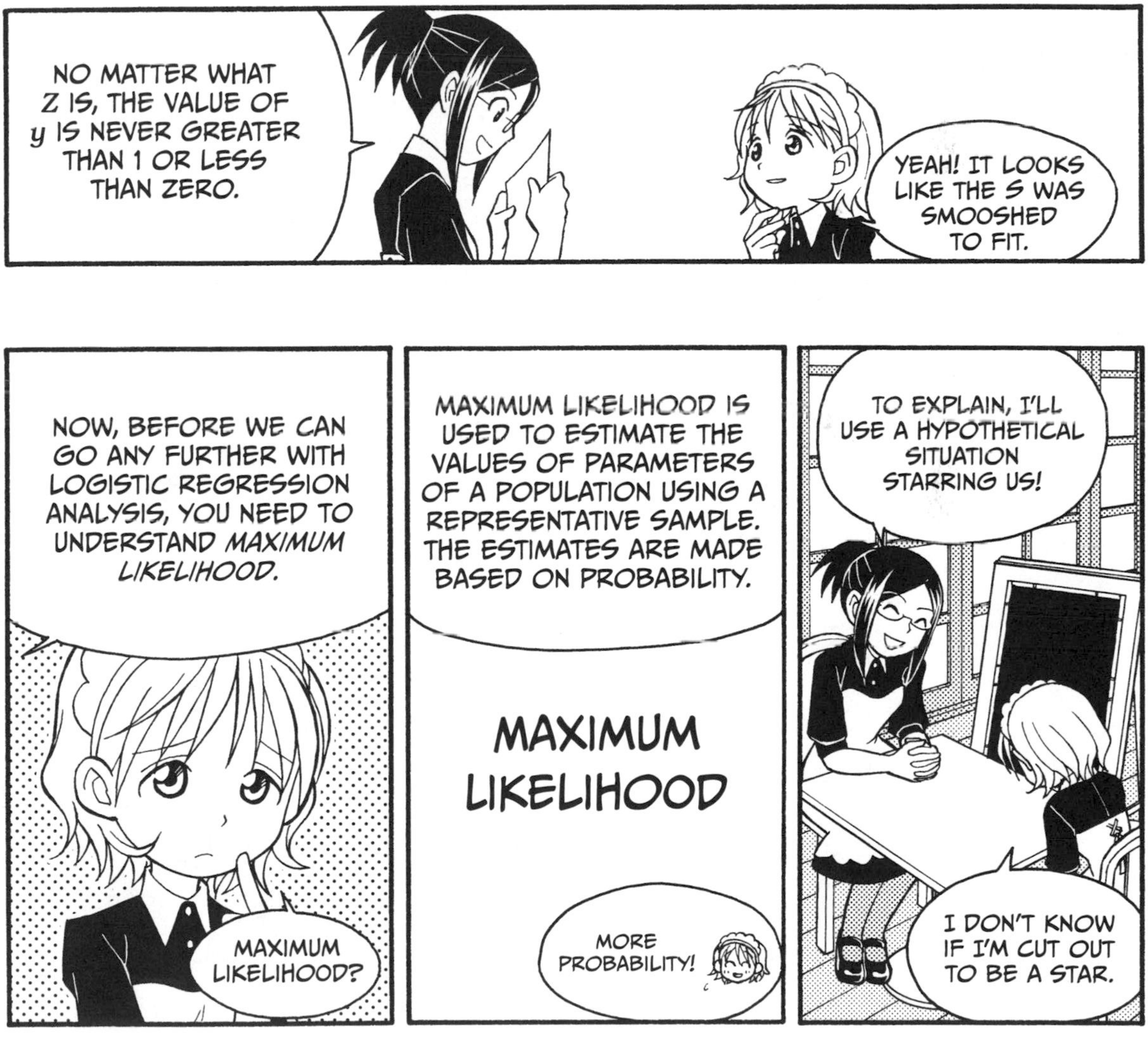
NO MATTER WHAT Z IS, THE VALUE OF y IS NEVER GREATER THAN 1 OR LESS THAN ZERO.
YEAH! IT LOOKS LIKE THE S WAS SMOOSHED TO FIT.
NOW, BEFORE WE CAN GO ANY FURTHER WITH LOGISTIC REGRESSION ANALYSIS, YOU NEED TO UNDERSTAND *MAXIMUM LIKELIHOOD.*
MAXIMUM LIKELIHOOD?
MAXIMUM LIKELIHOOD IS USED TO ESTIMATE THE VALUES OF PARAMETERS OF A POPULATION USING A REPRESENTATIVE SAMPLE. THE ESTIMATES ARE MADE BASED ON PROBABILITY.
MAXIMUM LIKELIHOOD
MORE PROBABILITY!
TO EXPLAIN, I'LL USE A HYPOTHETICAL SITUATION STARRING US!
I DON'T KNOW IF I'M CUT OUT TO BE A STAR.

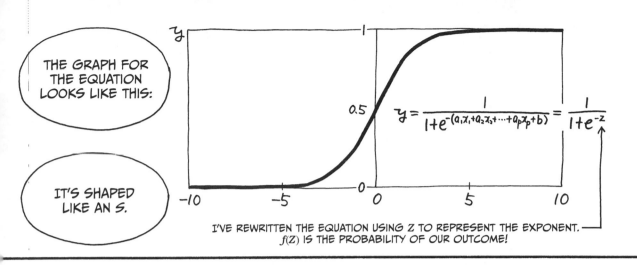

THE MAXIMUM LIKELIHOOD METHOD

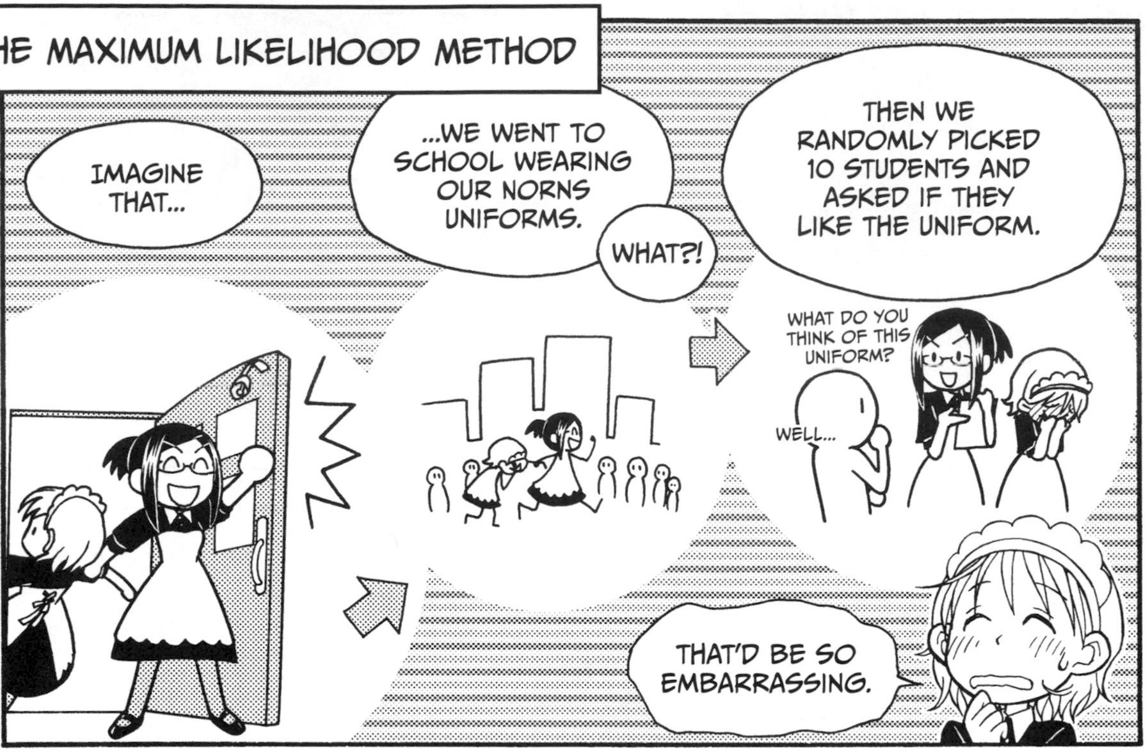

IMAGINE THAT...
...WE WENT TO SCHOOL WEARING OUR NORNS UNIFORMS.
WHAT?!
THEN WE RANDOMLY PICKED 10 STUDENTS AND ASKED IF THEY LIKE THE UNIFORM.
WHAT DO YOU THINK OF THIS UNIFORM?
WELL...
THAT'D BE SO EMBARRASSING.

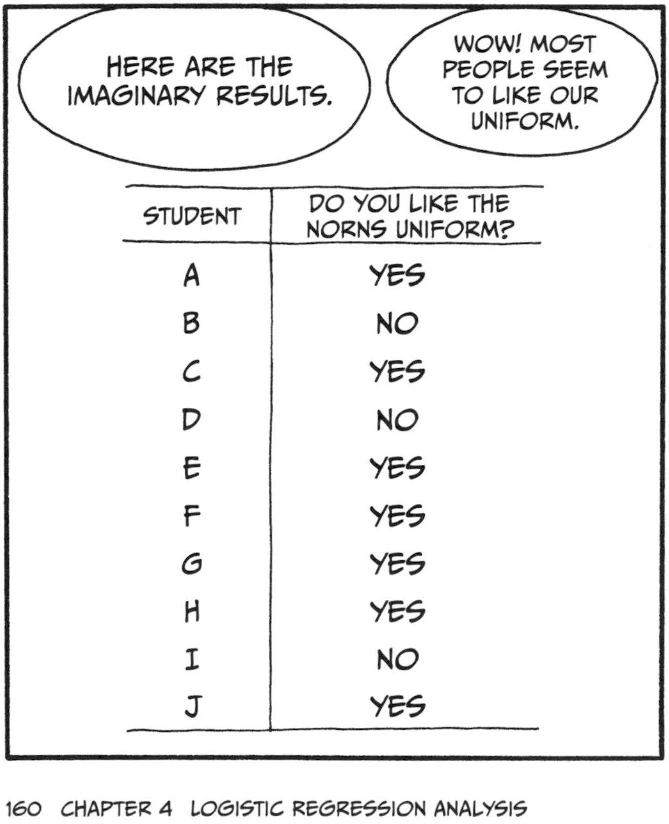

HERE ARE THE IMAGINARY RESULTS.
WOW! MOST PEOPLE SEEM TO LIKE OUR UNIFORM.

STUDENT	DO YOU LIKE THE NORNS UNIFORM?
A	YES
B	NO
C	YES
D	NO
E	YES
F	YES
G	YES
H	YES
I	NO
J	YES

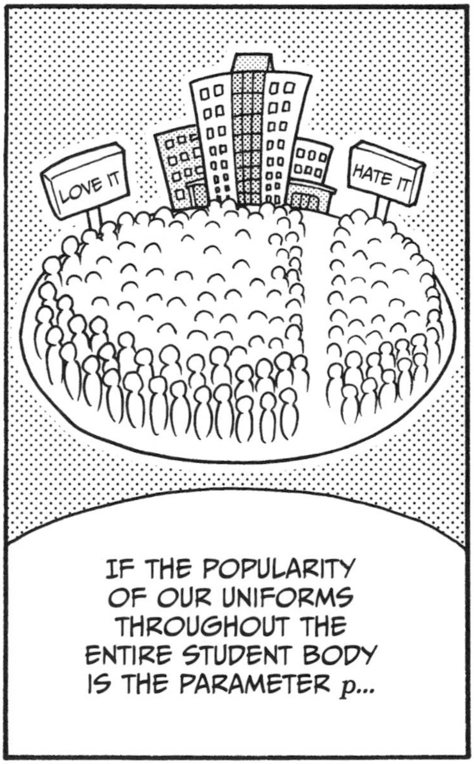

LOVE IT
HATE IT
IF THE POPULARITY OF OUR UNIFORMS THROUGHOUT THE ENTIRE STUDENT BODY IS THE PARAMETER *p*...

...THEN THE PROBABILITY BASED ON THE IMAGINARY SURVEY RESULTS IS THIS:

YES   NO   YES   NO   YES YES YES YES   NO   YES

$$p \times (1-p) \times p \times (1-p) \times p \times p \times p \times p \times (1-p) \times p$$

$$= p^7 (1-p)^3$$

IT'S AN EQUATION?

YES, WE SOLVE IT BY FINDING THE MOST LIKELY VALUE OF p.

$$p^7 (1-p)^3$$

OR

$$\log \{ p^7 (1-p)^3 \}^*$$

WE USE ONE OF THESE LIKELIHOOD FUNCTIONS.

EITHER WAY, THE ANSWER IS THE SAME.

* TAKING THE LOG OF THIS FUNCTION CAN MAKE LATER CALCULATIONS EASIER.

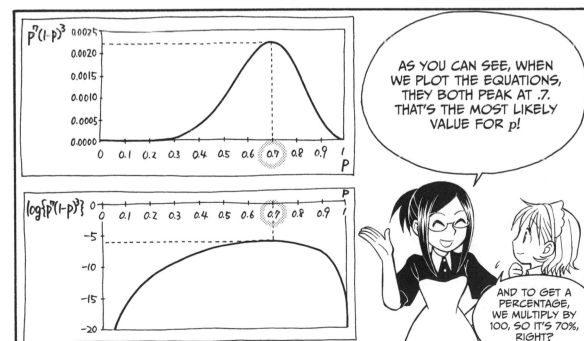

AS YOU CAN SEE, WHEN WE PLOT THE EQUATIONS, THEY BOTH PEAK AT .7. THAT'S THE MOST LIKELY VALUE FOR p!

AND TO GET A PERCENTAGE, WE MULTIPLY BY 100, SO IT'S 70%, RIGHT?

THAT'S RIGHT. WE TAKE THE LOG OF THIS FUNCTION BECAUSE IT MAKES IT EASIER TO CALCULATE THE DERIVATIVE, WHICH WE NEED TO FIND THE MAXIMUM LIKELIHOOD.

$$p^7(1-p)^3$$

→ LIKELIHOOD FUNCTION

$$\log\{p^7(1-p)^3\}$$

→ LOG-LIKELIHOOD FUNCTION

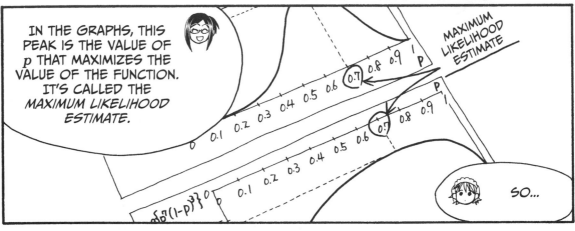

IN THE GRAPHS, THIS PEAK IS THE VALUE OF p THAT MAXIMIZES THE VALUE OF THE FUNCTION. IT'S CALLED THE *MAXIMUM LIKELIHOOD ESTIMATE.*

MAXIMUM LIKELIHOOD ESTIMATE

SO...

...SINCE THE FUNCTIONS PEAK AT THE SAME PLACE, EVEN THOUGH THEY HAVE A DIFFERENT SHAPE, THEY GIVE US THE SAME ANSWER.

EXACTLY!

NOW, LET'S REVIEW THE MAXIMUM LIKELIHOOD ESTIMATE FOR THE POPULARITY OF OUR UNIFORMS.

OKAY, RISA.

FINDING THE MAXIMUM LIKELIHOOD USING THE LIKELIHOOD FUNCTION

Step 1 Find the likelihood function. Here, p stands for *Yes*, and $1 - p$ stands for *No*. There were 7 *Yeses* and 3 *Nos*.

$$p \times (1-p) \times p \times (1-p) \times p \times p \times p \times p \times (1-p) \times p$$
$$= p^7 (1-p)^3$$

Step 2 Obtain the log-likelihood function and rearrange it.

$$L = \log\left\{p^7 (1-p)^3\right\}$$
$$= \log p^7 + \log(1-p)^3 \quad \longleftarrow \text{Take the log of each component.}$$
$$= 7\log p + 3\log(1-p) \quad \longleftarrow \text{Use the Exponentiation Rule from page 22.}$$

WE'LL USE L TO MEAN THE LOG-LIKELIHOOD FUNCTION FROM NOW ON.

Step 3 Differentiate L with respect to p and set the expression equal to 0. Remember that when a function's rate of change is 0, we're finding the maxima.

$$\frac{dL}{dp} = 7 \times \frac{1}{p} + 3 \times \frac{1}{1-p} \times (-1) = 7 \times \frac{1}{p} - 3 \times \frac{1}{1-p} = 0$$

Step 4 Rearrange the equation in Step 3 to solve for p.

$$7 \times \frac{1}{p} - 3 \times \frac{1}{1-p} = 0$$
$$\left(7 \times \frac{1}{p} - 3 \times \frac{1}{1-p}\right) \times p(1-p) = 0 \times p(1-p) \quad \longleftarrow \text{Multiply both sides of the equation by } p(1-p).$$
$$7(1-p) - 3p = 0$$
$$7 - 7p - 3p = 0$$
$$7 - 10p = 0$$
$$p = \frac{7}{10}$$

AND HERE'S THE MAXIMUM LIKELIHOOD ESTIMATE!
YEP, 70%.

CHOOSING PREDICTOR VARIABLES

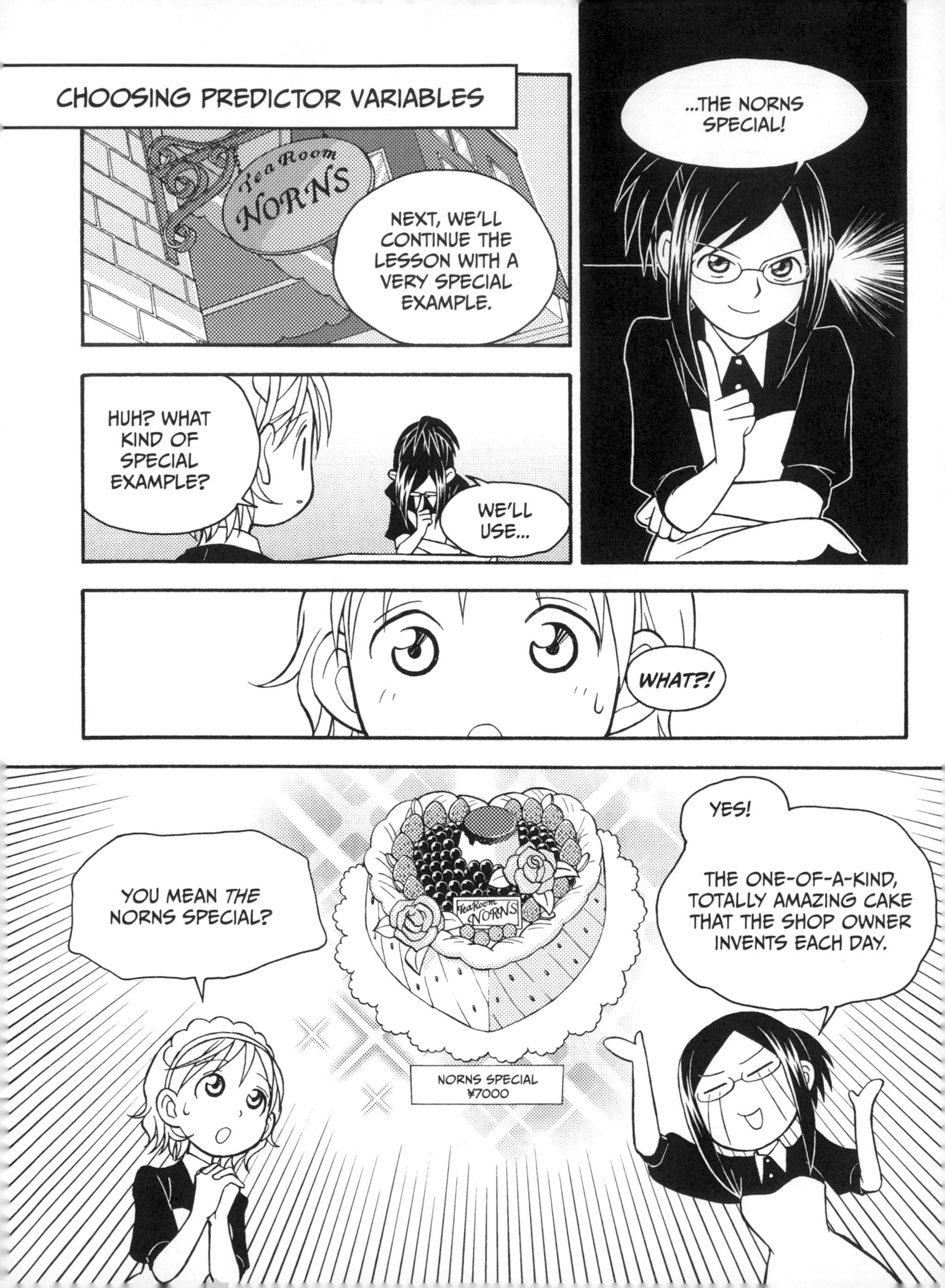
Tea Room NORNS
NEXT, WE'LL CONTINUE THE LESSON WITH A VERY SPECIAL EXAMPLE.
HUH? WHAT KIND OF SPECIAL EXAMPLE?
WE'LL USE...
...THE NORNS SPECIAL!
WHAT?!
YOU MEAN *THE* NORNS SPECIAL?
Tea Room NORNS
NORNS SPECIAL
¥7000
YES!
THE ONE-OF-A-KIND, TOTALLY AMAZING CAKE THAT THE SHOP OWNER INVENTS EACH DAY.

NORNS DOESN'T SELL A SPECIAL CAKE EVERY DAY. PEOPLE ONLY BUY IT FOR A REALLY SPECIAL OCCASION, LIKE A BIRTHDAY.

OR AN ANNIVERSARY...

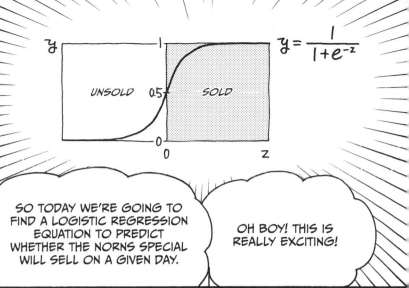

$$y = \frac{1}{1 + e^{-z}}$$

UNSOLD 0.5 SOLD

SO TODAY WE'RE GOING TO FIND A LOGISTIC REGRESSION EQUATION TO PREDICT WHETHER THE NORNS SPECIAL WILL SELL ON A GIVEN DAY.

OH BOY! THIS IS REALLY EXCITING!

I GUESS WE'LL NEED A DATA SAMPLE. BUT WHAT SHOULD WE MEASURE? WHAT AFFECTS IF IT SELLS OR NOT?

THAT'S A GREAT QUESTION.

I'VE BEEN TRYING TO FIGURE THAT OUT FOR A WHILE. I'VE NOTICED THAT MORE PEOPLE SEEM TO BUY THE NORNS SPECIAL WHEN THE TEMPERATURE IS HIGH, AND ON WEDNESDAYS, SATURDAYS, AND SUNDAYS.

REALLY?

YEP. THERE ARE WAY MORE CUSTOMERS ON THE WEEKEND...

PARTY TIME!

...AND ON WEDNESDAYS, A LARGE MANGA FAN CLUB MEETS HERE, AND THEY REALLY LIKE TO GO BIG.

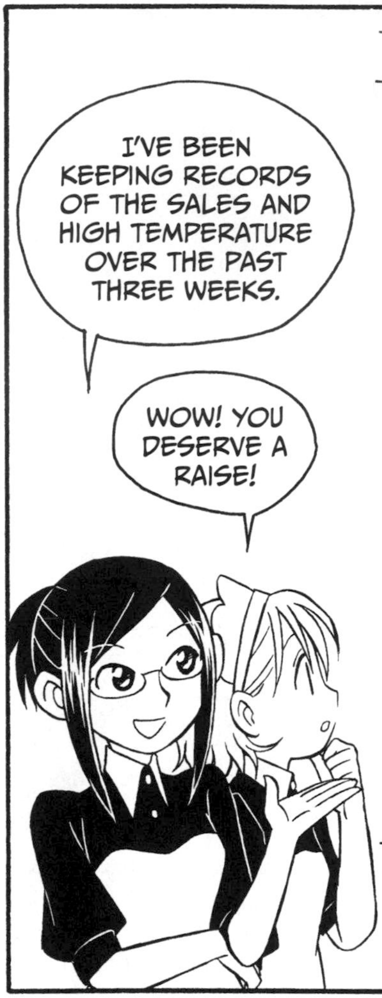
I'VE BEEN KEEPING RECORDS OF THE SALES AND HIGH TEMPERATURE OVER THE PAST THREE WEEKS.
WOW! YOU DESERVE A RAISE!

MARCH	WED, SAT, OR SUN	HIGH TEMP (°C)	SALES OF NORNS SPECIAL
5	0	28	1
6	0	24	0
7	1	26	0
8	0	24	0
9	0	23	0
10	1	28	1
11	1	24	0
12	0	26	1
13	0	25	0
14	1	28	1
15	0	21	0
16	0	22	0
17	1	27	1
18	1	26	1
19	0	26	0
20	0	21	0
21	1	21	1
22	0	27	0
23	0	23	0
24	1	22	0
25	1	24	1

1 MEANS WEDNESDAY, SATURDAY, OR SUNDAY. 0 MEANS OTHER DAYS.

1 MEANS IT WAS SOLD. 0 MEANS IT WAS NOT SOLD.

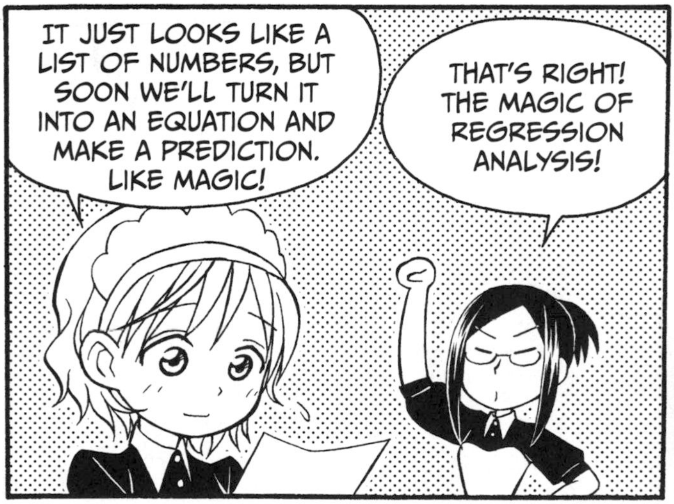
IT JUST LOOKS LIKE A LIST OF NUMBERS, BUT SOON WE'LL TURN IT INTO AN EQUATION AND MAKE A PREDICTION. LIKE MAGIC!
THAT'S RIGHT! THE MAGIC OF REGRESSION ANALYSIS!

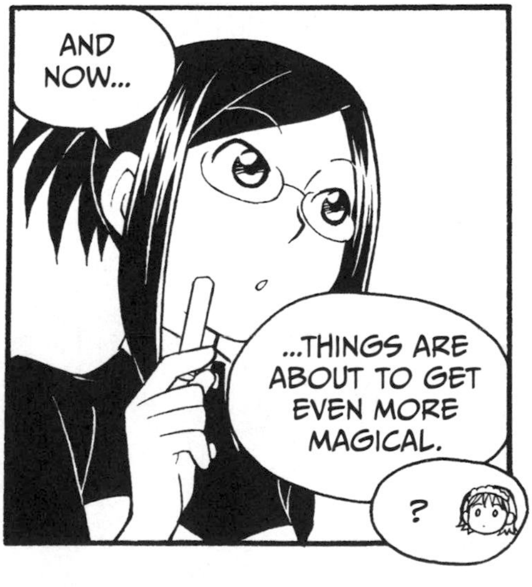
AND NOW...
...THINGS ARE ABOUT TO GET EVEN MORE MAGICAL.
?

WE USED 1 TO MEAN SOLD AND 0 TO MEAN UNSOLD...

1 = SOLD

0 = UNSOLD

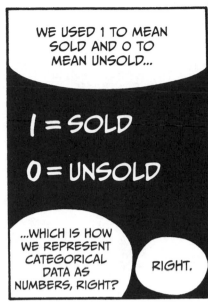

...WHICH IS HOW WE REPRESENT CATEGORICAL DATA AS NUMBERS, RIGHT?

RIGHT.

WELL, IN LOGISTIC REGRESSION ANALYSIS, THESE NUMBERS AREN'T JUST LABELS— THEY ACTUALLY MEASURE THE PROBABILITY THAT THE CAKE WAS SOLD. THAT'S BECAUSE 1 MEANS 100% AND 0 MEANS 0%.

OH! SINCE WE KNOW IT WAS SOLD, THERE'S A 100% PROBABILITY THAT IT WAS SOLD.

WE ALSO KNOW FOR SURE IF IT WAS WEDNESDAY, SATURDAY, OR SUNDAY.

WE SURE DO.

IN THIS CASE, HIGH TEMPERATURE IS MEASURABLE DATA, SO WE USE THE TEMPERATURE, JUST LIKE IN LINEAR REGRESSION ANALYSIS. CATEGORICAL DATA ALSO WORKS IN BASICALLY THE SAME WAY AS IN LINEAR REGRESSION ANALYSIS, AND ONCE AGAIN WE CAN USE ANY COMBINATION OF CATEGORICAL AND NUMERICAL DATA.

BUT CATEGORICAL DATA CAN HAVE MEASURABLE PROBABILITIES.

NOW, LET'S ANALYZE THE SALES OF THE NORNS SPECIAL. I TYPED UP THE STEPS.

WILL WE CALCULATE THE EQUATION AND THEN GET R^2? AND THEN FIND CONFIDENCE AND PREDICTION INTERVALS? OH, AND THE HYPOTHESIS THING?

YEAH, SOMETHING LIKE THAT.

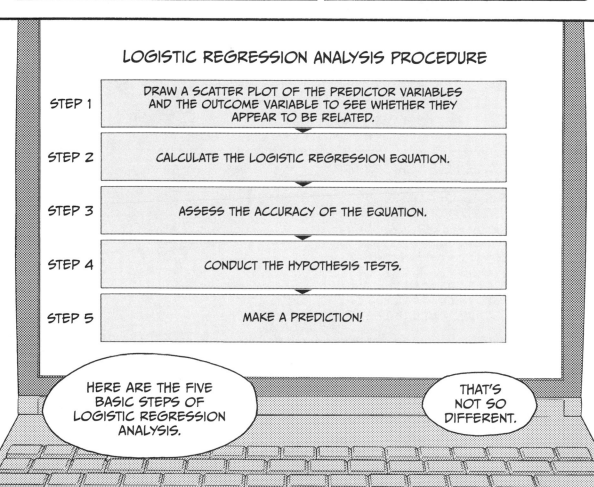

LOGISTIC REGRESSION ANALYSIS PROCEDURE

STEP 1	DRAW A SCATTER PLOT OF THE PREDICTOR VARIABLES AND THE OUTCOME VARIABLE TO SEE WHETHER THEY APPEAR TO BE RELATED.
STEP 2	CALCULATE THE LOGISTIC REGRESSION EQUATION.
STEP 3	ASSESS THE ACCURACY OF THE EQUATION.
STEP 4	CONDUCT THE HYPOTHESIS TESTS.
STEP 5	MAKE A PREDICTION!

HERE ARE THE FIVE BASIC STEPS OF LOGISTIC REGRESSION ANALYSIS.

THAT'S NOT SO DIFFERENT.

STEP 1: DRAW A SCATTER PLOT OF THE PREDICTOR VARIABLES AND THE OUTCOME VARIABLE TO SEE WHETHER THEY APPEAR TO BE RELATED.

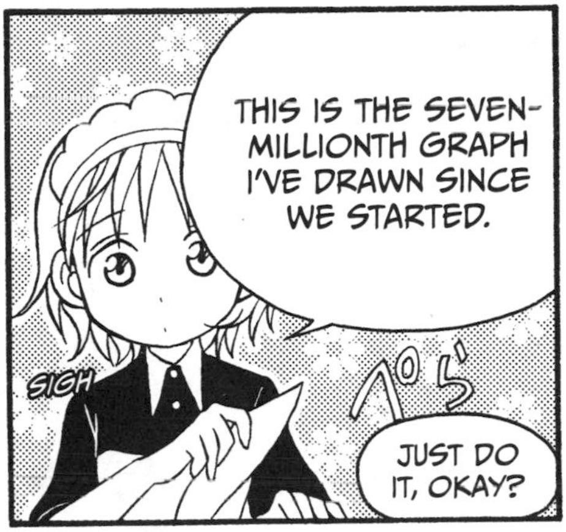
THIS IS THE SEVEN-MILLIONTH GRAPH I'VE DRAWN SINCE WE STARTED.
SIGH
JUST DO IT, OKAY?

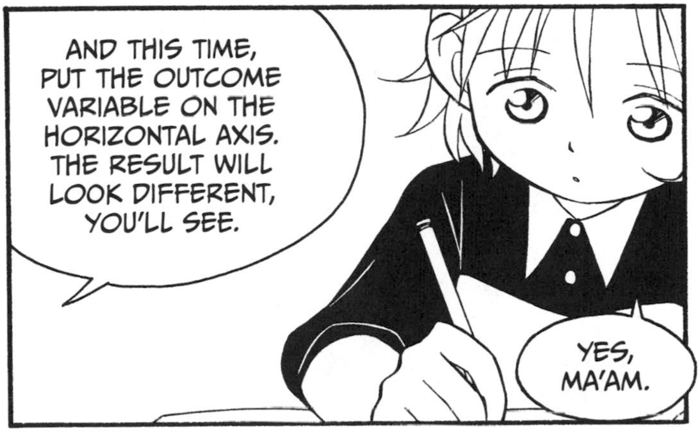
AND THIS TIME, PUT THE OUTCOME VARIABLE ON THE HORIZONTAL AXIS. THE RESULT WILL LOOK DIFFERENT, YOU'LL SEE.
YES, MA'AM.

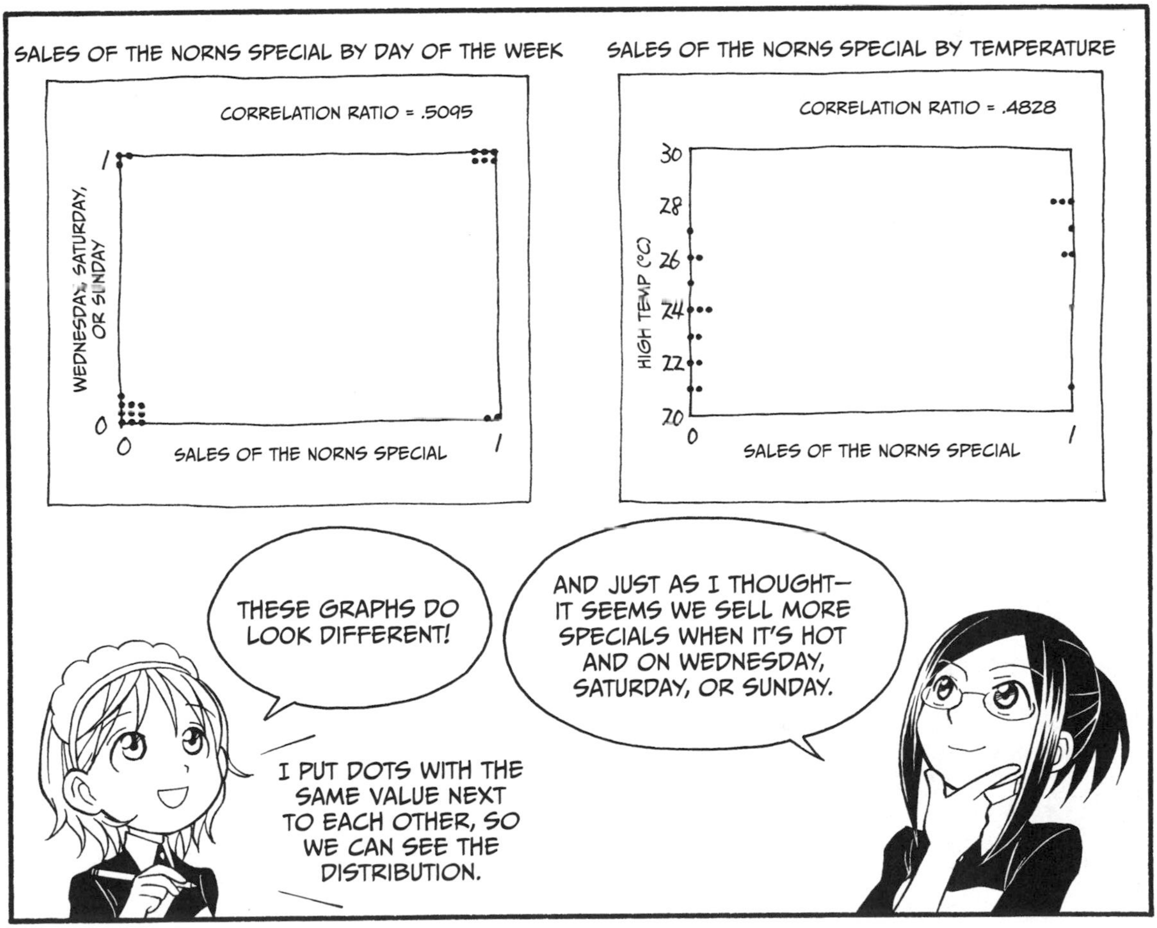
SALES OF THE NORNS SPECIAL BY DAY OF THE WEEK
CORRELATION RATIO = .5095
WEDNESDAY, SATURDAY, OR SUNDAY
1
0
0
SALES OF THE NORNS SPECIAL
1
SALES OF THE NORNS SPECIAL BY TEMPERATURE
CORRELATION RATIO = .4828
HIGH TEMP (°C)
30
28
26
24
22
20
0
SALES OF THE NORNS SPECIAL
1
THESE GRAPHS DO LOOK DIFFERENT!
AND JUST AS I THOUGHT—IT SEEMS WE SELL MORE SPECIALS WHEN IT'S HOT AND ON WEDNESDAY, SATURDAY, OR SUNDAY.
I PUT DOTS WITH THE SAME VALUE NEXT TO EACH OTHER, SO WE CAN SEE THE DISTRIBUTION.

STEP 2: CALCULATE THE LOGISTIC REGRESSION EQUATION.

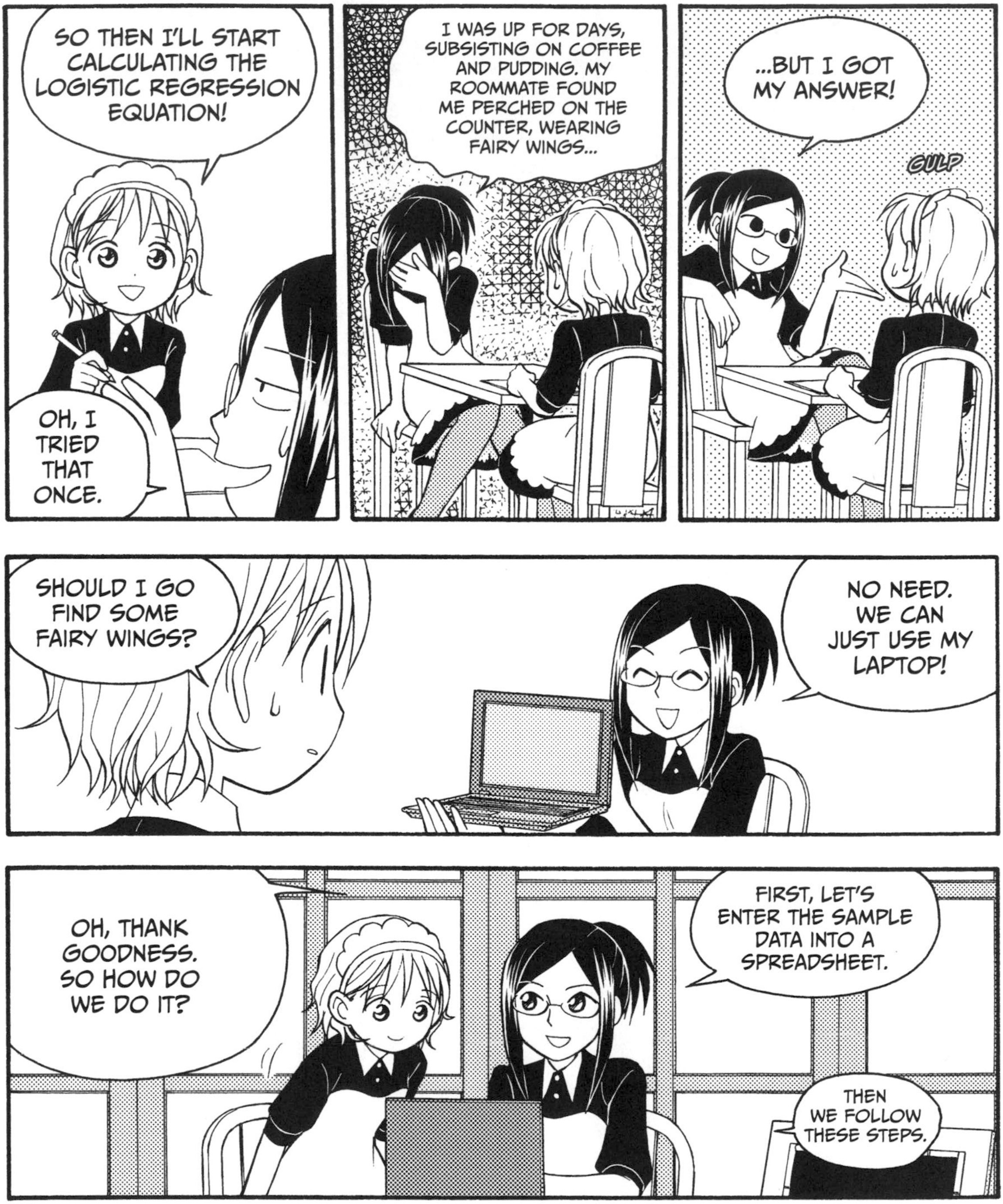
SO THEN I'LL START CALCULATING THE LOGISTIC REGRESSION EQUATION!
OH, I TRIED THAT ONCE.
I WAS UP FOR DAYS, SUBSISTING ON COFFEE AND PUDDING. MY ROOMMATE FOUND ME PERCHED ON THE COUNTER, WEARING FAIRY WINGS...
...BUT I GOT MY ANSWER!
GULP
SHOULD I GO FIND SOME FAIRY WINGS?
NO NEED. WE CAN JUST USE MY LAPTOP!
OH, THANK GOODNESS. SO HOW DO WE DO IT?
FIRST, LET'S ENTER THE SAMPLE DATA INTO A SPREADSHEET.
THEN WE FOLLOW THESE STEPS.

Step 1 Determine the binomial logistic equation for each sample.

Wednesday, Saturday, or Sunday x_1	High temperature x_2	Sales of the Norns special y	Sales of the Norns special $\hat{y} = \frac{1}{1+e^{-(a_1x_1+a_2x_2+b)}}$
0	28	1	$\frac{1}{1+e^{-(a_1\times 0+a_2\times 28+b)}}$
0	24	0	$\frac{1}{1+e^{-(a_1\times 0+a_2\times 24+b)}}$
⋮	⋮	⋮	⋮
1	24	0	$\frac{1}{1+e^{-(a_1\times 1+a_2\times 24+b)}}$

Step 2 Obtain the likelihood function. The equation from Step 1 represents a sold cake, and (1 – the equation) represents an unsold cake.

$$\underbrace{\frac{1}{1+e^{-(a_1\times 0+a_2\times 28+b)}}}_{\text{Sold}} \times \underbrace{\left(1-\frac{1}{1+e^{-(a_1\times 0+a_2\times 24+b)}}\right)}_{\text{Unsold}} \times \cdots \times \underbrace{\frac{1}{1+e^{-(a_1\times 1+a_2\times 24+b)}}}_{\text{Sold}}$$

Step 3 Take the natural log to find the log-likelihood function, L.

$$L = \log_e\left[\frac{1}{1+e^{-(a_1\times 0+a_2\times 28+b)}} \times \left(1-\frac{1}{1+e^{-(a_1\times 0+a_2\times 24+b)}}\right) \times \cdots \times \frac{1}{1+e^{-(a_1\times 1+a_2\times 24+b)}}\right]$$

$$= \log_e\left(\frac{1}{1+e^{-(a_1\times 0+a_2\times 28+b)}}\right) + \log_e\left(1-\frac{1}{1+e^{-(a_1\times 0+a_2\times 24+b)}}\right) + \cdots + \log_e\left(\frac{1}{1+e^{-(a_1\times 1+a_2\times 24+b)}}\right)$$

Step4 Find the maximum likelihood coefficients. These coefficients maximize log-likelihood function L.

The values are:*

$$\begin{cases} a_1 = 2.44 \\ a_2 = 0.54 \\ b = -15.20 \end{cases}$$

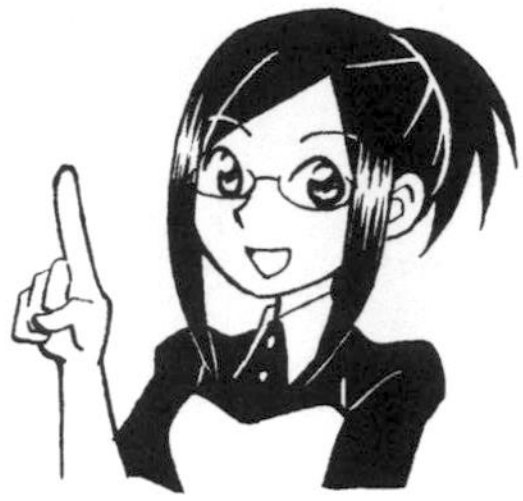

We can plug these values into the likelihood function to calculate L, which we'll use to calculate R^2.

$$L = \log_e\left(\frac{1}{1+e^{-(2.44\times0+0.54\times28-15.20)}}\right) + \log_e\left(1-\frac{1}{1+e^{-(2.44\times0+0.54\times24-15.20)}}\right) + \cdots + \log_e\left(\frac{1}{1+e^{-(2.44\times1+0.54\times24-15.20)}}\right)$$

$$= -8.9$$

*See page 210 for a more detailed explanation of these calculations.

Step5 Calculate the logistic regression equation.

We fill in the coefficients calculated in Step 4 to get the following logistic regression equation:

$$y = \frac{1}{1+e^{-(2.44x_1+0.54x_2-15.20)}}$$

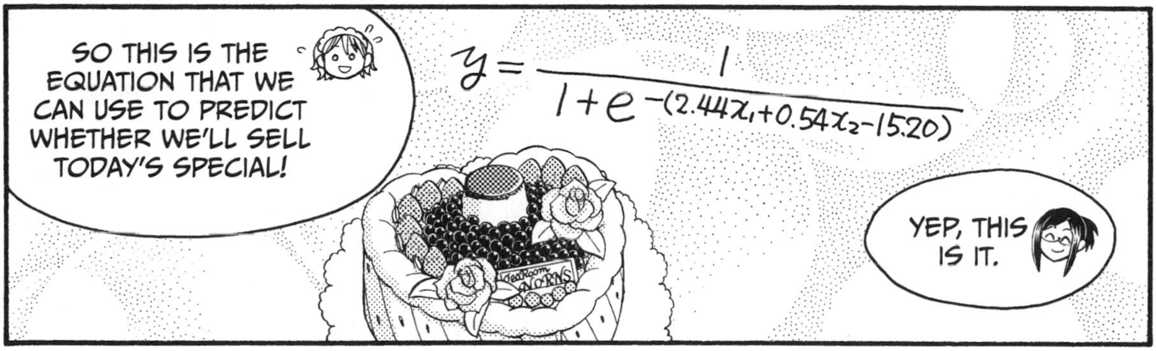
SO THIS IS THE EQUATION THAT WE CAN USE TO PREDICT WHETHER WE'LL SELL TODAY'S SPECIAL!
$y = \frac{1}{1+e^{-(2.44x_1+0.54x_2-15.20)}}$
YEP, THIS IS IT.

STEP 3: ASSESS THE ACCURACY OF THE EQUATION.

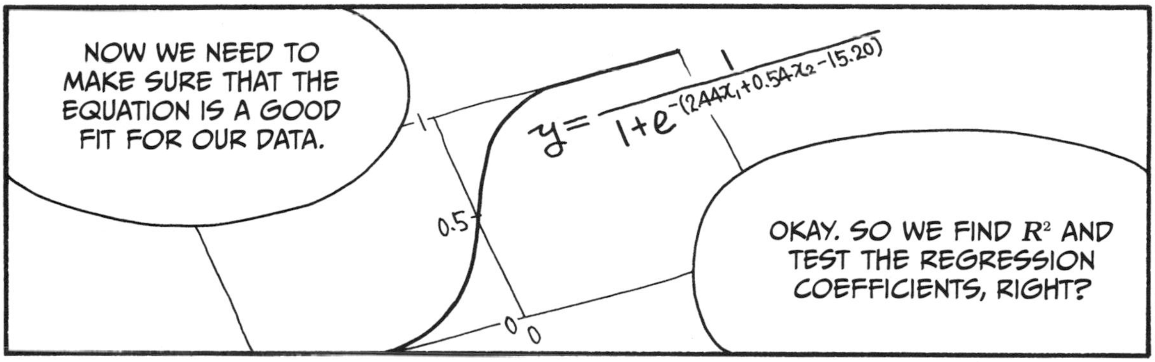
NOW WE NEED TO MAKE SURE THAT THE EQUATION IS A GOOD FIT FOR OUR DATA.
$y = \frac{1}{1+e^{-(2.44x_1+0.54x_2-15.20)}}$
1
0.5
0
OKAY. SO WE FIND R^2 AND TEST THE REGRESSION COEFFICIENTS, RIGHT?

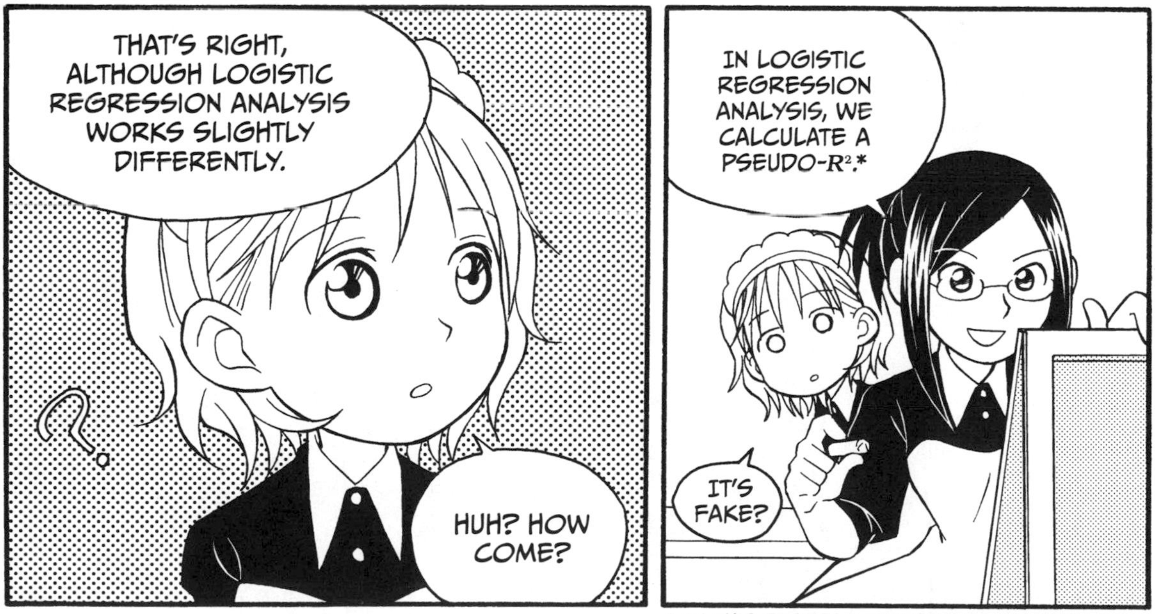
THAT'S RIGHT, ALTHOUGH LOGISTIC REGRESSION ANALYSIS WORKS SLIGHTLY DIFFERENTLY.
HUH? HOW COME?
IN LOGISTIC REGRESSION ANALYSIS, WE CALCULATE A PSEUDO-R^2.*
IT'S FAKE?

* IN THIS EXAMPLE, WE USE MCFADDEN'S PSEUDO-R^2 FORMULA.

HERE'S THE EQUATION THAT WE USE TO CALCULATE R^2 IN LOGISTIC REGRESSION ANALYSIS.

$$R^2 = 1 - \frac{\text{MAXIMUM VALUE OF LOG-LIKELIHOOD FUNCTION } L}{n_1 \log n_1 + n_0 \log n_0 - (n_1 + n_0) \log (n_1 + n_0)}$$

ACK!!! IT'S SO LONG!

THE n VARIABLES ARE A TALLY OF THE CAKES THAT ARE SOLD (n_1) OR UNSOLD (n_0).

n_1	THE NUMBER OF DATA POINTS WHOSE OUTCOME VARIABLE'S VALUE IS 1
n_0	THE NUMBER OF DATA POINTS WHOSE OUTCOME VARIABLE'S VALUE IS 0

AND HERE'S A MORE GENERAL DEFINITION.

I'M STILL NOT SURE HOW TO USE THIS EQUATION WITH THE NORNS SPECIAL DATA.

DON'T WORRY, IT'S NOT THAT HARD.

WE JUST FILL IN THE NUMBERS FOR THE NORNS SPECIAL...

$$R^2 = 1 - \frac{\text{MAXIMUM VALUE OF LOG-LIKELIHOOD FUNCTION } L}{n_1 \log n_1 + n_0 \log n_0 - (n_1 + n_0) \log (n_1 + n_0)}$$

$$= 1 - \frac{-8.9}{8 \log 8 + 13 \log 13 - (8 + 13) \log (8 + 13)}$$

$$= .3622$$

WHOA, I WASN'T EXPECTING THAT.

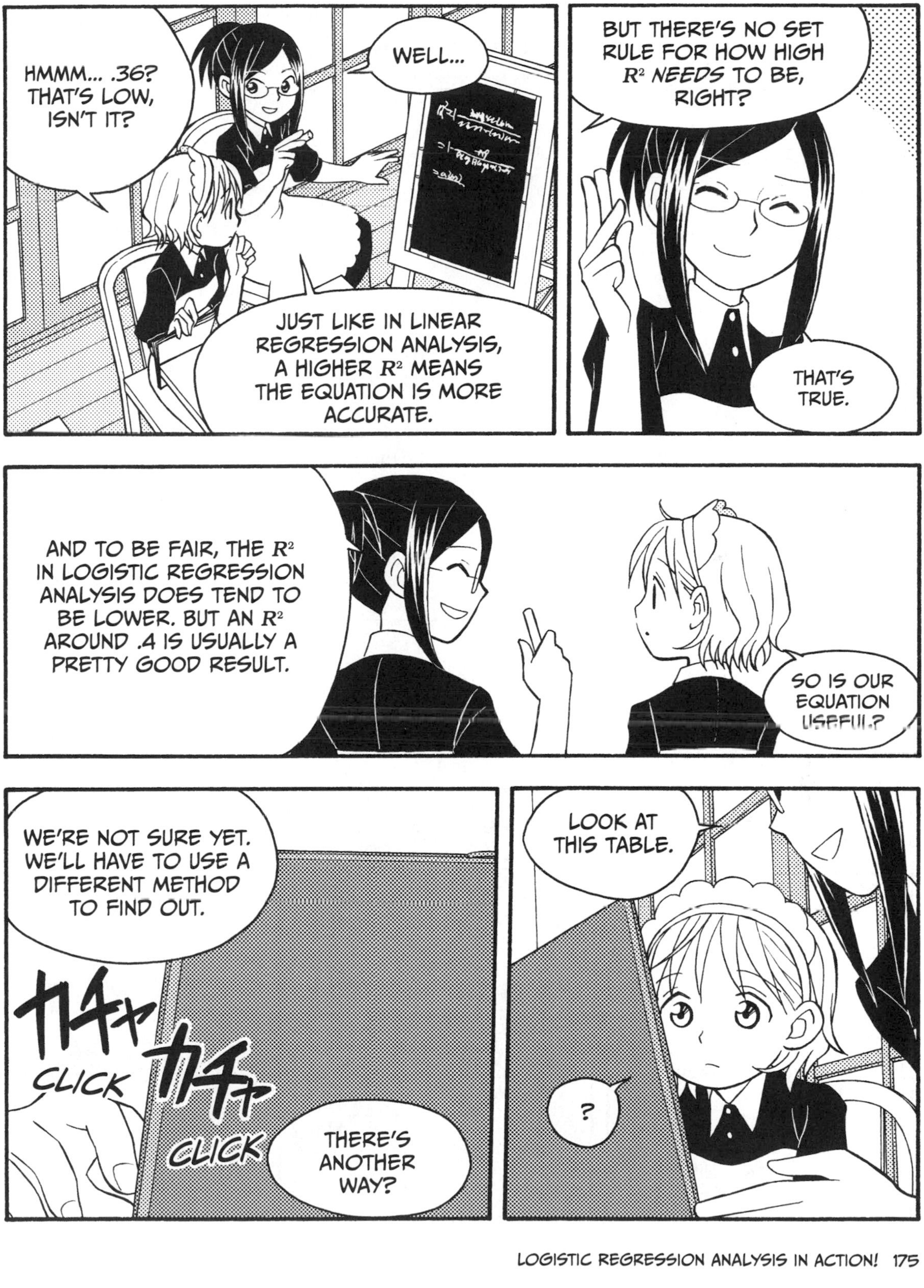
HMMM... .36? THAT'S LOW, ISN'T IT?
WELL...
JUST LIKE IN LINEAR REGRESSION ANALYSIS, A HIGHER R^2 MEANS THE EQUATION IS MORE ACCURATE.
BUT THERE'S NO SET RULE FOR HOW HIGH R^2 *NEEDS* TO BE, RIGHT?
THAT'S TRUE.
AND TO BE FAIR, THE R^2 IN LOGISTIC REGRESSION ANALYSIS DOES TEND TO BE LOWER. BUT AN R^2 AROUND .4 IS USUALLY A PRETTY GOOD RESULT.
SO IS OUR EQUATION USEFUL?
WE'RE NOT SURE YET. WE'LL HAVE TO USE A DIFFERENT METHOD TO FIND OUT.
カチャ
CLICK
カチャ
CLICK
THERE'S ANOTHER WAY?
LOOK AT THIS TABLE.
?

Day	Wednesday, Saturday, or Sunday x_1	High temp. (°C) x_2	Actual sales y	Predicted sales $\hat{y}$
5	0	28	1	.51 sold
6	0	24	0	.11 unsold
7	1	26	0	.80 sold
8	0	24	0	.11 unsold
9	0	23	0	.06 unsold
10	1	28	1	.92 sold
11	1	24	0	.58 sold
12	0	26	1	.26 unsold
13	0	25	0	.17 unsold
14	1	28	1	.92 sold
15	0	21	0	.02 unsold
16	0	22	0	.04 unsold
17	1	27	1	.87 sold
18	1	26	1	.80 sold
19	0	26	0	.26 unsold
20	0	21	0	.02 unsold
21	1	21	1	.21 unsold
22	0	27	0	.38 unsold
23	0	23	0	.06 unsold
24	1	22	0	.31 unsold
25	1	24	1	.58 sold

$$\frac{1}{1+e^{-(2.44\times1+0.54\times24-15.20)}} = .58$$

THIS TABLE SHOWS THE ACTUAL SALES DATA FOR THE NORNS SPECIAL AND OUR PREDICTION. IF THE PREDICTION IS GREATER THAN .50, WE SAY IT SOLD.

BUT THE TABLE SHOWS SOMETHING ELSE. CAN YOU SEE IT?

HMM...

WELL...

FOR ONE THING, THE NORNS SPECIAL DID NOT SELL ON THE 7TH AND THE 11TH, EVEN THOUGH WE PREDICTED THAT IT WOULD.

Day	y	$\hat{y}$
7	0	.80 sold
11	0	.58 sold

GREAT! ANYTHING ELSE?

ON THE 12TH AND THE 21ST, WE PREDICTED THAT IT WOULDN'T SELL, BUT IT DID! WE CAN SEE WHERE THE EQUATION WAS WRONG.

BRILLIANT!

BEST STUDENT EVER!

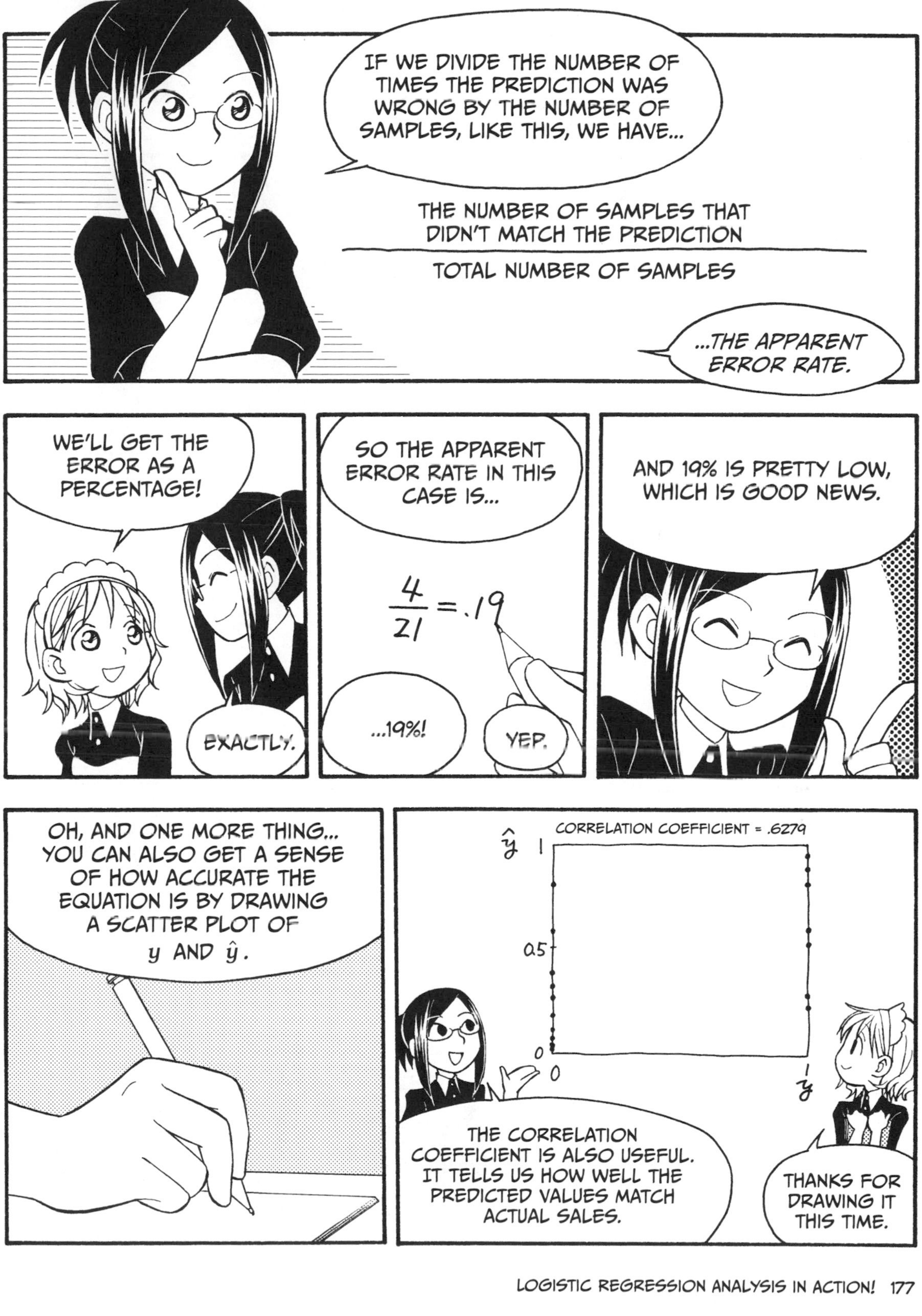
IF WE DIVIDE THE NUMBER OF TIMES THE PREDICTION WAS WRONG BY THE NUMBER OF SAMPLES, LIKE THIS, WE HAVE...
THE NUMBER OF SAMPLES THAT DIDN'T MATCH THE PREDICTION / TOTAL NUMBER OF SAMPLES
...THE APPARENT ERROR RATE.
WE'LL GET THE ERROR AS A PERCENTAGE!
EXACTLY.
SO THE APPARENT ERROR RATE IN THIS CASE IS...
4/21 = .19
...19%!
YEP.
AND 19% IS PRETTY LOW, WHICH IS GOOD NEWS.
OH, AND ONE MORE THING... YOU CAN ALSO GET A SENSE OF HOW ACCURATE THE EQUATION IS BY DRAWING A SCATTER PLOT OF y AND ŷ.
CORRELATION COEFFICIENT = .6279
ŷ
1
0.5
0
0
1
y
THE CORRELATION COEFFICIENT IS ALSO USEFUL. IT TELLS US HOW WELL THE PREDICTED VALUES MATCH ACTUAL SALES.
THANKS FOR DRAWING IT THIS TIME.

STEP 4: CONDUCT THE HYPOTHESIS TESTS.

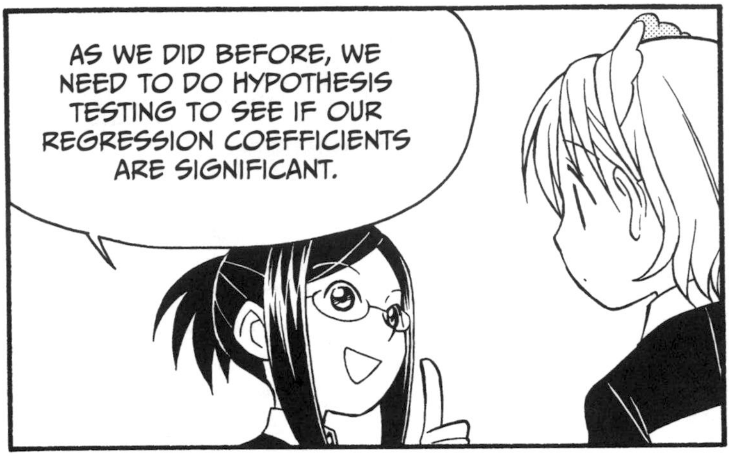
AS WE DID BEFORE, WE NEED TO DO HYPOTHESIS TESTING TO SEE IF OUR REGRESSION COEFFICIENTS ARE SIGNIFICANT.

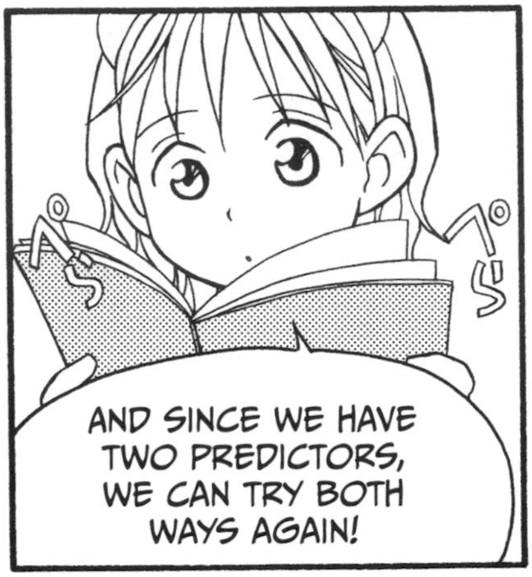
AND SINCE WE HAVE TWO PREDICTORS, WE CAN TRY BOTH WAYS AGAIN!

COMPREHENSIVE HYPOTHESIS TEST

NULL HYPOTHESIS	$A_1 = A_2 = 0$
ALTERNATIVE HYPOTHESIS	ONE OF THE FOLLOWING IS TRUE: • $A_1 \neq 0$ AND $A_2 \neq 0$ • $A_1 \neq 0$ AND $A_2 = 0$ • $A_1 = 0$ AND $A_2 \neq 0$

HYPOTHESIS TEST FOR AN INDIVIDUAL REGRESSION COEFFICIENT

NULL HYPOTHESIS	$A_i = 0$
ALTERNATIVE HYPOTHESIS	$A_i \neq 0$

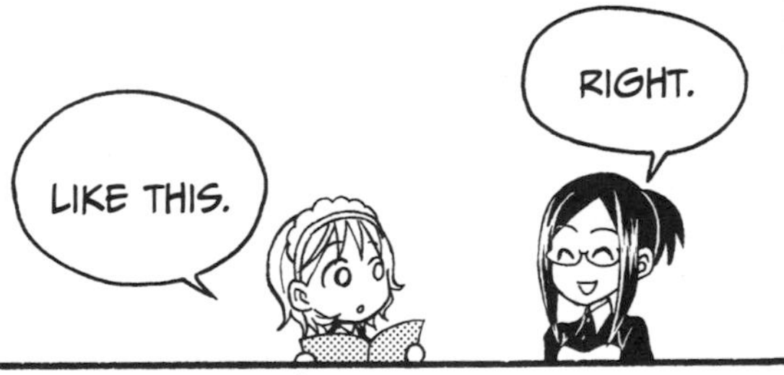
LIKE THIS.
RIGHT.

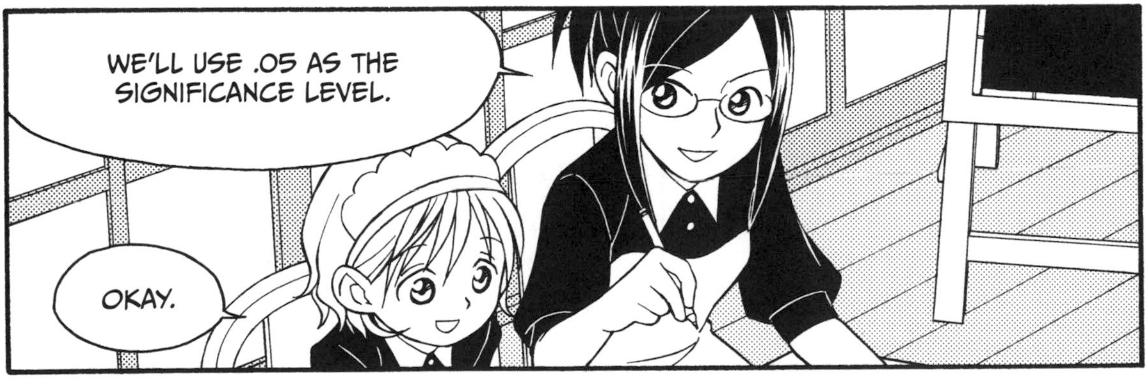
WE'LL USE .05 AS THE SIGNIFICANCE LEVEL.
OKAY.

WE'LL DO THE LIKELIHOOD RATIO TEST. THIS TEST LETS US EXAMINE ALL THE COEFFICIENTS AT ONCE AND ASSESS THE RELATIONSHIPS AMONG THE COEFFICIENTS.

THE STEPS OF THE LIKELIHOOD RATIO TEST

Step 1	Define the populations.	All days the Norns Special is sold, comparing Wednesdays, Saturdays, and Sundays against the remaining days, at each high temperature.
Step 2	Set up a null hypothesis and an alternative hypothesis.	Null hypothesis is $A_1 = 0$ and $A_2 = 0$. Alternative hypothesis is $A_1 \neq 0$ or $A_2 \neq 0$.
Step 3	Select which hypothesis test to conduct.	We'll perform the likelihood ratio test.
Step 4	Choose the significance level.	We'll use a significance level of .05.
Step 5	Calculate the test statistic from the sample data.	The test statistic is: $2[L - n_1\log_e(n_1) - n_0\log_e(n_0) + (n_1 + n_0)\log_e(n_1 + n_0)]$ When we plug in our data, we get: $2[-8.9010 - 8\log_e 8 - 13\log_e 13 + (8 + 13)\log_e(8 + 13)]$ $= 10.1$ The test statistic follows a chi-squared distribution with 2 degrees of freedom (the number of predictor variables), if the null hypothesis is true.
Step 6	Determine whether the p-value for the test statistic obtained in Step 5 is smaller than the significance level.	The significance level is .05. The value of the test statistic is 10.1, so the p-value is .006. Finally, .006 < .05.*
Step 7	Decide whether you can reject the null hypothesis.	Since the p-value is smaller than the significance level, we reject the null hypothesis.

* How to obtain the p-value in a chi-squared distribution is explained on page 205.

NEXT, WE'LL USE THE WALD TEST TO SEE WHETHER EACH OF OUR PREDICTOR VARIABLES HAS A SIGNIFICANT EFFECT ON THE SALE OF THE NORNS SPECIAL. WE'LL DEMONSTRATE USING DAYS OF THE WEEK.

THE STEPS OF THE WALD TEST

Step 1	Define the population.	All days the Norns Special is sold, comparing Wednesdays, Saturdays, and Sundays against the remaining days, at each high temperature.
Step 2	Set up a null hypothesis and an alternative hypothesis.	Null hypothesis is $A = 0$. Alternative hypothesis is $A \neq 0$.
Step 3	Select which hypothesis test to conduct.	Perform the Wald test.
Step 4	Choose the significance level.	We'll use a significance level of .05.
Step 5	Calculate the test statistic from the sample data.	The test statistic for the Wald test is $$\frac{a_1^2}{S_{11}}$$ In this example, the value of the test statistic is: $$\frac{2.44^2}{1.5475} = 3.9$$ The test statistic will follow a chi-squared distribution with 1 degree of freedom, if the null hypothesis is true.
Step 6	Determine whether the p-value for the test statistic obtained in Step 5 is smaller than the significance level.	The value of the test statistic is 3.9, so the p-value is .0489. You can see that .0489 < .05, so the p-value is smaller.
Step 7	Decide whether you can reject the null hypothesis.	Since the p-value is smaller than the significance level, we reject the null hypothesis.

IN SOME REFERENCES, THIS PROCESS IS EXPLAINED USING NORMAL DISTRIBUTION INSTEAD OF CHI-SQUARED DISTRIBUTION. THE FINAL RESULT WILL BE THE SAME NO MATTER WHICH METHOD YOU CHOOSE.

This is how we calculate the standard error matrix. The values of this matrix are used to calculate the Wald test statistic in Step 5 on page 180.

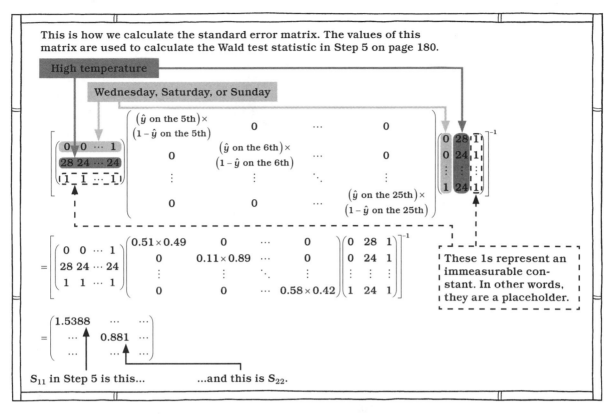

High temperature

Wednesday, Saturday, or Sunday

$$\begin{bmatrix} 0 & 0 & \cdots & 1 \\ 28 & 24 & \cdots & 24 \\ 1 & 1 & \cdots & 1 \end{bmatrix} \begin{pmatrix} \begin{pmatrix}\hat{y}\ \text{on the 5th}\end{pmatrix}\times \\ \begin{pmatrix}1-\hat{y}\ \text{on the 5th}\end{pmatrix} & 0 & \cdots & 0 \\ 0 & \begin{pmatrix}\hat{y}\ \text{on the 6th}\end{pmatrix}\times \\ \begin{pmatrix}1-\hat{y}\ \text{on the 6th}\end{pmatrix} & \cdots & 0 \\ \vdots & \vdots & \ddots & \vdots \\ 0 & 0 & \cdots & \begin{pmatrix}\hat{y}\ \text{on the 25th}\end{pmatrix}\times \\ \begin{pmatrix}1-\hat{y}\ \text{on the 25th}\end{pmatrix} \end{pmatrix} \begin{bmatrix} 0 & 28 & 1 \\ 0 & 24 & 1 \\ \vdots & \vdots & \vdots \\ 1 & 24 & 1 \end{bmatrix}^{-1}$$

$$= \begin{bmatrix} 0 & 0 & \cdots & 1 \\ 28 & 24 & \cdots & 24 \\ 1 & 1 & \cdots & 1 \end{bmatrix} \begin{pmatrix} 0.51\times0.49 & 0 & \cdots & 0 \\ 0 & 0.11\times0.89 & \cdots & 0 \\ \vdots & \vdots & \ddots & \vdots \\ 0 & 0 & \cdots & 0.58\times0.42 \end{pmatrix} \begin{pmatrix} 0 & 28 & 1 \\ 0 & 24 & 1 \\ \vdots & \vdots & \vdots \\ 1 & 24 & 1 \end{pmatrix}^{-1}$$

These 1s represent an immeasurable constant. In other words, they are a placeholder.

$$= \begin{pmatrix} 1.5388 & \cdots & \cdots \\ \cdots & 0.881 & \cdots \\ \cdots & \cdots & \cdots \end{pmatrix}$$

S_{11} in Step 5 is this...   ...and this is S_{22}.

SO A ≠ 0. WE CAN REJECT THE NULL!

YES, WE CAN.

AND NOW...

RATTLE

...THE MOST IMPORTANT PART.

STEP 5: PREDICT WHETHER THE NORNS SPECIAL WILL SELL.

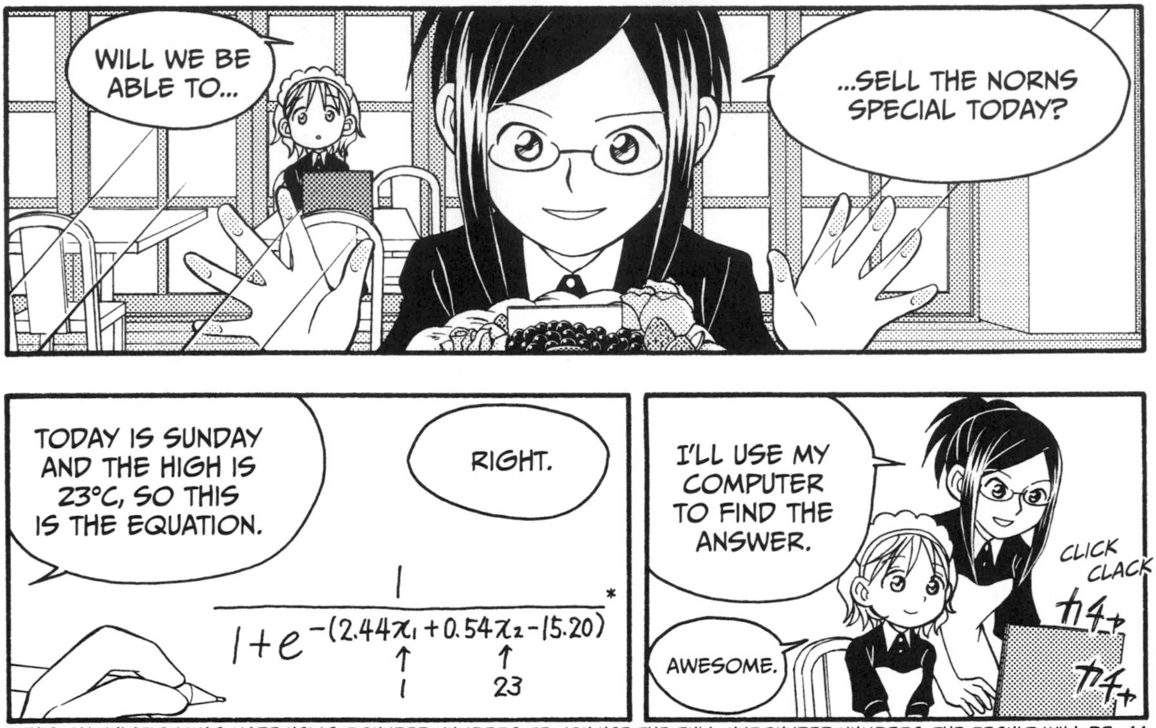

WILL WE BE ABLE TO...
...SELL THE NORNS SPECIAL TODAY?
TODAY IS SUNDAY AND THE HIGH IS 23°C, SO THIS IS THE EQUATION.
RIGHT.
$\frac{1}{1+e^{-(2.44x_1+0.54x_2-15.20)}}$*
1
23
I'LL USE MY COMPUTER TO FIND THE ANSWER.
AWESOME.
CLICK CLACK

* THIS CALCULATION WAS MADE USING ROUNDED NUMBERS. IF YOU USE THE FULL, UNROUNDED NUMBERS, THE RESULT WILL BE .44.

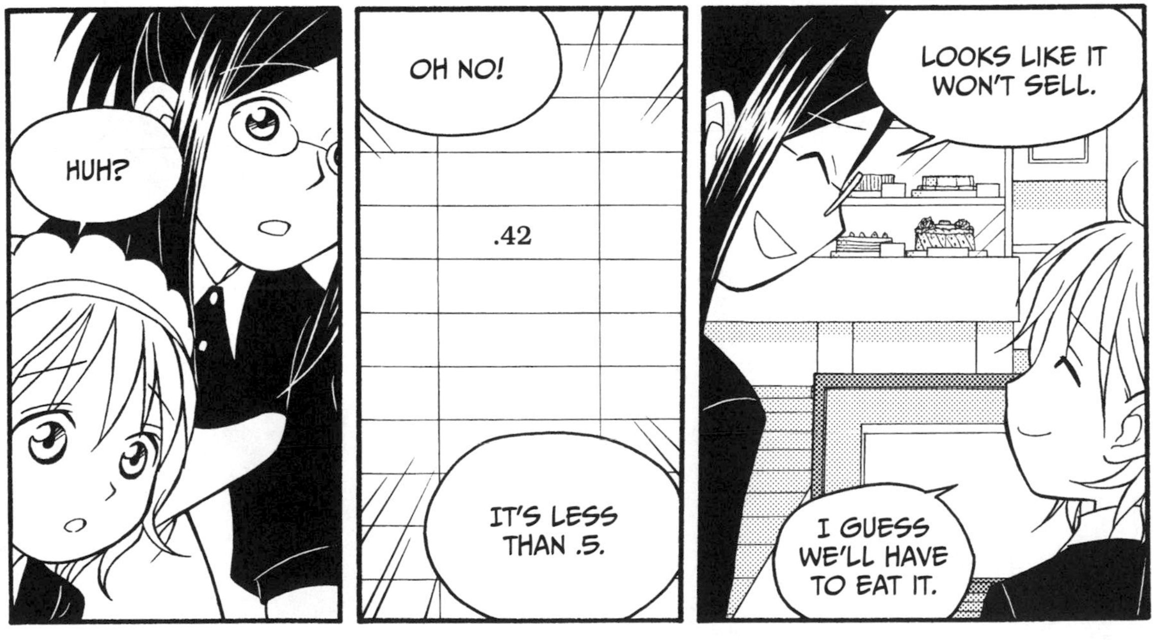

HUH?
OH NO!
.42
IT'S LESS THAN .5.
LOOKS LIKE IT WON'T SELL.
I GUESS WE'LL HAVE TO EAT IT.

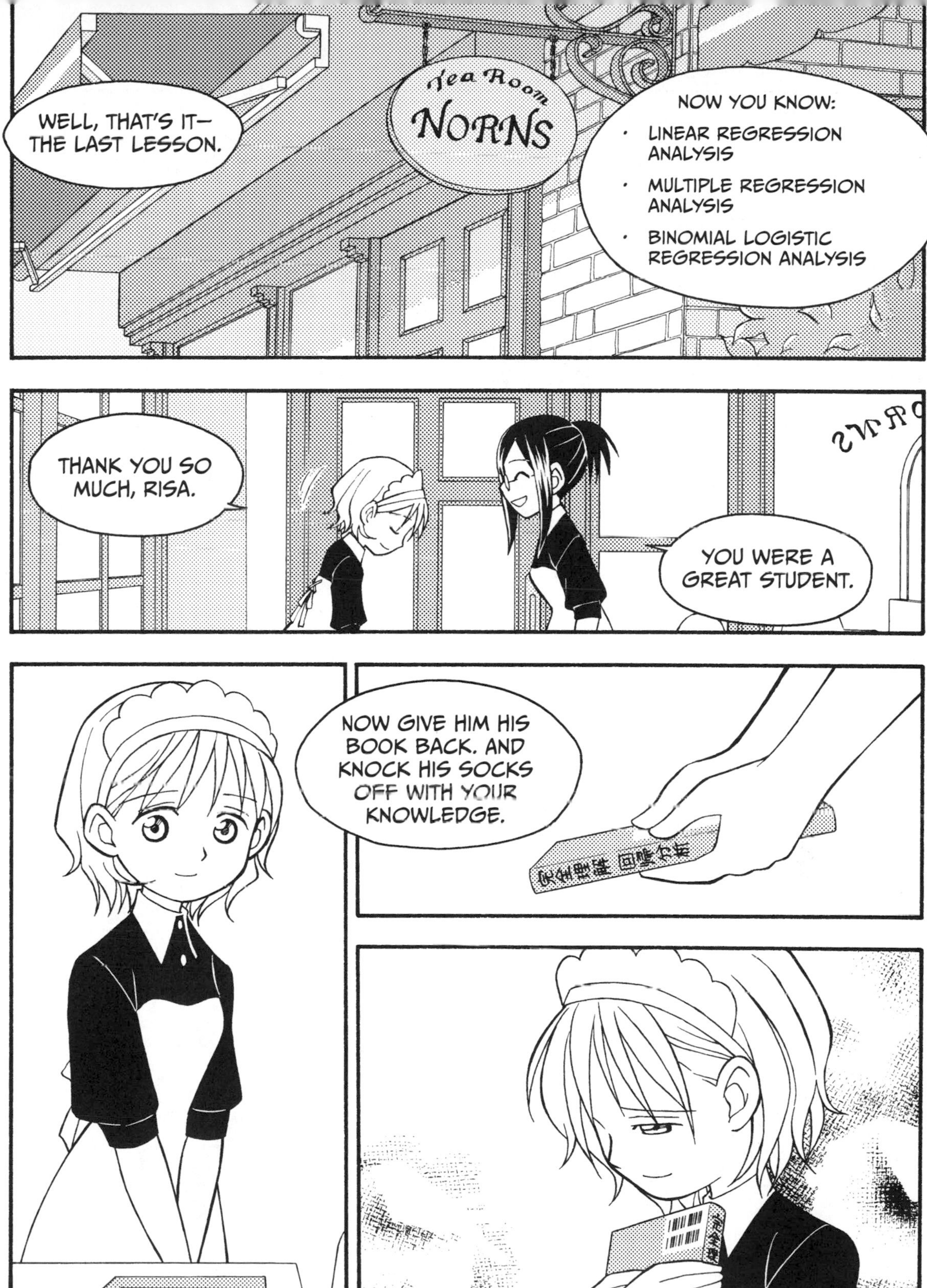
WELL, THAT'S IT—THE LAST LESSON.
Tea Room NORNS
NOW YOU KNOW:
· LINEAR REGRESSION ANALYSIS
· MULTIPLE REGRESSION ANALYSIS
· BINOMIAL LOGISTIC REGRESSION ANALYSIS
THANK YOU SO MUCH, RISA.
YOU WERE A GREAT STUDENT.
NOW GIVE HIM HIS BOOK BACK. AND KNOCK HIS SOCKS OFF WITH YOUR KNOWLEDGE.

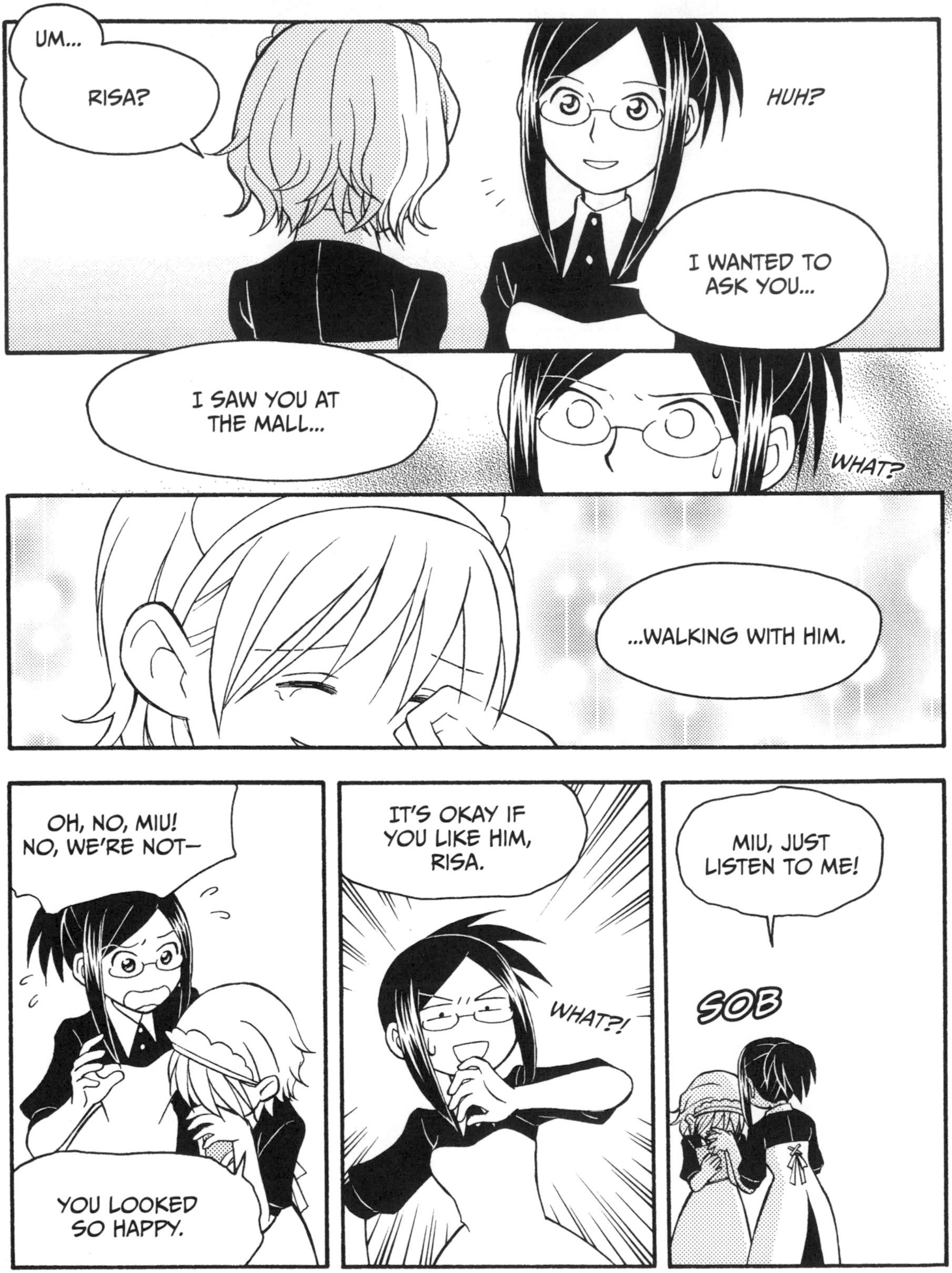
UM...
RISA?
HUH?
I WANTED TO ASK YOU...
I SAW YOU AT THE MALL...
WHAT?
...WALKING WITH HIM.
OH, NO, MIU! NO, WE'RE NOT—
YOU LOOKED SO HAPPY.
IT'S OKAY IF YOU LIKE HIM, RISA.
WHAT?!
MIU, JUST LISTEN TO ME!
SOB

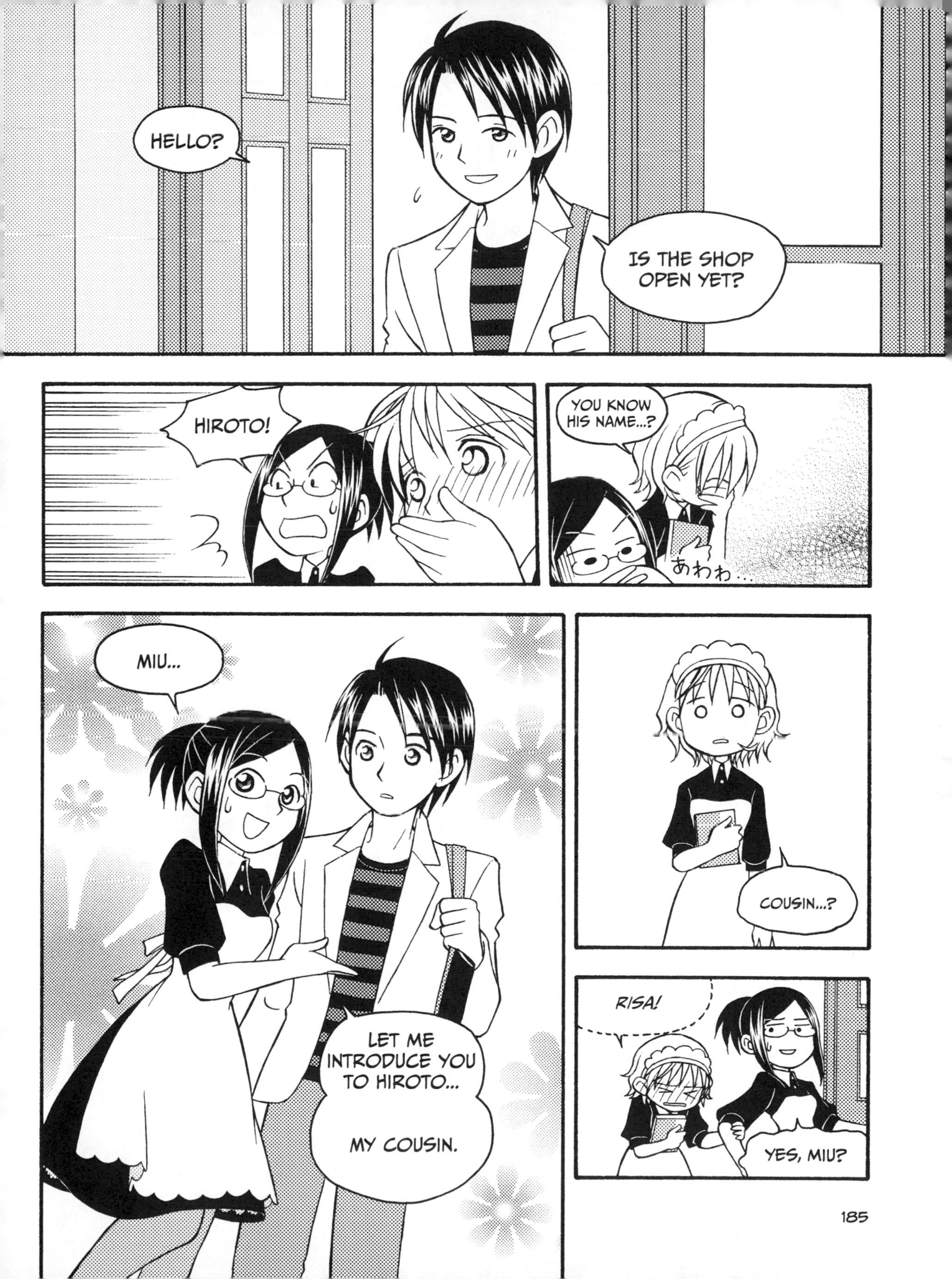
HELLO?
IS THE SHOP OPEN YET?
HIROTO!
YOU KNOW HIS NAME...?
あわわ...
MIU...
LET ME INTRODUCE YOU TO HIROTO...
MY COUSIN.
COUSIN...?
RISA!
YES, MIU?

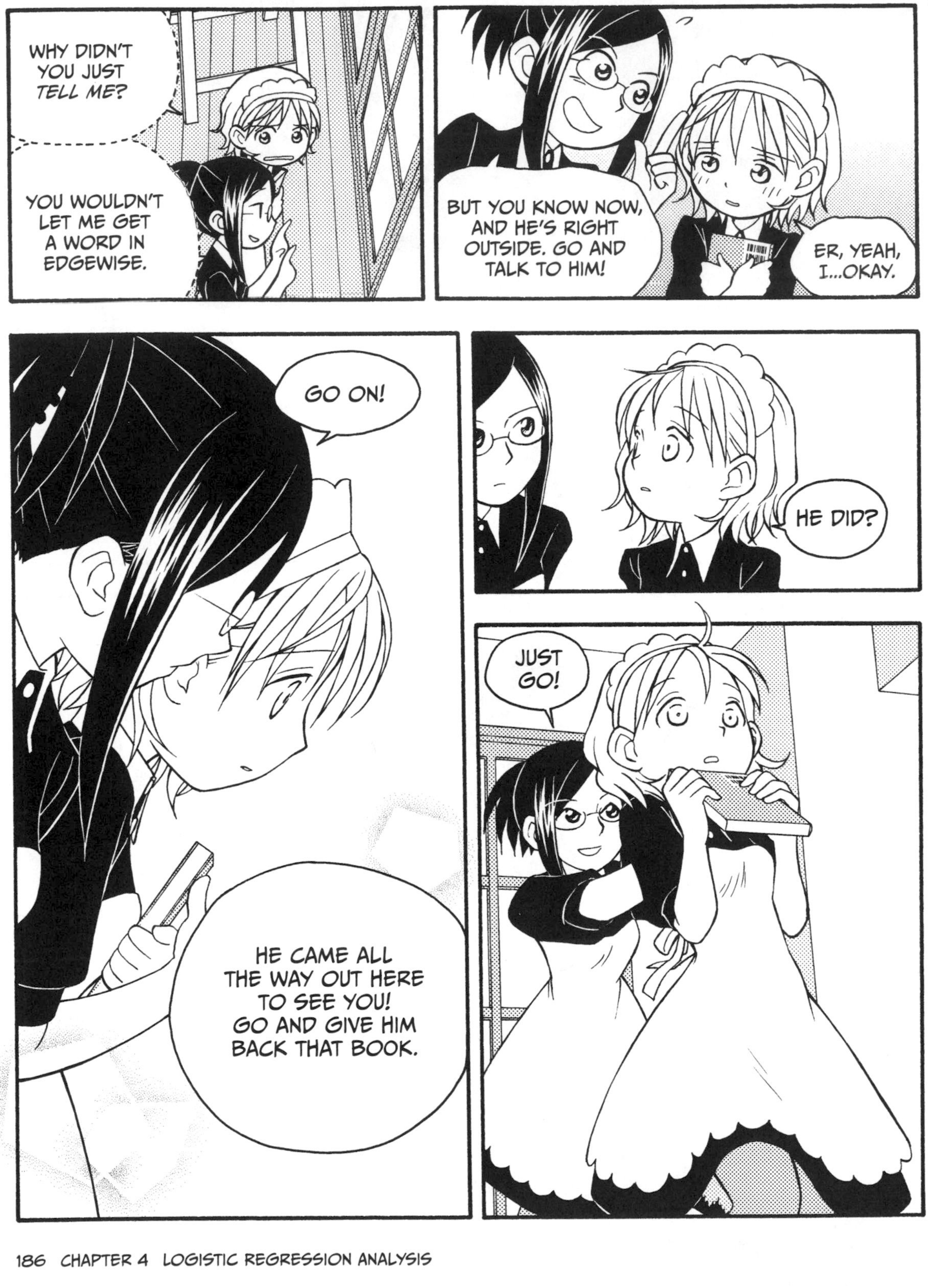
WHY DIDN'T YOU JUST *TELL ME?*
YOU WOULDN'T LET ME GET A WORD IN EDGEWISE.
BUT YOU KNOW NOW, AND HE'S RIGHT OUTSIDE. GO AND TALK TO HIM!
ER, YEAH, I...OKAY.
GO ON!
HE CAME ALL THE WAY OUT HERE TO SEE YOU! GO AND GIVE HIM BACK THAT BOOK.
HE DID?
JUST GO!

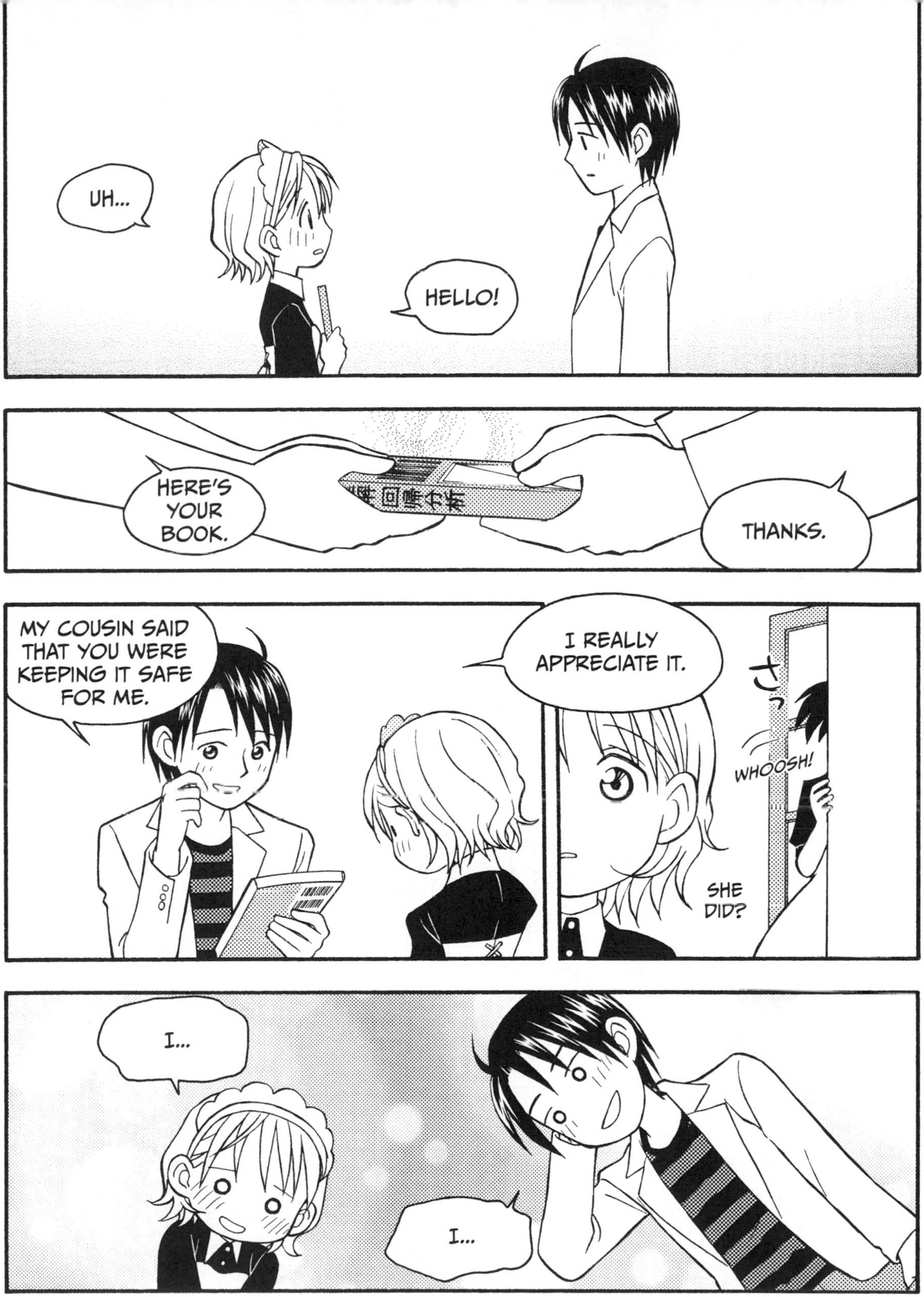
UH...
HELLO!
HERE'S YOUR BOOK.
THANKS.
MY COUSIN SAID THAT YOU WERE KEEPING IT SAFE FOR ME.
I REALLY APPRECIATE IT.
さっ
WHOOSH!
SHE DID?
I...
I...

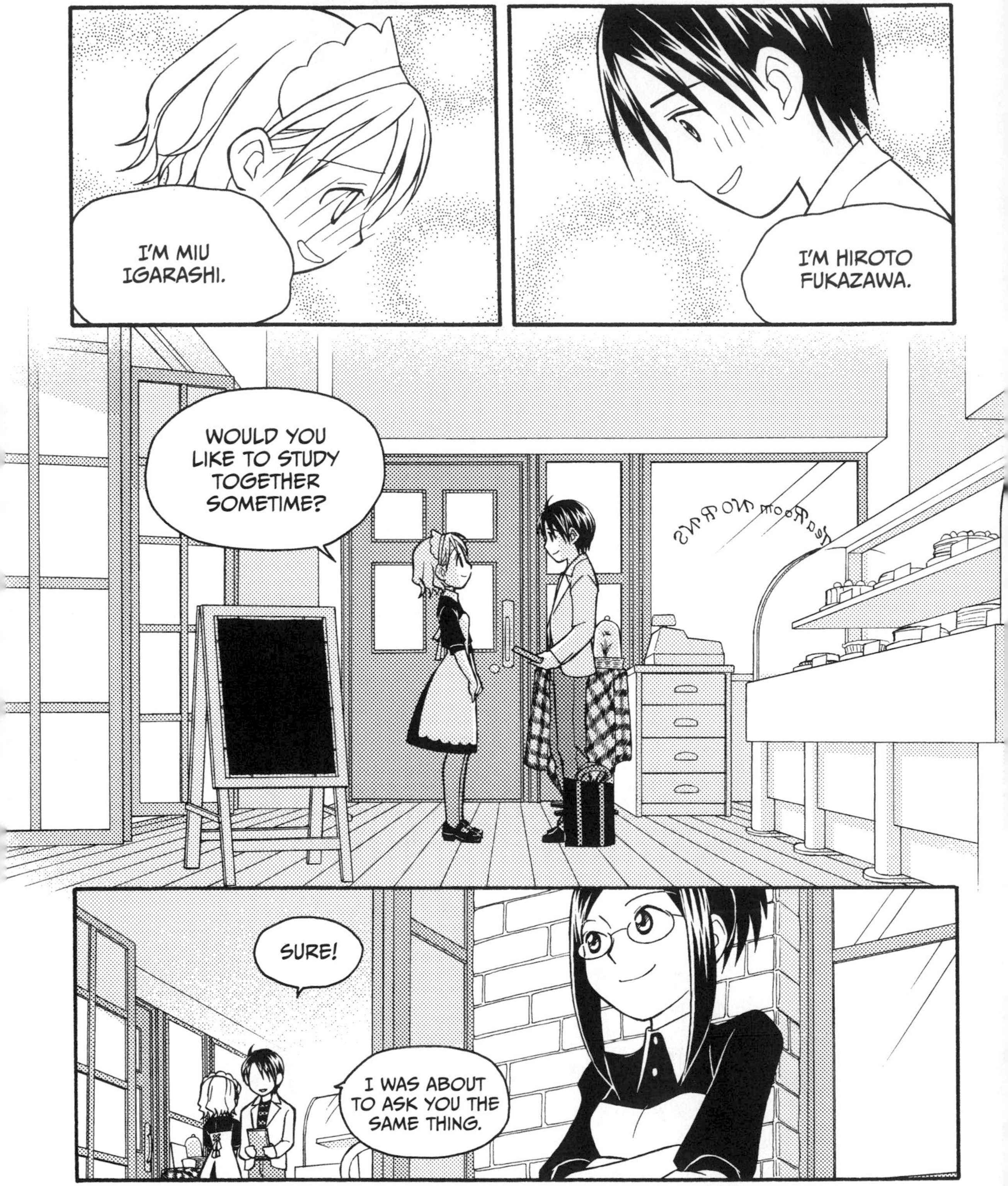
I'M MIU IGARASHI.
I'M HIROTO FUKAZAWA.
WOULD YOU LIKE TO STUDY TOGETHER SOMETIME?
SURE!
I WAS ABOUT TO ASK YOU THE SAME THING.

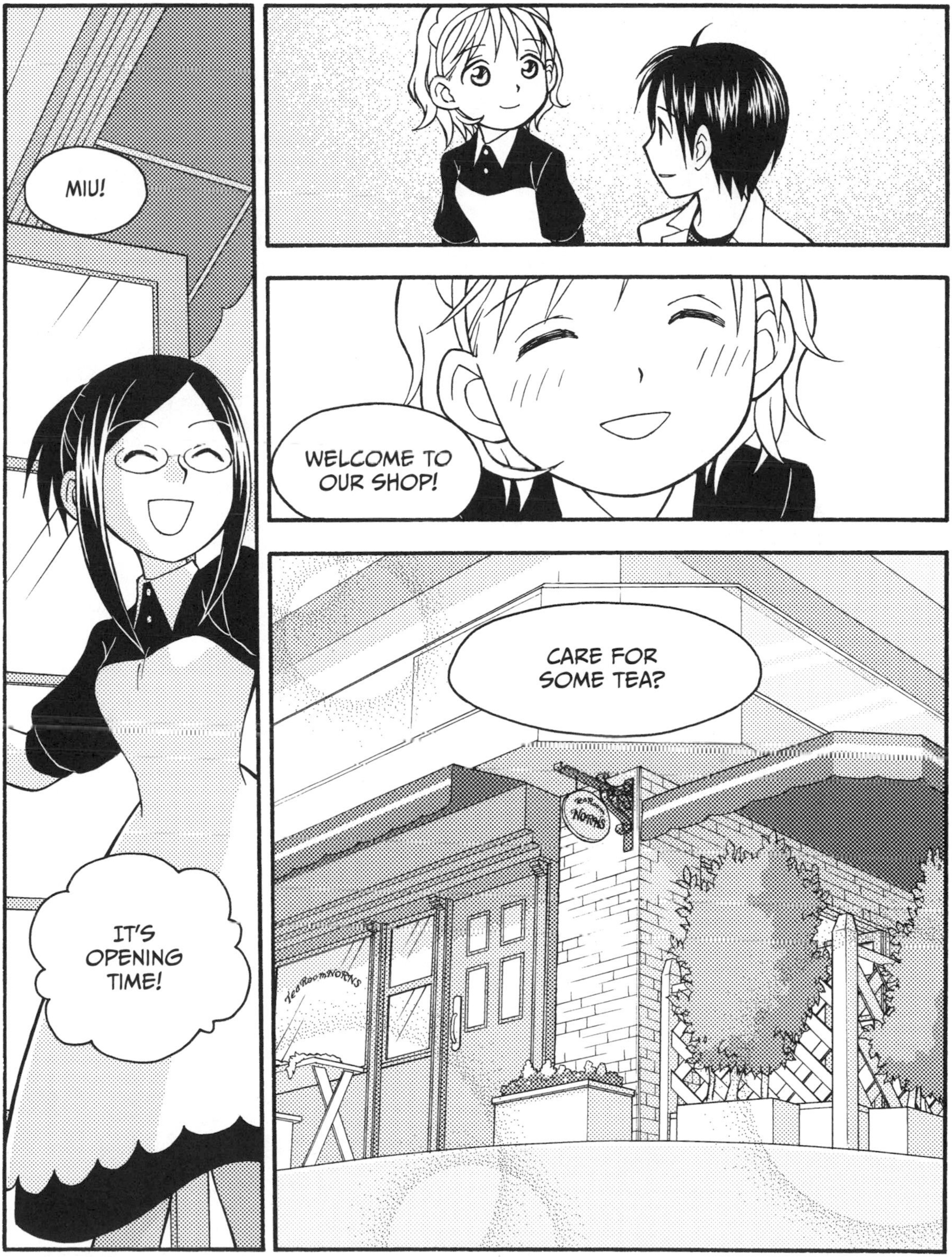
MIU!
IT'S OPENING TIME!
WELCOME TO OUR SHOP!
CARE FOR SOME TEA?

LOGISTIC REGRESSION ANALYSIS IN THE REAL WORLD

On page 68, Risa made a list of the all the steps of regression analysis, but later it was noted that it's not always necessary to perform each of the steps. For example, if we're analyzing Miu's height over time, there's just one Miu, and she was just one height at a given age. There's no population of Miu heights at age 6, so analyzing the "population" wouldn't make sense.

In the real world too, it's not uncommon to skip Step 1, drawing the scatter plots—especially when there are thousands of data points to consider. For example, in a clinical trial with many participants, researchers may choose to start at Step 2 to save time, especially if they have software that can do the calculations quickly for them.

Furthermore, when you do statistics in the real world, don't just dive in and apply tests. Think about your data and the purpose of the test. Without context, the numbers are just numbers and signify nothing.

LOGIT, ODDS RATIO, AND RELATIVE RISK

Odds are a measure that suggests how closely a predictor and an outcome are associated. They are defined as the ratio of the probability of an outcome happening in a given situation (y) to the probability of the outcome not happening ($1 - y$):

$$\frac{y}{1-y}$$

LOGIT

The *logit* is the log of the odds. The logistic function is its inverse, taking a log-odds and turning it into a probability. The logit is mathematically related to the regression coefficients: for every unit of increase in the predictor, the logit of the outcome increases by the value of the regression coefficient.

The equation for the logistic function, which you saw earlier when we calculated that logistic regression equation on page 170, is as follows:

$$y = \frac{1}{1+e^{-z}}$$

where z is the logit and y is the probability.

To find the logit, we invert the logistic equation like this:

$$\log \frac{y}{1-y} = z.$$

This inverse function gives the logit based on the original logistic regression equation. The process of finding the logit is like finding any other mathematical inverse:

$$y = \frac{1}{1+e^{-z}} = \frac{1}{1+e^{-z}} \times \frac{e^z}{e^z} = \frac{e^z}{e^z+1}$$

$$y \times (e^z + 1) = \frac{e^z}{e^z+1} \times (e^z + 1) \quad \boxed{\text{MULTIPLY BOTH SIDE OF THE EQUATION BY } (e^z + 1).}$$

$$y \times e^z + y = e^z$$

$$y = e^z - y \times e^z \quad \boxed{\text{TRANSPOSE TERMS.}}$$

$$y = (1-y)e^z$$

$$y \times \frac{1}{1 \times y} = (1-y)e^z \times \frac{1}{1-y} \quad \boxed{\text{MULTIPLY BOTH SIDE OF THE EQUATION BY } \frac{1}{1-y}.}$$

$$\frac{y}{1-y} = e^z$$

$$\log \frac{y}{1-y} = \log e^z = z$$

Therefore, the logistic regression equation for selling the Norns Special (obtained on page 172),

$$y = \frac{1}{1+e^{-(2.44x_1 + 0.54x_2 - 15.20)}},$$

can be rewritten as

$$\log \frac{y}{1-y} = 2.44x_1 + 0.54x_2 - 15.20.$$

So the odds of selling the Norns Special on a given day, at a given temperature are $e^{2.44x_1 + 0.54x_2 - 15.20}$, and the logit is $2.44x_1 + 0.54x_2 - 15.20$.

ODDS RATIO

Another way to quantify the association between a predictor and an outcome is the *odds ratio (OR)*. The odds ratio compares two sets of odds for different conditions of the same variable.

Let's calculate the odds ratio for selling the Norns Special on Wednesday, Saturday, or Sunday versus other days of the week:

$$\frac{\left(\dfrac{\text{sales rate of Wed, Sat, or Sun}}{1 - \text{sales rate of Wed, Sat, or Sun}}\right)}{\left(\dfrac{\text{sales rate of days other than Wed, Sat, or Sun}}{1 - \text{sales rate of days other than Wed, Sat, or Sun}}\right)} = \frac{\left[\dfrac{(6/9)}{1 - (6/9)}\right]}{\left[\dfrac{(2/12)}{1 - (2/12)}\right]} = \frac{\left[\dfrac{(6/9)}{(3/9)}\right]}{\left[\dfrac{(2/12)}{(10/12)}\right]} =$$

$$\frac{(6/3)}{(2/10)} = \frac{6}{3} \div \frac{2}{10} = \frac{6}{3} \times \frac{10}{2} = 2 \times 5 = 10$$

This shows that the odds of selling the Norns special on one of those three days are 10 times higher than on the other days of the week.

ADJUSTED ODDS RATIO

So far, we've used only the odds based on the day of the week. If we want to find the truest representation of the odds ratio, we would need to calculate the odds ratio of each variable in turn and then combine the ratios. This is called the *adjusted odds ratio*. For the data collected by Risa on page 176, this means finding the odds ratio for two variables—day of the week and temperature—at the same time.

Table 4-1 shows the logistic regression equations and odds when considering each variable separately and when considering them together, which we'll need to calculate the adjusted odds ratios.

TABLE 4-1: THE LOGISTIC REGRESSION EQUATIONS AND ODDS FOR THE DATA ON PAGE 176

Predictor variable	Logistic regression equation	Odds
"Wed, Sat, or Sun" only	$y = \dfrac{1}{1 + e^{-(2.30x_1 - 1.61)}}$	$e^{(2.30x_1 - 1.61)}$
"High temperature" only	$y = \dfrac{1}{1 + e^{-(0.52x_2 - 13.44)}}$	$e^{(0.52x_2 - 13.44)}$
"Wed, Sat, or Sun" and "High temperature"	$y = \dfrac{1}{1 + e^{-(2.44x_1 + 0.54x_2 - 15.20)}}$	$e^{(2.44x_1 + 0.54x_2 - 15.20)}$

The odds of a sale based only on the day of the week are calculated as follows:

$$\frac{\text{odds of a sale on Wed, Sat, or Sun}}{\text{odds of a sale on days other than Wed, Sat, or Sun}} = \frac{e^{2.30 \times 1 - 1.61}}{e^{2.30 \times 0 - 1.61}} =$$

$$e^{2.30 \times 1 - 1.61 - (2.30 \times 0 - 1.61)} = e^{2.30}$$

This is the unadjusted odds ratio for "Wednesday, Saturday, or Sunday." If we evaluate that, we get $e^{2.30} = 10$, the same value we got for the odds ratio on page 192, as you would expect!

To find the odds of a sale based only on temperature, we look at the effect a change in temperature has. We therefore find the odds of making a sale with a temperature difference of 1 degree calculated as follows:

$$\frac{\text{odds of a sale with high temp of } (k+1) \text{ degrees}}{\text{odds of a sale with high temp of } k \text{ degrees}} = \frac{e^{0.52 \times (k+1) - 13.44}}{e^{0.52 \times k \; 13.44}} =$$

$$e^{0.52 \times (k+1) - 13.44 - (0.52 \times k - 13.44)} = e^{0.52}$$

This is the unadjusted odds ratio for a one degree increase in temperature.

However, the logistic regression equation that was calculated from this data considered both of these variables together, so the regression coefficients (and thus the odds ratios) have to be adjusted to account for multiple variables.

In this case, when the regression equation is calculated using both day of the week and temperature, we see that both exponents and the constant have changed. For day of the week, the coefficient has increased from 2.30 to 2.44, temperature increased from 0.52 to 0.54, and the constant is now –15.20. These changes are due to *interactions* between variables—when changes in one variable alter the effects of another variable, for example if the day being a Saturday changes the effect that a rise in temperature has on sales. With these new numbers, the same calculations are performed, first varying the day of the week:

$$\frac{e^{2.44 \times 1 + 0.54 \times k - 15.20}}{e^{2.44 \times 0 + 0.54 \times k - 15.20}} = e^{2.44 \times 1 + 0.54 \times k - 15.20 - (2.44 \times 0 + 0.54 \times k - 15.20)} = e^{2.44}$$

This is the adjusted odds ratio for "Wednesday, Saturday, or Sunday." In other words, the day-of-the-week odds have been adjusted to account for any combined effects that may be seen when temperature is also considered.

Likewise, after adjusting the coefficients, the odds ratio for temperature can be recalculated:

$$\frac{e^{2.44\times1+0.54\times(k+1)-15.20}}{e^{2.44\times1+0.54\times k-15.20}} = \frac{e^{2.44\times0+0.54\times(k+1)-15.20}}{e^{2.44\times0+0.54\times k-15.20}} = e^{0.54\times(k+1)-15.20-(0.54\times k-15.20)} = e^{0.54}$$

This is the adjusted odds ratio for "high temperature." In this case, the temperature odds ratio has been adjusted to account for possible effects of the day of the week.

HYPOTHESIS TESTING WITH ODDS

As you'll remember, in linear regression analysis, we perform a hypothesis test by asking whether A is equal to zero, like this:

Null hypothesis	$A_i = 0$
Alternative hypothesis	$A_i \neq 0$

In logistic regression analysis, we perform a hypothesis test by evaluating whether coefficient A as a power of e equals e^0:

Null hypothesis	$e^{A_i} = e^0 = 1$
Alternative hypothesis	$e^{A_i} \neq e^0 = 1$

Remember from Table 4-1 that $e^{(2.30x_1-1.61)}$ is the odds of selling the Norns Special based on the day of the week. If, instead, the odds were found to be $e^{0x_1-1.61}$, it would mean the odds of selling the special were the same every day of the week. Therefore, the null hypothesis would be true: day of the week has no effect on sales. Checking whether $A_i = 0$ and whether $e^{A_i} = e^0 = 1$ are effectively the same thing, but because logistic regression analysis is about odds and probabilities, it is more relevant to write the hypothesis test in terms of odds.

CONFIDENCE INTERVAL FOR AN ODDS RATIO

Odds ratios are often used in clinical studies, and they're generally presented with a confidence interval. For example, if medical researchers were trying to determine whether ginger helps to alleviate an upset stomach, they might separate people with stomach ailments into two groups and then give one group ginger pills and the other a placebo. The scientists would then measure the discomfort of the people after taking the pills and calculate an odds ratio. If the odds ratio showed that people given ginger felt better than people given a placebo, the researchers could use a confidence interval to get a sense of the standard error and the accuracy of the result.

We can also calculate a confidence interval for the Norns Special data. Below, we calculate the interval with a 95% confidence rate.

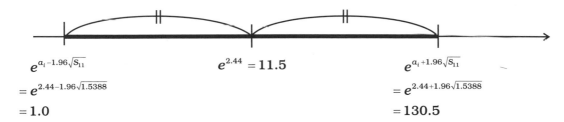

$$e^{a_i - 1.96\sqrt{S_{11}}}$$

$$= e^{2.44 - 1.96\sqrt{1.5388}}$$

$$= 1.0$$

$$e^{2.44} = 11.5$$

$$e^{a_i + 1.96\sqrt{S_{11}}}$$

$$= e^{2.44 + 1.96\sqrt{1.5388}}$$

$$= 130.5$$

If we look at a population of all days that a Norns Special was on sale, we can be sure the odds ratio is somewhere between 1 and 130.5. In other words, at worst, there is no difference in sales based on day of the week (when the odds ratio = 1), and at best, there is a very large difference based on the day of the week. If we chose a confidence rate of 99%, we would change the 1.96 above to 2.58, which makes the interval 0.5 to 281.6. As you can see, a higher confidence rate leads to a larger interval.

RELATIVE RISK

The *relative risk (RR)*, another type of ratio, compares the probability of an event occurring in a group exposed to a particular factor to the probability of the same event occurring in a nonexposed group. This ratio is often used in statistics when a researcher wants to compare two outcomes and the outcome of interest is relatively rare. For example, it's often used to study factors associated with contracting a disease or the side effects of a medication.

You can also use relative risk to study something less serious (and less rare), namely whether day of the week increases the chances that the Norns Special will sell. We'll use the data from page 166.

First, we make a table like Table 4-2 with the condition on one side and the outcome on the other. In this case, the condition is the day of the week. The condition must be binary (yes or no), so since Risa thinks the Norns special sells best on Wednesday, Saturday, and Sunday, we consider the condition present on one of those three days and absent on any other day. As for the outcome, either the cake sold or it didn't.

TABLE 4-2: CROSS-TABULATION TABLE OF "WEDNESDAY, SATURDAY, OR SUNDAY" AND "SALES OF NORNS SPECIAL"

		Sales of Norns Special		Sum
		Yes	No	
Wed, Sat, or Sun	Yes	6	3	9
	No	2	10	12
Sum		8	13	21

To find the relative risk, we need to find the ratio of Norns Specials sold on Wednesday, Saturday, or Sunday to the total number offered for sale on those days. In our sample data, the number sold was 6, and the number offered for sale was 9 (3 were not sold). Thus, the ratio is 6:9.

Next, we need the ratio of the number sold on any other day to the total number offered for sale on any other day. This ratio is 2:12.

Finally, we divide these ratios to find the relative risk:

$$\frac{\text{sales rate of Wed, Sat, or Sun}}{\text{the sales rate of days other than Wed, Sat, or Sun}} = \frac{(6/9)}{(2/12)} = \frac{6}{9} \div \frac{2}{12} = \frac{6}{9} \times \frac{12}{2} = \frac{2}{3} \times 6 = 4$$

So the Norns Special is 4 times more likely to sell on Wednesday, Saturday or Sunday. It looks like Risa was right!

It's important to note that often researchers will report the odds ratio in lieu of the relative risk because the odds ratio is more closely associated with the results of logistic regression analysis and because sometimes you aren't able to calculate the relative risk; for example, if you didn't have complete data for sales rates on all days other than Wednesday, Saturday, and Sunday. However, relative risk is more useful in some situations and is often easier to understand because it deals with probabilities and not odds.

APPENDIX
REGRESSION CALCULATIONS WITH EXCEL

This appendix will show you how to use Excel functions to calculate the following:

- Euler's number (*e*)
- Powers
- Natural logarithms
- Matrix multiplication
- Matrix inverses
- Chi-squared statistic from a *p*-value
- *p*-value from a chi-squared statistic
- *F* statistic from a *p*-value
- *p*-value from an *F* statistic
- Partial regression coefficient of a multiple regression analysis
- Regression coefficient of a logistic regression equation

We'll use a spreadsheet that already includes the data for the examples in this appendix. Download the Excel spreadsheet from *http://www.nostarch.com/regression/*.

EULER'S NUMBER

Euler's number (*e*), introduced on page 19, is the base number of the natural logarithm. This function will allow you to raise Euler's number to a power. In this example, we'll calculate e^1.

1. Go to the *Euler's Number* sheet in the spreadsheet.
2. Select cell **B1**.

B1
fx
A
B
C
D
1
e1
2

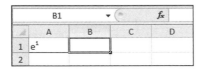

3. Click **Formulas** in the top menu bar and select **Insert Function**.

File
Home
Insert
Page Layout
Formulas
Data
Review
View
Insert Function
AutoSum
Recently Used
Financial
Logical
Text
Date & Time
Lookup & Reference
Math & Trig
More Functions
Function Library

4. From the category drop-down menu, select **Math & Trig**. Select the **EXP** function and then click **OK**.

Insert Function
Search for a function:
Type a brief description of what you want to do and then click Go
Go
Or select a category: Math & Trig
Select a function:
COSH
DEGREES
EVEN
EXP
FACT
FACTDOUBLE
FLOOR
EXP(number)
Returns e raised to the power of a given number.
Help on this function
OK
Cancel

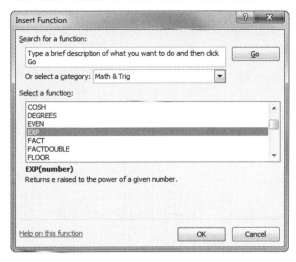

5. You'll now see a dialog where you can enter the power to which you want to raise *e*. Enter **1** and then click **OK**.

Function Arguments
EXP
Number 1 = 1
= 2.718281828
Returns e raised to the power of a given number.
Number is the exponent applied to the base e. The constant e equals 2.71828182845904, the base of the natural logarithm.
Formula result = 2.718281828
Help on this function
OK
Cancel

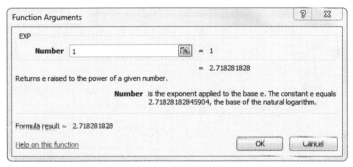

Because we've calculated Euler's number to the power of 1, you'll just get the value of *e* (to a few decimal places), but you can raise *e* to any power using the EXP function.

B1
=EXP(1)
A B C D E
1 e1 2.718282
2

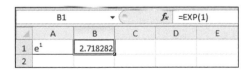

NOTE *You can avoid using the Insert Function menu by entering =EXP(X) into the cell. For example, entering =EXP(1) will also give you the value of e. This is the case for any function: after using the Insert Function menu, simply look at the formula bar for the function you can enter directly into the cell.*

POWERS

This function can be used to raise any number to any power. We'll use the example question from page 14: "What's 2 cubed?"

1. Go to the *Power* sheet in the spreadsheet.
2. Select cell **B1** and type **=2^3**. Press ENTER.

B1
=2^3
A
B
C
D
1
2^3
8
2

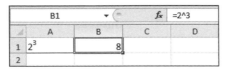

In Excel, we use the ^ symbol to mean "to the power of," so 2^3 is 2^3, and the result is 8. Make sure to include the equal sign (=) at the start or Excel will not calculate the answer for you.

NATURAL LOGARITHMS

This function will perform a natural log transformation (see page 20).

1. Go to the *Natural Log* sheet in the spreadsheet.
2. Select cell **B1**. Click **Formulas** in the top menu bar and select **Insert Function**.
3. From the category drop-down menu, select **Math & Trig**. Select the **LN** function and then click **OK**.

Insert Function
Search for a function:
Type a brief description of what you want to do and then click Go
Go
Or select a category: Math & Trig
Select a function:
GCD
INT
LCM
LN
LOG
LOG10
MDETERM
LN(number)
Returns the natural logarithm of a number.
Help on this function
OK
Cancel

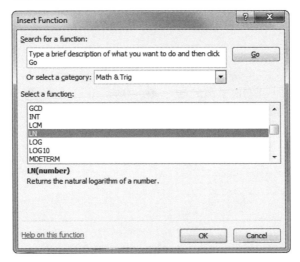

4. Enter **exp(3)** and click **OK**.

Function Arguments
LN
Number exp(3) = 20.08553692
= 3
Returns the natural logarithm of a number.
Number is the positive real number for which you want the natural logarithm.
Formula result = 3
Help on this function
OK
Cancel

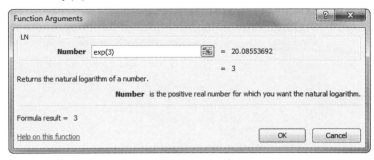

You should get the natural logarithm of e^3, which, according to Rule 3 on page 22, will of course be 3. You can enter any number here, with a base of e or not, to find its natural log. For example, entering exp(2) would produce 2, while entering just 2 would give 0.6931.

MATRIX MULTIPLICATION

This function is used to multiply matrices—we'll calculate the multiplication example shown in Example Problem 1 on page 41.

1. Go to the *Matrix Multiplication* sheet in the spreadsheet.
2. Select cell **G1**. Click **Formulas** in the top menu bar and select **Insert Function**.
3. From the category drop-down menu, select **Math & Trig**. Select the **MMULT** function and then click **OK**.

Insert Function
Search for a function:
Type a brief description of what you want to do and then click Go
Go
Or select a category: Math & Trig
Select a function:
LOG10
MDETERM
MINVERSE
MMULT
MOD
MROUND
MULTINOMIAL
MMULT(array1,array2)
Returns the matrix product of two arrays, an array with the same number of rows as array1 and columns as array2.
Help on this function
OK
Cancel

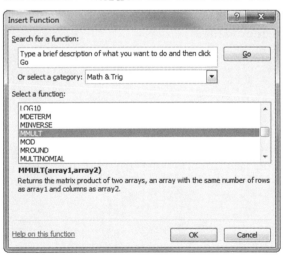

4. Click in the **Array1** field and highlight all the cells of the first matrix in the spreadsheet. Then click in the **Array2** field and highlight the cells containing the second matrix. Click **OK**.

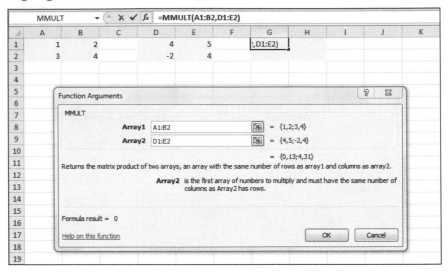

5. Starting with **G1**, highlight a matrix of cells with the same dimensions as the matrices you are multiplying—G1 to H2 in this example. Then click in the formula bar.

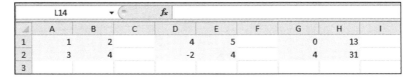

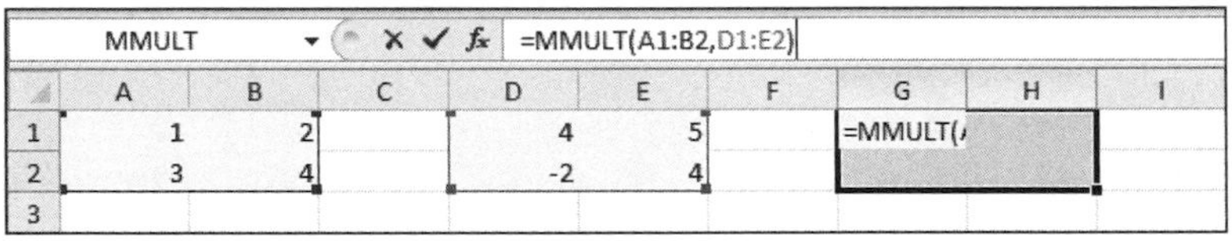

6. Press CTRL-SHIFT-ENTER. The fields in your matrix should fill with the correct values.

	L14		▼		f_x			
◢	A	B	C	D	E	F	G	H	I
1	1	2		4	5		0	13	
2	3	4		-2	4		4	31	
3									

You should get the same results as Risa gets at the bottom of page 41. You can do this with any matrices that share the same dimensions.

MATRIX INVERSION

This function calculates matrix inverses—we'll use the example shown on page 44.

1. Go to the *Matrix Inversion* sheet in the spreadsheet.
2. Select cell **D1**. Click **Formulas** in the top menu bar and select **Insert Function**.
3. From the category drop-down menu, select **Math & Trig**. Select the **MINVERSE** function and then click **OK**.

Insert Function
Search for a function:
Type a brief description of what you want to do and then click Go
Go
Or select a category: Math & Trig
Select a function:
LN
LOG
LOG10
MDETERM
MINVERSE
MMULT
MOD
MINVERSE(array)
Returns the inverse matrix for the matrix stored in an array.
Help on this function
OK
Cancel

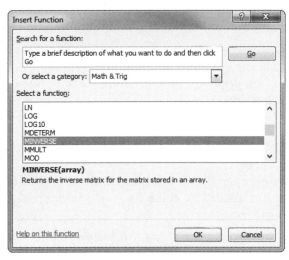

4. Select and highlight the matrix in the sheet—that's cells A1 to B2—and click **OK**.

MINVERSE
=MINVERSE(A1:B2)
(A1:B2)
Function Arguments
MINVERSE
Array A1:B2 = {1,2;3,4}
= {-2,1;1.5,-0.5}
Returns the inverse matrix for the matrix stored in an array.
Array is a numeric array with an equal number of rows and columns, either a cell range or an array constant.
Formula result = -2
Help on this function
OK
Cancel

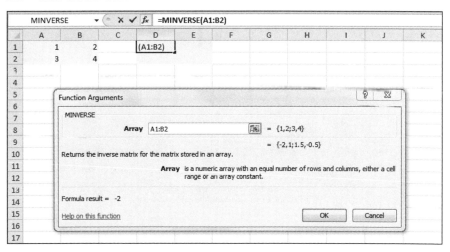

5. Starting with **D1**, select and highlight a matrix of cells with the same dimensions as the first matrix—in this case, D1 to E2. Then click in the formula bar.

MINVERSE
=MINVERSE(A1:B2)
=MINVERS

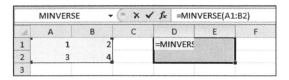

6. Press CTRL-SHIFT-ENTER. The fields in your matrix should fill with the correct values.

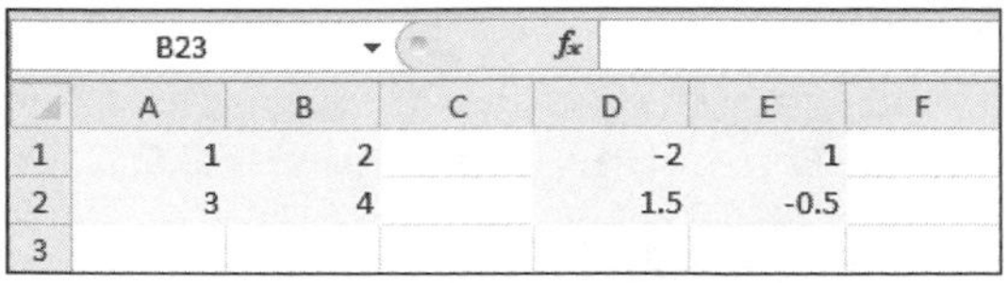

	A	B	C	D	E	F
1	1	2		-2	1	
2	3	4		1.5	-0.5	
3						

You should get the same result as Risa does on page 44. You can use this on any matrix you want to invert; just make sure the matrix of cells you choose for the results has the same dimensions as the matrix you're inverting.

CALCULATING A CHI-SQUARED STATISTIC FROM A P-VALUE

This function calculates a test statistic from a chi-squared distribution, as discussed on page 54. We'll use a *p*-value of .05 and 2 degrees of freedom.

1. Go to the *Chi-Squared from p-Value* sheet in the spreadsheet.
2. Select cell **B3**. Click **Formulas** in the top menu bar and then select **Insert Function**.
3. From the category drop-down menu, select **Statistical**. Select the **CHISQ.INV.RT** function and then click **OK**.

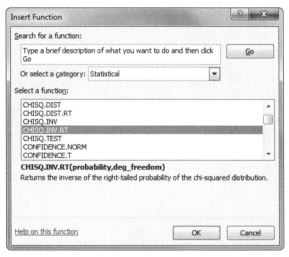

4. Click in the **Probability** field and enter **B1** to select the probability value in that cell. Then click in the **Deg_freedom** field and enter **B2** to select the degrees of freedom value. When (B1,B2) appears in cell B3, click **OK**.

Function Arguments
CHISQ.INV.RT
Probability B1 = 0.05
Deg_freedom B2 = 2
= 5.991464547
Returns the inverse of the right-tailed probability of the chi-squared distribution.
Deg_freedom is the number of degrees of freedom, a number between 1 and 10^10, excluding 10^10.
Formula result = 5.991464547
Help on this function
OK
Cancel

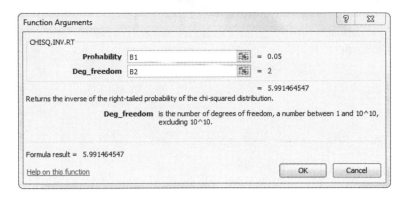

You can check this calculation against Table 1-6 on page 56.

CALCULATING A P-VALUE FROM A CHI-SQUARED STATISTIC

This function is used on page 179 in the likelihood ratio test to obtain a *p*-value. We're using a test statistic value of 10.1 and 2 degrees of freedom.

1. Go to the *p-Value from Chi-Squared* sheet in the spreadsheet.
2. Select cell **B3**. Click **Formulas** in the top menu bar and select **Insert Function**.
3. From the category drop-down menu, select **Statistical**. Select the **CHISQ.DIST.RT** function and then click **OK**.

Insert Function
Search for a function:
Type a brief description of what you want to do and then click Go
Go
Or select a category: Statistical
Select a function:
BETA.INV
BINOM.DIST
BINOM.INV
CHISQ.DIST
CHISQ.DIST.RT
CHISQ.INV
CHISQ.INV.RT
CHISQ.DIST.RT(x,deg_freedom)
Returns the right-tailed probability of the chi-squared distribution.
Help on this function
OK
Cancel

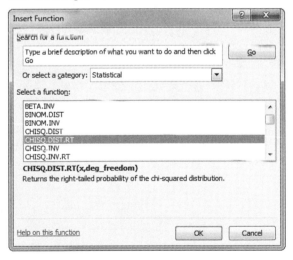

4. Click in the **X** field and enter **B1** to select the chi-squared value in that cell. Then click the **Deg_freedom** field and enter **B2** to select the degrees of freedom value. When (B1,B2) appears in cell B3, click **OK**.

Function Arguments
CHISQ.DIST.RT
X B1 = 10.1
Deg_freedom B2 = 2
= 0.006409333
Returns the right-tailed probability of the chi-squared distribution.
Deg_freedom is the number of degrees of freedom, a number between 1 and 10^10, excluding 10^10.
Formula result = 0.006409333
Help on this function
OK
Cancel

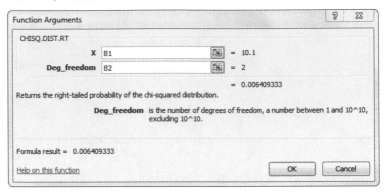

We get 0.006409, which on page 179 has been rounded down to 0.006.

B3
=CHISQ.DIST.RT(B1,B2)
A B C D E
1 Chi-squared 10.1
2 Freedom 2
3 Probability 0.006409
4

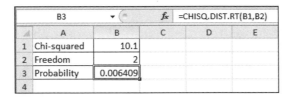

CALCULATING AN F STATISTIC FROM A P-VALUE

This function gives us the *F* statistic we calculated on page 58.

1. Go to the *F Statistic from p-Value* sheet in the spreadsheet.
2. Select cell **B4**. Click **Formulas** in the top menu bar and select **Insert Function**.
3. From the category drop-down menu, select **Statistical**. Select the **F.INV.RT** function and then click **OK**.

Insert Function
Search for a function:
Type a brief description of what you want to do and then click Go
Go
Or select a category: Statistical
Select a function:
F.DIST
F.DIST.RT
F.INV
F.INV.RT
F.TEST
FISHER
FISHERINV
F.INV.RT(probability,deg_freedom1,deg_freedom2)
Returns the inverse of the (right-tailed) F probability distribution: if p = F.DIST.RT(x,...), then F.INV.RT(p,...) = x.
Help on this function
OK
Cancel

4. Click in the **Probability** field and enter **B1** to select the probability value in that cell. Click in the **Deg_freedom1** field and enter **B2** and then select the **Deg_freedom2** field and enter **B3**. When (B1,B2,B3) appears in cell B3, click **OK**.

Function Arguments
F.INV.RT
Probability B1 = 0.05
Deg_freedom1 B2 = 1
Deg_freedom2 B3 = 12
= 4.747225347
Returns the inverse of the (right-tailed) F probability distribution: if p = F.DIST.RT(x,...), then F.INV.RT(p,...) = x.
Deg_freedom2 is the denominator degrees of freedom, a number between 1 and 10^10, excluding 10^10.
Formula result = 4.747225347
Help on this function
OK
Cancel

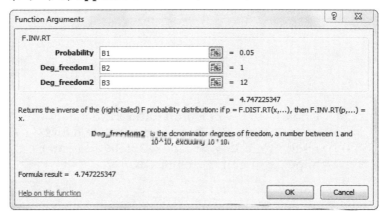

We get 4.747225, which has been rounded down to 4.7 in Table 1-7 on page 58.

B4 f_x =F.INV.RT(B1,B2,B3)

	A	B	C	D
1	Probability	0.05		
2	1 degree of freedom	1		
3	2 degrees of freedom	12		
4	F	4.747225		
5				

CALCULATING A P-VALUE FROM AN F STATISTIC

This function is used on page 90 to calculate the *p*-value in an ANOVA.

1. Go to the *p-Value for F Statistic* sheet in the spreadsheet.
2. Select cell **B4**. Click **Formulas** in the top menu bar and select **Insert Function**.
3. From the category drop-down menu, select **Statistical**. Select the **F.DIST.RT** function and then click **OK**.

Insert Function
Search for a function:
Type a brief description of what you want to do and then click Go
Go
Or select a category: Statistical
Select a function:
DEVSQ
EXPON.DIST
F.DIST
F.DIST.RT
F.INV
F.INV.RT
F.TEST
F.DIST.RT(x,deg_freedom1,deg_freedom2)
Returns the (right-tailed) F probability distribution (degree of diversity) for two data sets.
Help on this function
OK
Cancel

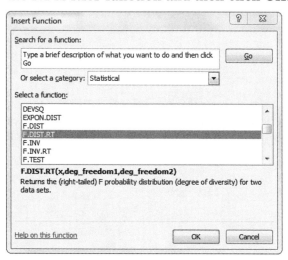

4. Click in the **X** field and enter **B1** to select the *F* value in that cell. Click in the **Deg_freedom1** field and enter **B2**, and then click in the **Deg_freedom2** field and enter **B3**. When (B1,B2,B3) appears in cell B3, click **OK**.

Function Arguments
F.DIST.RT
X B1 = 55.6
Deg_freedom1 B2 = 1
Deg_freedom2 B3 = 12
= 7.66775E-06
Returns the (right-tailed) F probability distribution (degree of diversity) for two data sets.
Deg_freedom2 is the denominator degrees of freedom, a number between 1 and 10^10, excluding 10^10.
Formula result = 7.66775E-06
Help on this function
OK
Cancel

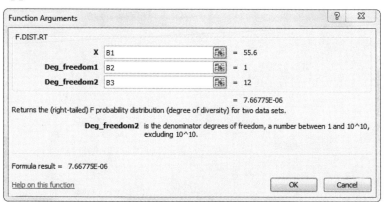

The result, 7.66775E-06, is the way Excel presents the value 7.66775×10^{-6}. If we were testing at the $p = .05$ level, this would be a significant result because it is less than .05.

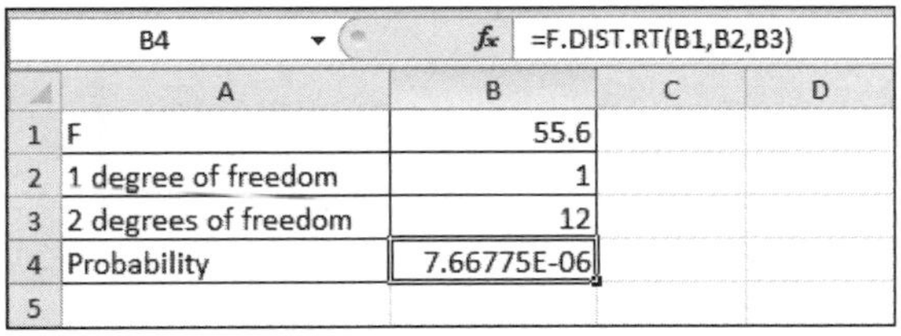

B4 =F.DIST.RT(B1,B2,B3)

	A	B	C	D
1	F	55.6		
2	1 degree of freedom	1		
3	2 degrees of freedom	12		
4	Probability	7.66775E-06		
5				

PARTIAL REGRESSION COEFFICIENT OF A MULTIPLE REGRESSION ANALYSIS

This function calculates the partial regression coefficients for the data on page 113, giving the results that Risa gets on page 118.

1. Go to the *Partial Regression Coefficient* sheet in the spreadsheet.
2. Select cell **G2**. Click **Formulas** in the top menu bar and select **Insert Function**.
3. From the category drop-down menu, select **Statistical**. Select the **LINEST** function and then click **OK**.

Insert Function
Search for a function:
Type a brief description of what you want to do and then click Go
Go
Or select a category: Statistical
Select a function:
INTERCEPT
KURT
LARGE
LINEST
LOGEST
LOGNORM.DIST
LOGNORM.INV
LINEST(known_y's,known_x's,const,stats)
Returns statistics that describe a linear trend matching known data points, by fitting a straight line using the least squares method.
Help on this function
OK
Cancel

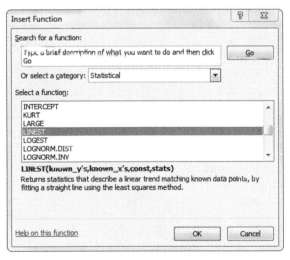

4. Click in the **Known_y's** field and highlight the data cells for your outcome variable—here it's D2 to D11. Click in the **Known_x's** field and highlight the data cells for your predictor variables—here B2 to C11. You don't need any values for Const and Stats, so click **OK**.

Function Arguments
LINEST
Known_y's D2:D11 = {469;366;371;208;246;297;363;436...
Known_x's B2:C11 = {10,80;8,0;8,200;5,200;7,300;8,23...
Const = logical
Stats = logical
= {-0.340882685663619,41.51347825...
Returns statistics that describe a linear trend matching known data points, by fitting a straight line using the least squares method.
Known_y's is the set of y-values you already know in the relationship y = mx + b.
Formula result = -0.340882686
Help on this function
OK
Cancel

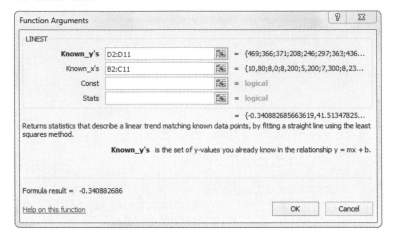

5. The full function gives you three values, so highlight G1 to I1 and click the function bar. Press CTRL-SHIFT-ENTER, and the highlighted fields should fill with the correct values.

G2 | {=LINEST(D2:D11,B2:C11)}

	A	B	C	D	E	F	G	H	I	J
1		Floor space (tsubo)	Distiance to nearest station (meters)	Monthly sales			Distance to nearest station (meters)	Floor space (tsubo)	Constant term	
2	Yumenooka Shop	10	80	469		Partial regression coefficient	-0.3409	41.5135	65.3239	
3	Terai Station Shop	8	0	366						
4	Sone Shop	8	200	371						
5	Hashimoto Station Shop	5	200	208						
6	Kikyou Town Shop	7	300	246						
7	Post Office Shop	8	230	297						
8	Suidobashi Station Shop	7	40	363						
9	Rokujo Station Shop	9	0	436						
10	Wakaba Riverside Shop	6	330	198						
11	Misato Shop	9	180	364						
12										

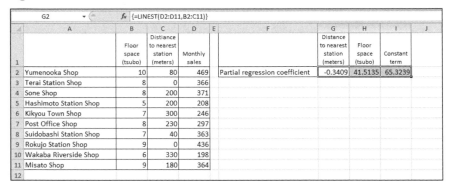

You can see that the results are the same as Risa's results on page 118 (in the text, they have been rounded).

REGRESSION COEFFICIENT OF A LOGISTIC REGRESSION EQUATION

There is no Excel function that calculates the logistic regression coefficient, but you can use Excel's Solver tool. This example calculates the maximum likelihood coefficients for the logistic regression equation using the data on page 166.

1. Go to the *Logistic Regression Coefficient* sheet in the spreadsheet.
2. First you'll need to check whether Excel has Solver loaded. When you select **Data** in the top menu bar, you should see a button to the far right named *Solver*. If it is there, skip ahead to Step 4; otherwise, continue on to Step 3.

File
Home
Insert
Page Layout
Formulas
Data
Review
View
Acrobat
Get External Data
Refresh All
Connections
Properties
Edit Links
Sort
Filter
Clear
Reapply
Advanced
Text to Columns
Remove Duplicates
Data Validation
Consolidate
What-If Analysis
Group
Ungroup
Subtotal
Solver
Connections
Sort & Filter
Data Tools
Outline
Analysis

3. If the Solver button isn't there, go to **File ▸ Options ▸ Add-Ins** and select the **Solver Add-in**. Click **Go**, select **Solver Add-in** in the Add-Ins dialog, and then click **OK**. Now when you select Data in the top menu bar, the Solver button should be there.

Excel Options
General
Formulas
Proofing
Save
Language
Advanced
Customize Ribbon
Quick Access Toolbar
Add-Ins
Trust Center
View and manage Microsoft Office Add-ins.
Add-ins
Name
Location
Type
Active Application Add-ins
Acrobat PDFMaker Office COM Addin
C:\...FMaker\Office\x64\PDFMOfficeAddin.dll
COM Add-in
Inactive Application Add-ins
Analysis ToolPak
C:\...e\Office14\Library\Analysis\ANALYS32.XLL
Excel Add-in
Analysis ToolPak - VBA
C:\...ffice14\Library\Analysis\ATPVBAEN.XLAM
Excel Add-in
Custom XML Data
C:\...\Microsoft Office\Office14\OFFRHD.DLL
Document Inspector
Date (XML)
C:\...es\Microsoft Shared\Smart Tag\MOFL.DLL
Action
Euro Currency Tools
C:\... Office\Office14\Library\EUROTOOL.XLAM
Excel Add-in
Financial Symbol (XML)
C:\...es\Microsoft Shared\Smart Tag\MOFL.DLL
Action
Headers and Footers
C:\...\Microsoft Office\Office14\OFFRHD.DLL
Document Inspector
Hidden Rows and Columns
C:\...\Microsoft Office\Office14\OFFRHD.DLL
Document Inspector
Hidden Worksheets
C:\...\Microsoft Office\Office14\OFFRHD.DLL
Document Inspector
Invisible Content
C:\...\Microsoft Office\Office14\OFFRHD.DLL
Document Inspector
Microsoft Actions Pane 3
XML Expansion Pack
Solver Add-in
C:\...e\Office14\Library\SOLVER\SOLVER.XLAM
Excel Add-in
Document Related Add-ins
No Document Related Add-ins
Disabled Application Add-ins
No Disabled Application Add-ins
Add-in: Solver Add-in
Publisher:
Compatibility: No compatibility information available
Location: C:\Program Files\Microsoft Office\Office14\Library\SOLVER\SOLVER.XLAM
Description: Tool for optimization and equation solving
Manage: Excel Add-ins
Go...
OK
Cancel

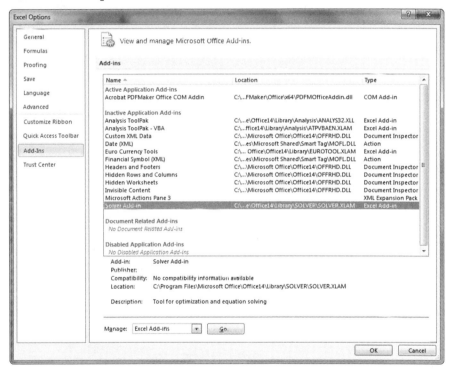

4. Click the **Solver** button. Click in the **Set Objective** field and select cell L3 to select the log likelihood data. Click in the **By Changing Variable Cells** field and select the cells where you want your results to appear—in this case L5 to L7. Click **Solve**.

Solver Parameters
Set Objective:
$L$3
To:
Max
Min
Value Of:
0
By Changing Variable Cells:
$L$5:$L$7
Subject to the Constraints:
Add
Change
Delete
Reset All
Load/Save
Make Unconstrained Variables Non-Negative
Select a Solving Method:
GRG Nonlinear
Options
Solving Method
Select the GRG Nonlinear engine for Solver Problems that are smooth nonlinear. Select the LP Simplex engine for linear Solver Problems, and select the Evolutionary engine for Solver problems that are non-smooth.
Help
Solve
Close

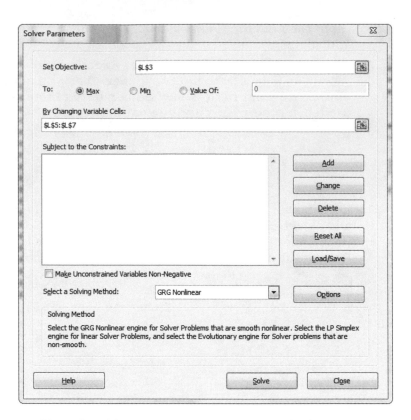

You should get the same answers as in Step 4 on page 172 (in the text, they've been rounded).

INDEX

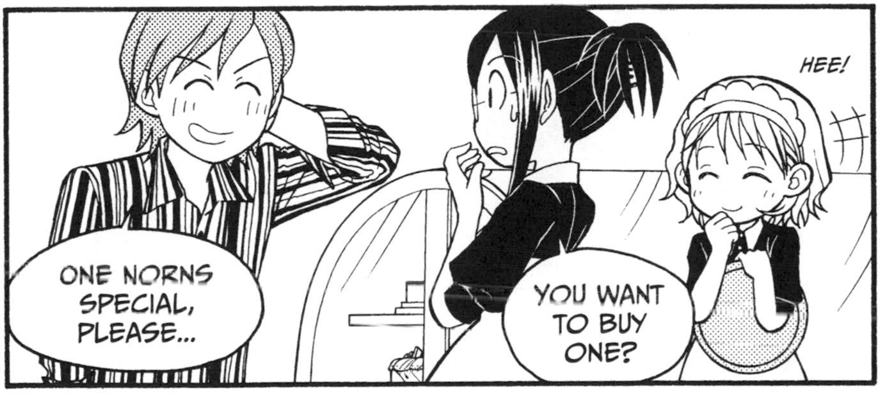
ONE NORNS SPECIAL, PLEASE...
YOU WANT TO BUY ONE?
HEE!

ABOUT THE AUTHOR

Shin Takahashi was born in 1972 in Niigata. He received a master's degree from Kyushu Institute of Design (known as Kyushu University today). Having previously worked both as an analyst and as a seminar leader, he is now an author specializing in technical literature.

Homepage: *http://www.takahashishin.jp/*

PRODUCTION TEAM FOR THE JAPANESE EDITION

SCENARIO: re_akino

ARTIST: Iroha Inoue

HOW THIS BOOK WAS MADE

The *Manga Guide* series is a co-publication of No Starch Press and Ohmsha, Ltd. of Tokyo, Japan, one of Japan's oldest and most respected scientific and technical book publishers. Each title in the best-selling *Manga Guide* series is the product of the combined work of a manga illustrator, scenario writer, and expert scientist or mathematician. Once each title is translated into English, we rewrite and edit the translation as necessary and have an expert review each volume. The result is the English version you hold in your hands.

MORE MANGA GUIDES

Find more *Manga Guides* at your favorite bookstore, and learn more about the series at *http://www.nostarch.com/manga/*.

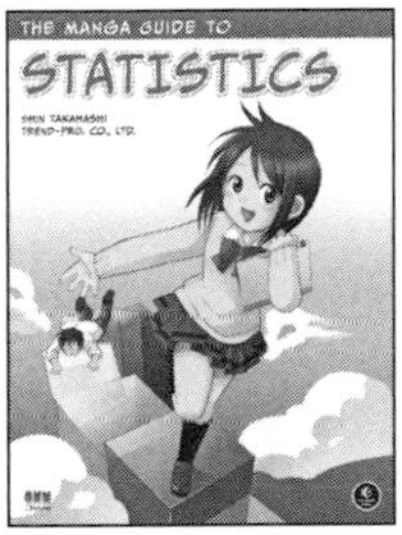
THE MANGA GUIDE TO
STATISTICS

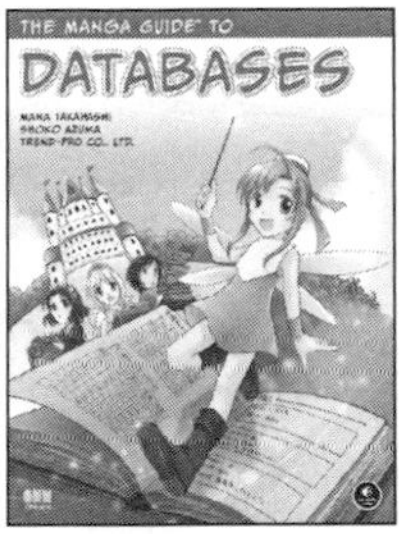
THE MANGA GUIDE TO
DATABASES

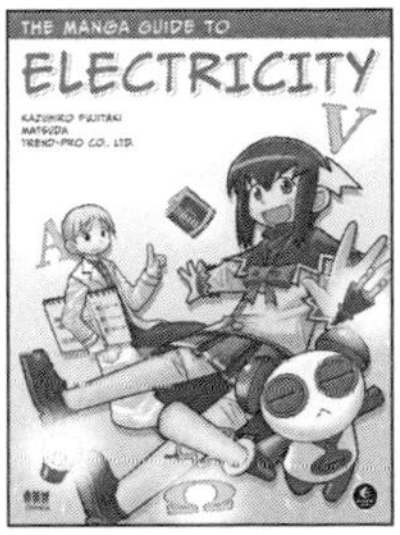
THE MANGA GUIDE TO
ELECTRICITY

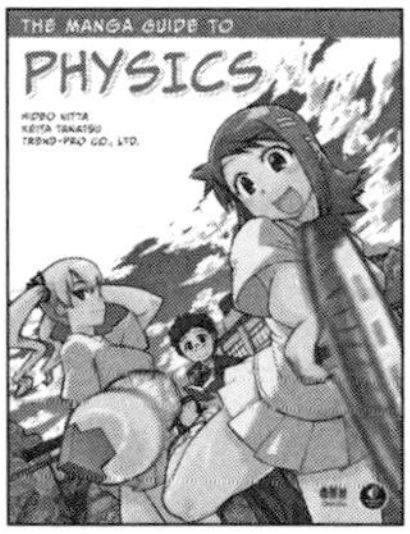
THE MANGA GUIDE TO
PHYSICS

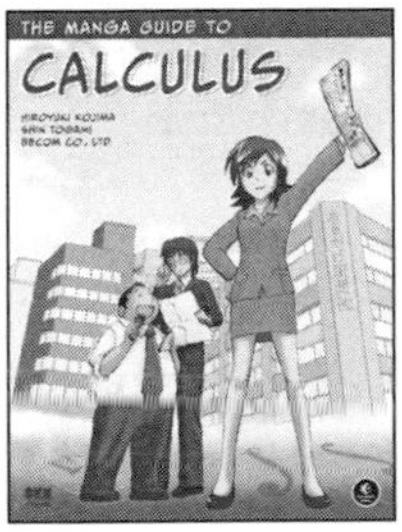
THE MANGA GUIDE TO
CALCULUS

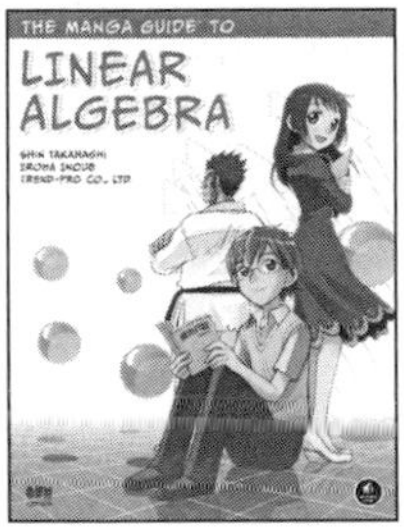
THE MANGA GUIDE TO
LINEAR ALGEBRA

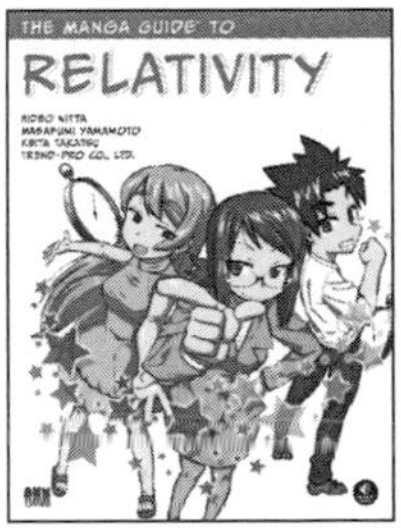
THE MANGA GUIDE TO
RELATIVITY

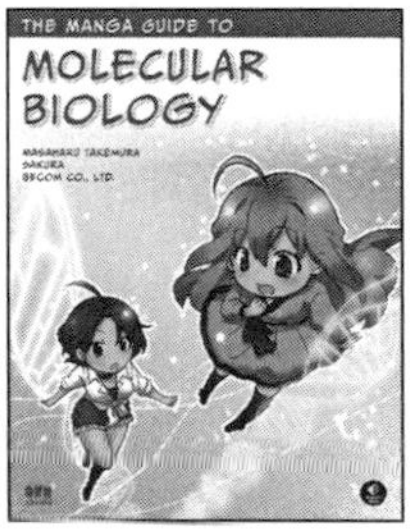
THE MANGA GUIDE TO
MOLECULAR BIOLOGY

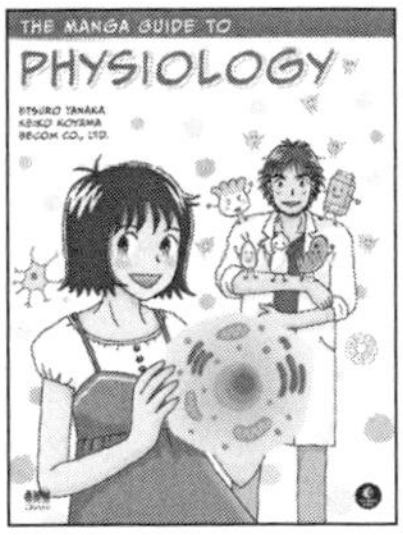
THE MANGA GUIDE TO
PHYSIOLOGY

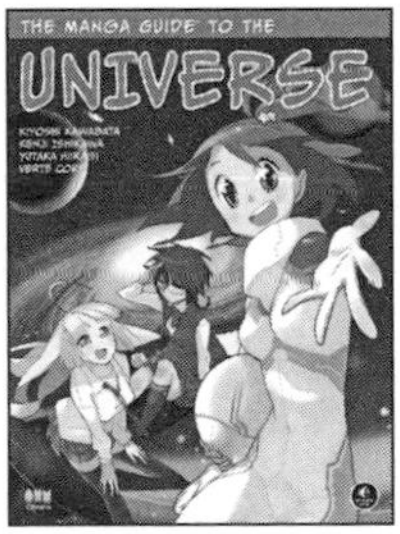
THE MANGA GUIDE TO THE
UNIVERSE

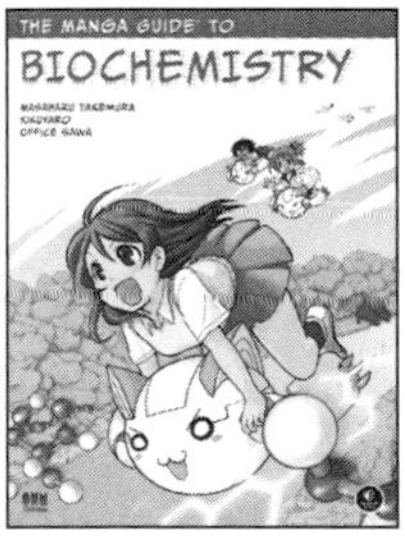
THE MANGA GUIDE TO
BIOCHEMISTRY

UPDATES

Visit *http://www.nostarch.com/regression/* for updates, errata, and other information.

COLOPHON

The Manga Guide to Regression Analysis is set in CCMeanwhile and Bookman. This book was printed and bound by Sheridan Books, Inc. in Chelsea, Michigan. The paper is 60# Finch Offset, which is certified by the Forest Stewardship Council (FSC).